C. West Churchman and Related Works Series
Series Editor – John P. Van Gigch

Cover design by Susanne Bagnato, a Graphic Designer in Adelaide, Australia.
Contact details are: sbagnato@picknowl.com.au

RESCUING THE ENLIGHTENMENT FROM ITSELF:

Critical and Systemic Implications for Democracy

Editor: Janet McIntyre-Mills

Series Editor: John P. van Gigch

VOLUME 1: RESCUING THE ENLIGHTENMENT FROM ITSELF:

Critical and Systemic Implications for Democracy

Editor: Janet McIntyre-Mills
Flinders Institute of Public Policy and Management
Flinders University
Adelaide, Australia

Series Editor: John P. van Gigch
Emeritus Professor
California State University,
Sacramento, CA., USA.

Springer

Janet McIntyre-Mills
Flinders Institute of Public Policy and Management
Flinders University, Adelaide, Australia

Library of Congress Cataloging-in-Publication Data

A C.I.P. Catalogue record for this book is available
from the Library of Congress.

ISBN-10: 0-387-27587-8 ISBN-13: 978-0387-27587-1
ISBN-10: 0-387-27589-4 (e-book) ISBN-13: 978-0387-27589-5 (e-book)

Printed in the United States of America. Printed on acid-free paper.

9 8 7 6 5 4 3 2 1 SPIN 11320708

springeronline.com

CONTENTS

BIOGRAPHY OF THE CONTRIBUTORS

Ackoff, Russell L., a colleague of Churchman is the founder of INTERACT, USA and is based at Ackoff Centre for Advanced Systems Approaches, University of Pennsylvania, USA.

Kenneth Bausch, Ph.D. Executive Director, Institute for 21st Century Agoras; Renaissance scholar; one-time priest, teacher, organizer, counselor, social service administrator, real estate agent, homebuilder, contractor, university professor, research director, and organizational consultant. Author of *The Emerging Consensus in Social Systems Theory* and *Body Wisdom: Faith in Chaos, Catching the Tune*; Co-Chair of 47th annual Conference of the International Society for the Systems Sciences, Crete 2003.

Carugati, Andrea, is based at IESEG School of Management in Lille (France) where he is Assistant Professor of Information Systems. His research interests focus on the role of knowledge creation and learning in information systems development.

Christakis, Alexander N. Christakis, CWA Ltd, USA, a founder member of the Club of Rome and developer of the CogniScope system approach and software for managing complexity, President of the International Society for the Systems Sciences, 2002.

Ferris, Tim, University of South Australia is director of Electronic Engineering, his research interests are in the theory of measurement, design, development and function of systems.

Forsgren, Olov, a colleague of C. West Churchman is based at the University College of Borås, Sweden.

Gharajedaghi, Jamshid is based at Ackoff Centre for Advanced Systems Approaches, University of Pennsylvania, USA.

Matthews, David is a Defence Scientist with the Defence Science and Technology Organisation (DSTO) of Australia. He has an Honours Degree in Applied Mathematics and has recently completed a Ph.D. in Systems Thinking, both from the University of Adelaide. His main research interests involve problems at the interface of philosophy, social theory, mathematics and the management and planning sciences.

McIntyre, Janet, is a Senior Lecturer at Flinders University in the Flinders Institute of Public Policy and Management. Her recent publications include *Critical Systemic Praxis for Social and Environmental Justice*, 2003 and *Global Citizenship and Social Movements: Creating Transcultural Webs of Meaning for the New Millennium*" in 2000. She currently facilitates research with a transdiciplinary design team to address social inclusion with Aboriginal Australians.

Metcalfe, Mike, is Associate Research Professor, University of South Australia (City West), Adelaide, South Australia. His PhD is from Adelaide University, on group problem solving. He has published 5 books and over 50 lead-author refereed academic articles on different aspects of perspectival systems thinking and argumentative inquiry in journals that include Systems Research and Behavioral Science, IT & People, Informal Logic and the European Journal Of Information Systems.

Morgan, D., is a senior research fellow at the University of South Australia and Chair of Neporendi Forum Inc. His PhD thesis was on Aboriginal Health: *The role of Cultural Antecedents in Training Health Professionals to Work with Difference*, The School of Education, The Flinders University of South Australia and he has written widely in the area of health, wellbeing, education and social wellbeing.

Mitroff, Ian, is Harold Quinton Distinguished Professor of Business Policy, Marshall School of Business, University of Southern California, Founder and Director, USC Center for Crisis Management (1986 to 1996), President, Comprehensive Crisis Management. Department of Management and Organization author of *Managing Crises Before They Happen* with Gus Anagnos and numerous works on management, ethics and spirituality.

Romm, Norma's primary research focus is on exploring the way in which researchers can contribute to social development in an accountable way. She is a Research Consultant, ABET, University of South Africa. Her career spans university positions as Dean of the Faculty of Social Science in Swaziland and in Cyprus, Associate Professor of Sociology, University of South Africa and Senior Researcher at universities in United Kingdom and Europe. She has published several books. For example: *People's Education in Theoretical Perspective* (with V.I. McKay, 1992, Cape Town: Longman); and *Diversity Management* (with R.L. Flood, 1996, Chichester: Wiley) and *Critical Systems Thinking* (edited with R.L. Flood, 1996, New York: Plenum). *The Methodologies of Positivism and Marxism*, 1991, Macmillan,

London; *Accountability in Social Research*, 2001, Kluwer/Plenum, New York.

van Gigch, John, Professor Emeritus, Califonia State University, is series editor for the C. West Churchman's Legacy Series. He is editor of the forthcoming Volume Two of the Series in collaboration with Janet McIntyre Mills. van Gigch has published widely, for instance *Applied General Systems Theory* (1974; 1978), *Decision Making about Decision Making* (1987), *System Modeling and Metamodeling* (1991). His most recent publication is *Metadecisions: Rehabilitating Epsistemology*, Kluwer, published in 2003.

Young, Janette, a student at the Flinders Institute of Public Policy and Management is a winner of the Catherine Spence Scholarship for research on social justice and the Sir Geoffrey Vickers Award in 2004, for her paper on narrative and empowerment, is based at University of South Australia where she is a researcher.

PROLOGUE

Why should humanity care if the powerful choose to ignore and exclude the powerless when making governance decisions? We argue that caring for the other and the environment is both idealistic and pragmatic. Idealism and pragmatism are one. Moving from pragmatism to idealism is not difficult once we appreciate the 'boomerang affect' (Beck 1992) of cruelty, wastefulness and greed.

This work on governance, democracy and international relations explores and creates opportunities for building capacity in not only weak states (as suggested by Fukuyama 2004), but instead identifies ways for strong states to address 'the enemy within', rather than merely 'dealing with the other' as nations or civilizations or imposing communitarian solutions (Etzioni, 2004, Huntington, 1996) that are not based on participatory design processes.

C. West Churchman (1913-2004).
(Photograph From the Churchman Family Collection).

PREFACE TO A NEW BOOK SERIES

C. West Churchman Legacy and Related Works

JOHN P. VAN GIGCH, Series Editor, Professor Emeritus, California State University,USA.
JANET MCINTYRE-MILLS, Flinders Institute of Public Policy and Management. Editor of Volume 1

This is the first volume in a series that aims to commemorate the work of C. West Churchman (1913-2004), one of the important thinkers within the last century. The defining characteristics of his writing and his work are his humility and compassion, expressed in the poetry of a human being who loved all that is beautiful in this world. He tries to be open to the diverse ideas of all people and respectful of all living things. Ackoff and Jamshid Gharajedaghi stress that the "world is a mess" and that we can draw on West Churchman for better designs to enhance governance. By working across conceptual boundaries through multidisciplinary work and conversation that is inclusive of differences, we can make steps towards rescuing the so-called 'Enlightenment' from itself through more democratic questioning by those who are to be at the receiving end of the decisions. Aleco Christakis, Norma Romm and Janet McIntyre stress in their contribution to this volume that he helps to liberate our thinking from working within the boundaries of one discipline and they make a plea for the practical ideal of working with ideas and working across disciplines, in order to enhance social and environmental justice.

The texts to be published in this series which is entitled: *C.W. Churchman's Legacy and Related Works,* will consist of volumes submitted by the members of the systems community who wish to honor West and his professional achievements. Churchman was a pioneer and his works continue to guide us, because he showed us the way to address problems in a manner that had not been done before. He contributed a great deal to pioneering the systemic approach.

For those who were fortunate to know him personally we remember him as a gentle man. He never volunteered his own opinion and asked questions to invite (and incite) you to think and to consider the issues carefully and with passion. Around 1978, in a Foreword to John van Gigch's book *(Applied General Systems Theory,* Harper and Row, 1974, 1978) he wrote:

> "Note that a 'good' question is not simply a query followed by a question mark. If the question does not ignite in you to the point of becoming a burning issue - one deeply felt - it is not a good question; on the other hand, if the ignition burns you up, the question is simply disastrous. The artistry of fire is to ignite and burn intensely without destruction."

His turn of phrase is poetic while right to the point. It is typical of all his writings. His work remains relevant and guides us as we tackle the issues of the world requiring attention. On behalf of the system community and beyond, we dedicate the first volume edited by Janet of Flinders University, Adelaide, Australia, as well as all the volumes to follow, to honor the Churchman Legacy. We invite prospective authors to submit their proposal to us and help to make this celebration vibrant. You can also contact the publisher.

We would like to ensure a range of authors is represented in the proposed volume and that many interpretations are included. In the spirit of collaboration we would like to ensure that the contributions that are drawn together are presented in an egalitarian manner and that anyone contributing will be attributed collectively[1].

The starting point for this volume is the challenge posed by Pierre (2000) in his edited volume on the potential for regional governance based on people working together across the boundaries of nation state governments[2].

The catastrophic tsunami that impacted directly on East African states, Maldives, Indonesia, Syri Lanka, Thailand, India and the many international tourists has created an international governance challenge. Early warning systems for prevention, disaster relief and reconstruction requires an international response by networks of concerned people and also co-ordination by the United Nations.

The desire for social and environmental justice is the motivation for shared rights and responsibility, rather than state power. Throughout history

[1] Any royalties for this volume will go to Streetwise, a project for homeless children in South Africa. The Pretoria Streetwise Branch started in 1988 as a Participatory Action Project with colleagues and many dedicated volunteers and supporters.

[2] The European Union is discussed in terms of its potential for another way of governance that is 'de-centred' and 'de-institutionalised' and based on a sense of shared policy direction, rather than on state power.

people have networked, but the scope of networking has expanded in a digital and global world. Where there is a will, people can network across organizations, they can advocate and bring pressure to bear through social movements. Thus ordinary people can play a role in making and re-making futures outside organizational contexts (McIntyre-Mills 2000, 2003). Nevertheless existing structures place enormous limitations on life chances and cannot be dismissed without the vision, creativity and determination of individuals and diverse groups of concerned people.

Acknowledgements

My thanks to John P. van Gigch, whose friendship made this series possible.

My thanks to all those who contributed to this volume and to those who have enthusiastically offered contributions to forthcoming volumes.

My thanks to Gloria Churchman for sharing the photographs of West Churchman.

My thanks to Kerrin Croft, Flinders Institute of Public Policy and Management whose intelligence, good humour and patience contributed to the technical editing process.

My thanks to the generosity of Susanne Bagnato who transformed the conceptual diagrams into graphics for this series and for my previous volume, *Critical Systemic Praxis for Social and Environmental Justice*.

Chapter 1

INTRODUCTION TO VOLUME 1: THE CONTRIBUTION OF WEST CHURCHMAN TO SUSTAINABLE GOVERNANCE AND INTERNATIONAL RELATIONS

JANET MCINTYRE-MILLS
Flinders Institute of Public Policy and Management, Volume Editor.

The central questions are: What is good governance[3]? How can governance support a sustainable future? What does this mean and what are the issues associated with striving for accountability to self, others (including sentient beings) and the environment?

The diverse knowledges and experiences of current and future generations of stakeholders need to be considered in making policy decisions. Testing out ideas is a task for all those who are to be at the receiving end of a decision, but so is advocacy and accountability to those who are voiceless. This is the basis of a reworked enlightenment and democracy that draws on the insights of many.

1.1 INTRODUCTION

This book provides many conceptual tools that can enhance our ability to design *inquiring systems* that are more mindful and more conscious of both our rights and responsibilities to humanity and to all sentient beings. At a certain point life becomes self-aware. As conscious caretakers we have a role of achieving balance between the freedoms of individuals and the need for collective responsibility for social and environmental justice that uses consciousness to address sustainability.

[3] Rhodes (in Pierre 2000) gives seven different definitions of governance, namely: corporate governance, New Public Sector Management, good governance, international interdependence, socio-cybernetic system, New Political Economy, networks.

In this collected works and in the volumes that follow, we explore the implications of Churchman's (and other related) ideas for better governance, risk management and systemic accountability by exploring the boundaries of our thinking and asking questions.

The purpose of this volume is to argue that ideally a revised enlightenment and democracy needs to be seen not so much as a static universal law, but as a dynamic structure and process for balancing the eternal paradox that:

On the one hand: openness to debate and to other ideas and possibilities is the basis for both the ideal form of revised enlightenment process of testing and for ideal forms of democracy and

On the other hand, for openness to occur there has to be some trust that voicing new ideas will not lead to subtle or overt marginalisation of oneself or one's associates.

The West and the East face the challenge of preserving this openness and trust and redressing the imbalances in wealth and power caused by centuries of brutal colonization, modernization and globalization that is largely based on the single bottom line (see Elkington's 1997 critique) of profits in competitive markets, that support hegemony, rather than on a multiple dynamic awareness of socio-cultural, political, economic and environmental factors that (when considered together) support a sustainable future.

Philosophy and communication to span the divides across Foundationalism, non-Foundationalism and anti-Foundationalism (Romm 2004, pers. comm., Romm, 2001a, b, 2002) is needed as a means to share thinking and to co-create responses. Communication based on respect (in the sense explored by Derrida and Habermas, 2003) is perhaps more important than ever as the basis for governance, international relations, peace studies and socio-environmental policy. The paradox, namely that openness and sharing ideas (the idealist version of democracy and a transformed Enlightenment approach) requires trust. Trust in turn requires openness and sharing ideas. The leap of faith required to address this paradox and to develop resonance and trust through respectful communication is where spirituality, religions and the sciences touch fingers across the divide across subjective, objective and intersubjective experiences of the world.

In 1971 Churchman wrote *The Design of Inquiring Systems* and developed tools for theoretical and methodological literacy. The five basic systems are as follows:

- Leibnizean logic based on rationalism and fact nets
- Lockean empiricism based on data
- Kantian idealism based on universal law
- Hegelian dialectic based on thesis, antithesis and synthesis

- Singerian[4] pragmatism based on contextual 'sweeping in' and unfolding – an application of the dialectic, which he tended to apply. But understanding the other approaches is useful for bridging the divides across our understanding of the world.

In 1979 he wrote *The Systems Approach and its Enemies* and developed reflexive thinking based on the dialectic of unfolding values and 'sweeping in' multiple variables. We need to recognize 'the enemies' within and without- these include: 'politics, religion, morality and aesthetics', to use West Churchman's (1979: 23) concepts, so as to examine the human potential for hubris, which means thinking we can behave like a God who has all the answers for other people. Instead we have to accept that we all have potential to make contributions.

West Churchman stresses that we can never find a total or absolute solution and for that matter we are unlikely to ever know the full extent of any system, but nevertheless we need to try to appreciate other ways of seeing (Vickers, 1983). The greater the use of critical reflection and participatory design for governance and international relations, the better the local and international policy outcomes will be.

In 1982 he wrote *Thought and Wisdom* in which he suggests that the system of interest should in fact start with 'unfolding' values and 'sweeping in' (in the sense used by Churchman, 1979 a, b, 1982 who drew on Singer) multiple variables (social, cultural, economic, political and environmental) within the context of the inquiry. Dialogue and conversation to achieve resonance and radiance of shared meaning was a focus for this work. Critical reflection is the basis for testing out ideas and finding out how they are perceived, because the closest we can get to truth is through dialogue (McIntyre-Mills, 2000, 2003, 2004).

1.2 DIALECTICAL PRAXIS

Our thinking and practice needs to be dialectical or open to many ideas. For this to occur we argue that we need to recognize 'the enemies' within - these include: "politics, religion, morality and aesthetics", to use West Churchman's (1979a, page 23) concepts, so as to examine the human potential for hubris, which means thinking we can behave like a god who has all the answers. Instead we have to accept that we all have potential to make

[4] Edgar Singer's work provided a source of ideas for West Churchman who drew on his notion of unfolding ideas and sweeping in the factors that shape our thinking and practice. Peter Singer's (2002) work: One World: the ethics of globalization, is influential in shaping the work of a forthcoming volume: Working and Reworking the Conceptual and Geographical Boundaries of International Relations and Governance. McIntyre-Mills forthcoming (2005).

contributions and that we are all gods, in so far as we have some potential to construct our futures with others. We can construct our futures through dialogue and praxis (McIntyre-Mills 2000, Rhodes (in Pierre ed. 2000, Rhodes 2003). Governments competing with globalization have argued for the inevitability of markets. But people have diverse ideas and narratives about the nature of the state, the nation, market and society.

Our starting point is that people can make choices within social, cultural, political economic and environmental contexts. Shaping life chances is never unidirectional or linear. Complexity can be recognized as we work and re-work the relationships across self-other and the environment. Improving governance requires accountable thinking and praxis to ensure policy decisions that are mindful of structures and processes within specific contexts. We need to be able to make generalizations, based on statistical (nomothetic) patterns, but we also need 'thick description' (Geertz 1975) (ideographic details) based on rich case studies and stories.

West Churchman argues we can never find a total or absolute solution based on a total understanding of the system, but we do need to try to see wider horizons. Systems thinking, like all thinking, can be used creatively or destructively if it leads to the hubris of certainty, hence our approach is critical, reflexive and thus systemic. Space, time, mindfulness, willfulness and human creativity provide contextual variations to policy decisions. Critical reflection can help if we look both inwards and outwards. The appreciation of a system requires understanding or at least being prepared to strive for a glimpse of the expanse of the system. Current thinking that is limited within the enlightenment framework can cast a shadow of hegemony.

This work on governance, democracy and international relations explores and creates opportunities for building the capacity of people to think and practice systemically, in terms of open systems. International relations and governance need to be based on openness and transparency. Closure in terms of conceptual and geographical boundaries can lead to entropy (see Flood and Carson, 1997) that means the dissipation of energy. This is as important for policy makers and administrators in weak states (as suggested by Fukuyama, 2004), as it is for policy makers and administrators in strong states. We try to identify ways for both so-called strong states and weak states to look at 'the enemies within'. This requires looking inwards, not merely outwards, in order to 'deal with the other' as nations or civilizations or to impose communitarian solutions (Etzioni, 2004, Huntington, 1996).

Our system of interest starts with 'unfolding' values and 'sweeping in' (in the sense used by Churchman, 1979a, b, 1982 who drew on Singer) multiple variables (social, cultural, economic, political and environmental) within the context of the inquiry. Critical reflection is the basis for testing out ideas, because the closest we can get to truth is through dialogue (McIntyre-Mills, 2000, 2003).

This volume draws on C. West Churchman to re-work 'a design of inquiring systems' to promote systemic international relations and policy for socio-cultural and environmental justice. In so doing it celebrates the contribution of West Churchman by working with some of his ideas developed in *The Design of Inquiring Systems* (1971), *The Systems Approach and its Enemies* (1979) and *Thought and Wisdom* (1982). These works can help rescue the Enlightenment from itself through mindfulness. He stressed that dialogue with many stakeholders helps greater self-reflection and self-knowledge. Limited consciousness can make us focus on just the waves (continuums) or just the particles (categories) of energy, to use a well worn natural science analogy. We need to be aware of both in physics and even more so as policy makers and managers.

Rescuing the Enlightenment agenda from its failings through open debate to try to achieve greater transparency (and awareness) ought to be the ideal for democracy, based on working on our conceptual and geographical boundaries. Moving from pragmatism to idealism is not difficult once we appreciate the continuums or (at the very least) the feedback loop or 'boomerang affect' (Beck, 1992) and realize that democracy needs be reconsidered as being about human and environmental rights- not merely about bounded or categorical citizenship and nationalism. The UN Aarhus convention of 2001 makes a useful step according to Florini (2003) towards vigilance that spans borders of any one state, nation or region. This is a far cry from the present situation of war, cover-ups and toadying to the powerful - nationally and internationally - within the so-called coalition of the willing warring groups.

We need to do more than merely balance the state, market and society. This can only occur if people redraw conceptual and geographical boundaries of identity and politics[5].

[5] This requires a capacity to think critically and systemically. This goes beyond the mere 'tolerance' of diversity in a societal sense (see Derrida in his discussion with Borradori, 2003 entitled: Philosophy in a Time of Terror). Derrida (in conversation with Borradori, 2003) argues that international relations need to be based on an engaged and respectful dialogue and 'hospitality', not merely cold tolerance, because this is insufficient for building rapport that is vital for 'generative communication' (see Banathy 2000, Freire 1982). Singer (2002: 192-197 taking issue with Rawls, 1999) argues for a sense of responsibility for individuals, not merely social tolerance of groups. Peoples living in developed nations need to take responsibility for what war, poverty and pollution does to individuals and they cannot say that it is the responsibility only of elected leaders. Power is recognized as being central to (what passes for) knowledge when decisions are taken that affect ordinary and less powerful people. Ordinary people need to take on their rights and responsibilities through participatory governance.

1.3 SYSTEMIC APPROACH

A sense of self-assurance, self importance and then arrogance based on knowledge rooted in power can result in physical, structural and emotional violence to others (Foucault & Gordon, 1980, Foucault 1967, Foucault, 1979) that will impact on internal politics and international politics in the long term. West Churchman's (1971) *Design of Inquiring Systems* together with his later works on systemic thinking (1979 a, b) and conversation (*Thought and Wisdom* 1982) contribute to building our capacity to explore the shortcomings of the Enlightenment.

Most human beings have values, whether they are aware of them or not. 'Religious, political, moral and aesthetic' values can limit our vision. Churchman calls these values 'the enemies within'[6]. Seeing the enemy as 'the other', without seeing 'the enemy within' is the challenge posed by West Churchman. It can lead to worsening international relations and environmental management. Psychological and socio-cultural stereotyping can be based on misunderstanding and projections of our own 'shadow', to use a Jungian concept[7]. Occidentalism is the projection of the idea of fundamentalist West ruled by greed and the market. Orientalism is the projection of the fundamentalist east ruled by religion (see Baruma and Margalit, 2004). 'Unfolding' our own values and 'sweeping in' (West Churchman, 1982) the social, cultural, political and environmental context can be helpful as a means to enhance governance capacity.

We need to strive to co-create (see Reason and Bradbury, 2000) frameworks with others. We work with, rather than within West Churchman's Design for *Inquiring Systems*[8], in order to 'address the enemies within', to borrow the poetic phrase from 'The Trojan Women', a play by Euripides. West Churchman like many other Western scholars was a white, middle class American male. My own experience is also shaped by my place of birth and life chances as a white, middle class Australian female, born in Bulawayo Zimbabwe[9]. I strive systemically (within the limits of my experiences in a range of contexts) to include experiential learning and the radical working of the hyphens across self, other and the environment[10].

6 In Thought and Wisdom, West Churchman refers to religion, morality, politics, and aesthetics as 'the enemies within'.

7 For a discussion on the shadow within, see for example Jung, C.J., 1972, Translated by Hull, R., Mandala Symbolism from the Collected Works of Jung, C.G. Bollingen Series, Princeton University Press.

8 The contributions from colleagues who are critical and systemic thinkers such as Bob Flood, Norma Romm, Gerald Midgley, and John van Gigch have helped with this task.

9 See McIntyre- Mills, 2003, for a discussion on identity and meaning

10 An attempt is made to discuss the implications of working across self-other and the environment in more detail in a manuscript entitled: Working and reworking the

The potential value of West Churchman's work is that he addresses and combines multiple frameworks, in order to create a more systemic or ecological approach by working with frameworks, rather than within any one Enlightenment framework. Nevertheless he is aware of the limitations of any framework and strives to both recognize and address 'the enemies within' as they are 'the vipers within the nest'[11] that can paradoxically liberate the paradigm from its own shortcomings. The potential danger of his work is that it is limited to Western Enlightenment thinking. His *Design of Inquiring Systems* draws on Leibnizean logic, Lockean notions of representation, Kantian idealism, Hegelian dialectics and the pragmatism of Edgar Singer. A greater capacity to work beyond conceptual boundaries is needed, in order to enhance governance and international relations. Bateson (1972) in his work "*Steps to an Ecology of Mind: A Revolutionary Approach to Man's (sic) understanding of himself)* stressed that sustainable futures rested on an ability to think beyond the limits of one framework. Being able to move from Level 1 thinking within one paradigm to making comparisons across paradigms, at Level 2 and then being able 'to leap beyond' (see Banathy 1996,2000) the confines of our learning at Level 3, is a precursor to better governance and international relations.

Limited frameworks can and do lead to hegemonic thinking and practice, as history demonstrates. Questioning and openness (based on both hospitable and respectful conversation) holds the key to sustainable governance and international relations (Borradori in conversation with Habermas and Derrida, 2003). Discursive and respectful communication is needed for sustainable governance and international relations. We need to make connections to heal the divides within and across the self, the other and the environment. This work draws on the work of West Churchman and the related works of other thinkers and practitioners[12].

According to a Rawlsian utopia (1999), 'reasonable' and 'decent' people are able to think beyond the boundaries of the nation state in terms of human rights. In his utopia immigration issues, the creation of nuclear weapons,

---

conceptual and geographical boundaries of governance and international relations. In this chapter I work with the ideas of Donna Haraway and Michelle Fine, the cultural lessons of Bhaba, the transformative potential of Deleuze and Guattari's fixed and fluid notions of identity, the dialectical interpretations of Derrida and Habermas (in conversation with Borradori 2003) to take into account race, gender, identity and politics. Other key ideas are drawn from working with The Law of Peoples by Rawls (1999) and reference to Peter Singer's (2002) One World that encompasses both human and animal rights and Beck (1992, 1994) who links social and environmental justice.

11 Gerald Midgley used this phrase over lunch at the C.West Churchman workshop in Adelaide, October 2004.

12 Such as Fine (1992), Bhabha (1994), Deleuze and Guattari (Bogue 1989, Buchanan, and Colebrook. 2000), Haraway (1991) and Beck (1992, 1994).

catastrophic natural disasters and waging war simply would not occur. But practical political and social philosophy (after September 11th 2002 and the tsunami disaster of 26th of December 2004) needs to work with the reality of terror and hegemony and the possible dangers of trying to understand it in terms of limited social, cultural, political, economic or environmental frameworks.

1.4 CONSCIOUSNESS IS ABOUT CONNECTIVITY

Greenfield (2000) stresses that consciousness is about connectivity across multiple neurons within the highly plastic human brain. If consciousness is emergent from the complex interplay of matter, body and brain within our environment, then perhaps we could argue that *'The Private Life of The Brain'* is not so private after all. In some ways all consciousness is individual and unique, but consciousness is a product of an open, not a closed system of perception. Consciousness is a recursive experience that no two people can experience in exactly the same way argues Greenfield (2000). But paradoxically, it is only by both our shared understanding of language, experiences and our willingness to converse (with warmth and respect) that we can make meaning with one another.

Our individuality and perceptions are a product of personal experiences, life chances and our emotions. To the extent that we can be mindful of the many factors that shape us, we can think about our thinking and design options for better governance for the future. Our emotions are part of what make up our experience. We can be ourselves when we are conscious and aware of our emotions. Emotions are part of sentience or feeling. Feeling is not the preserve of the human animal. Consciousness is a continuum, according to Greenfield (2000) from inorganic to organic life.

> "Unlike machines, young children and nonhuman animals have very poor cognitive abilities, yet we attribute to them sentience…a portfolio of feelings…"(Greenfield, 2000: 32)

Haraway (1991) however argues that we could extend our ability to think through using technology, and thus we are in the process of extending the continuum of consciousness to include the human– machine interface, namely the cyborg. Our values will determine the extent to which social and environmental justice is achieved.

1.5 AN OVERVIEW OF CHAPTERS

The two introductory papers by John van Gigch and Ken Bausch talk specifically about 'the enemies within' and their implications for our thinking. They make us human and my paper stresses that these need to be owned and acknowledged and not merely projected onto others.

John van Gigch addresses *The Systems Approach And Its Enemies*, one of the most important works written by C.W. Churchman. At the same time, we - the contributors to this volume re-interpret CWC's ideas to search for the way to rescue democracy from its failings. Hegemony is a characteristic of western democracy of which we are not proud. Current events Post 9/11 underline this. We encounter countervailing forces (enemies and adversaries) that are bent in disrupting our efforts and contradict our premises. We will expand CWC's list of enemies to show that the possibility of conflict(s) is ever present and must be anticipated. We suggest reaching consensus on the basis of the use of the logic of an inquiring system (a CWC original idea) where agreement instead of conflict can be developed.

Most importantly West Churchman stresses that questions are as important as answers to a revised Enlightenment agenda. It is certainly vital for rescuing democracy.

It is these characteristics that enable his thinking to stand the test of time. Questioning is important for working the boundaries of our own maps of reality. Questioning helps us to consider our own ideas and to rework them in the light of ideas from others. Drawing a line between them and us needs to be considered in terms of differences that are couched in terms of categories such as age, gender, language, nationality or religion. Questioning or unfolding our own values and 'sweeping in' a wide range of social, cultural, political, economic and environmental considerations can enhance understanding others and ourselves. Whether we engage with differences as educators, in the course of business or as managers or policy makers in the public or private sector we need to find a way to strive for better understanding and for compassion.

This revised Enlightenment agenda provides a basis for a democracy that does not "cast a shadow of hegemony over the powerless", because it is "first and foremost open to all - irrespective of age, gender, level of education, income, culture or any other category that can be used to justify inclusion or exclusion or competition for resources. Most importantly Churchman is mindful of our rights and our responsibilities - not merely as citizens, but as compassionate caretakers of the environment of all living things and all sentient beings".

Ian Mitroff discusses practical ideas to make us more accountable in all walks of life.

Churchman talks of ethics, beauty and compassionate praxis. He is a practical man with suggestions that can make a difference if they are applied in a world that is greatly in need of better decision-making (see David Mathews's chapter) at a local, national and international level.

To our mind, his more lasting legacy will relate to his interest in important the philosophical issues that have been neglected. We are referring to the underlying morality (or lack thereof) of management, social and international affairs. The Enron collapses of the world would not have occurred if the business world were less based of unbridled greed, individualism and the destructive mentality of 'survival of the fittest.' If the business world were to embrace a design that is inclusive and based on many different stakeholders, it would be able to do better risk management (see van Gigch, Christakis, Ferris, McIntyre-Mills and Carugati's chapters).

We are also referring to the proclivity of Western nations to settle conflicts through violence, intimidation and bullying. Van Gigch stresses that Churchman was interested in the poverty of the disadvantaged and of the dispossessed wherever they may live, the inequality of social classes and, in general the morality of our actions toward each other and other fellow human beings (See John van Gigch, Ackoff and Gharajedaghi).

Decision-making based on looking inwards at ourselves and exploring the scope of problems through good communication are vital for better governance (see Bausch, McIntyre-Mills, Morgan and Romm).

1.6 ADDRESSING COMPLEXITY

The world comprises many different life chances and economies. For some it is eking out a subsistence existence and coming to terms with giving up a nomadic existence, as land is lost. For some it is fleeing war and living in camps/or incarcerated with or without refugee status. For some it is surviving the trauma and loss caused by natural disasters. For some it is entering into the world of commercial agriculture and trying to maintain a foothold in the market based economy that is controlled by the WTO. For others it is entering the postindustrial age of being knowledge workers. Life chances vary within and across national boundaries.

The reality of the boundaries caused by lack of education, citizenship and discrimination cannot be denied in a world where the global markets know few limits and where information technology can empower the literate, the numerate and those with access to electricity or at the very least access to mobile phones.

To talk of the "*borderless information economy*" (Long, 2002) makes sense for some, for others it is still a dream. The challenge for good governance and international relations is not only to try to redress the divides between those who can access digital economy and those who cannot. But to consider the challenge of different ways of knowing, different ways of being and earning a living that are now presented simultaneously as options. Globalization has changed the speed with which information flows and it has changed our sense of who we are in space and time. Some have increasing access to information through small radios, battery operated, and televisions or through mobile phones. Long (2002, 6), an advocate of borderless information, talks about the advantages to coffee farmers on the Ivory Coast who can use mobile phones to decide when to market their commodities. This is advantageous to them. Similarly he talks about Sri Lankan tea growers who learned a new and more efficient method to dry their tea by using the Internet. This information was translated and shared through local radio. Nanadorn (2004, pers. comm.) reporting on his doctoral research into the role of communication technology in Thailand rural communities talks of the potential and also the challenges. On the one hand, local teachers receive Internet training, but they do not have phone lines, so they are unable to practice their skills.

The challenge in the future will be to provide affordable phone lines and power supply to enable people to access resources, not only water and sanitation (Long, 2002). These tools for learning and communication whilst useful for supporting and deepening democracy are insufficient for good governance. The Internet does not prevent totalitarianism, but it can aid transparency even if its participation potential remains limited. Florini (2003) talks optimistically about the new options e-governance provides. Conceptually however information is not the same as knowledge as Long (2002) argues in his paper *Beyond Traditional Boundaries: Government in the Information Age*. If it were there would be more cause for optimism. Greenfield has argued that the challenge is to work with difference and different ways of knowing (Thinker in Residence in Adelaide, Australia, public lecture 2004), in order to enhance governance. We also need to add international relations. Making links across bits of data, creating information and then accepting that information will be perceived differently (because we are human) is a useful starting point for good governance and good international relations[13]. For ordinary citizens it means learning to live with difference and realising the value of diversity. For public policy makers at the local, national and international level it means learning to think about thinking and translating this into practice.

[13] Different ways of knowing and working towards shared knowledge can be about finding ways to work with different domains. Habermas and Derrida (in Borradori, 2003) talk of the challenges of communicating and making knowledge by thinking and practising in ways that are underpinned by philosophical rigour.

Unfortunately we need to do more than respond to "*the borderless nature of information*" (Long, 2002: 8). Firstly, borders do exist for information flow - they are poverty and lack of literacy and numeracy and powerlessness. Secondly information and knowledge cannot be conflated (as Long suggests on page 7). The boundedness of knowledge is an issue for governance. It requires more than so-called knowledge management to enhance connections. In fact the term suggests that knowledge and information are the same and that directing the flow of the right information for the task is the challenge. Secondly, we need to acknowledge that many kinds of knowledge exist and finding ways to communicate across knowledge domains is the challenge. Habermas (1984, 2003 in conversation with Borradori) suggests respectful communication is the way forward. Derrida (in conversation with Borradori, 2003) suggests that not only do we need respectful communication, in order to co-create shared domains of understanding, we also need to accept that people will see things differently. Spaces for conceptual difference need to be respected to the extent that they do not undermine the rights of others.

1.7 REWORKING DEMOCRACY AND THE ENLIGHTENMENT

Our starting point is the need for better communication- not just as a means to an end, namely closer representation of people and their ideas, but also because communication is the very essence of life. The book provides examples and discusses the learnings for re-working governance. Examples and their implications for social and environmental justice are discussed. Conversation that is discursive is essential for democracy, governance and better international relations.

> "Democracy, in its most basic sense of majority decision making, requires that those who decide be sufficiently alike that they will respect the will of the majority. Global- level decisions will inevitably have a highly restricted agenda, set by what the majority of the richest nations will tolerate, and a very reduced role for the world's publics. The difficulties of global democracy should make us pause when considering the rhetoric about democracy at the national level, for that too, despite the belief that nations are relatively homogenous political communities, is subject to similar limitations as to both agenda and participation. The question is whether one needs to rethink some of the assumptions of democratic theory in order to find ways to widen the scope of accountable government by consent." (Hist in Pierre 2000, 17).

To be classified as a non-citizen makes one aware of the benefits of citizenship. Being an asylum seeker in 2003-4 in Australia or a black South African in 1988 would make one realise that human dignity is given only to citizens, not to non-citizens. The nation state protects only some not others. Perhaps democratic rights based on rights and responsibilities of citizens need to be expanded to all human being who can be free to the extent that they do not undermine the rights of others (including sentient beings).

The Enlightenment thinkers such as Locke and Kant can be perceived by some as racists, because of the support of slavery, in the case of the former, or the belief in their own superiority, in the case of the latter (Bernasconi, 2003). We need to see that liberal thinking has coiled within it a serpent, which is appropriate, if the Enlightenment is to be rescued or wrested from itself.

For the dialectic to become a means of liberation, it needed to be reworked first by Marx and the members of the Frankfurt School who emphasized the systemic importance of both culture and economics in shaping life chances. It now needs to be reworked to add the environment, identity and consciousness.

Rawls (1999: 61, 71) talks of 'tolerance' for diverse societies, provided they are based on consultation and are thus '*decent hierarchical*' societies, even if they are not democratic in the ideal sense that he uses the word[14]. His ideal utopia only embraces societies in good standing who accept the principles of decency, defined in terms of health care, equality of opportunity in education and training, decent distribution of wealth, long term opportunities for work and support (as a last resort from government) and public financing of fair and open electoral system (Rawls, 1999: 50).

These decent hierarchical societies are prepared in his scheme to accept the principles of human rights; because they are defined in minimalist terms and they are not equivalent to citizenship rights in so-called liberal democracies. These human rights according to his schema include:

> "1. Their fulfillment is a necessary condition of the decency of a society's political institutions and of its legal order. 2. Their fulfillment

[14] Rawls (1999, 73) points out "Of course, a decent hierarchical society has never had the concept of one person, one vote, which is associated with a liberal democratic tradition of thought that is foreign to it, and perhaps would think as Hegel did that such as idea mistakenly expresses an individualistic idea that each person, as an atomistic unit, has the basic right to participate equally in political deliberation". Rawls cites the following from Hegel's Philosophy of Right (1821) as follows: "...The citizens come to the scene as isolated atoms, and the electoral assemblies as unordered inorganic aggregates, the people as a whole are dissolved into a heap. This is a form in which the community should never have appeared at all in undertaking any enterprise; it is a form most unworthy of the community and most in contradiction with its concepts as a spiritual order. Age and property are qualities affecting only the individual himself, not characteristically constituting his worth in the civil order".

is sufficient to exclude justified and forceful intervention by other peoples, for example, by diplomatic and economic sanctions, or in grave cases by military force. 3. They set a limit to the pluralism among peoples." (Rawls, 1999: 80).

In his revised work Rawls (1999: 30-34) argues that decisions should not always be made on the basis of even a 'thick veil of ignorance', because:

> "Whether our conjecture is borne out will depend on whether you and I, here and now, can, on due reflection, endorse the principles adopted. Even if the conjecture is intuitively plausible, there are different ways of interpreting the reasonable and the rational, and of specifying restrictions on reasons and explaining the primary goods. There is no a priori guarantee that we have matters right" (Rawls, 1999: 30).

The notion of a 'veil of ignorance' is based on the biblical notion that a law should be conceptualised in such as way that the makers of the law would be prepared to abide by it and he also argues that all those who make the law should be party to the process. This is the basis of liberal law. Rawls argues that it could lead to problems, because it does not provide space for decent hierarchical peoples, who have decent, but hierarchical consultation structures. He uses an example of an ideal Muslim society that has different ideas. His argument is a useful step in rescuing the Enlightenment from itself. The next step needs to be to extend his notion of tolerance to include a sense of hospitality to all (see Derrida in Borradori 2003) and to expand the notion of human rights to notions of social and environmental justice, because these policies are both idealistic and pragmatic. Thus they are sustainable. Also his work is not extended to individuals who suffer at the hands of nation states and his work also denies the need for a welfare net, other than an open opportunity and basic health care. He does not link the environment with sustainable development. *A Law of Peoples* (Rawls, 1999), according to Singer (2002) needs to be reworked contextually in such a way that discursive democracy takes into account what it means in the everyday life of ordinary ethical people who live in a world where nationally defined rights and responsibilities need to be reshaped to take into account international human rights and responsibilities of individuals[15].

In a volume 3 of this Book Series, a model of sustainable, systemic governance and international relations is developed through working and reworking the boundaries (both conceptual and geographical) across:

[15] I am grateful for a comment by Shaun Young, 2004 , pers comm. agreeing that Rawls is reshaping his argument in line with changed ideas of governance and boundaries.

1. Community governance
2. Third Way governance partnerships across public, private and volunteer sectors and developing community involvement
3. Corporate governance
4. Federal European Union models for regional and international governance. 1-4 are all interconnected and complementary.

We stress that diversity and creativity is fostered to the extent that it does not undermine the diversity of others. The central argument is that those at the receiving end of a decision need to be part of the process. British Third Way approaches pertain to citizenship rights. This means finding a way that is in between complete reliance on the state and complete reliance on the individual to address needs at the national level such as challenges of health, education and employment. This model emphasizes democratic solutions within the boundaries of the state. European federalism for the European Union emphasizes a balance between local and wider governance (national and international complementarity). It potentially incorporates United Nations policy and human rights. This model expands governance to the regional and international context. Some of the essential characteristics of good governance are:

- Openness to many people and many ideas,
- Representation of all the stakeholders in decision making,
- Advocacy for the voiceless, sentient beings based on compassionate observation, signaling and/or sign language[16],
- Listening to and responding to needs,
- Respectful communication,
- Working across organisations and across sectors (health, education, employment, for example)
- Forming responsive team approaches in response to issues,
- Holding in mind many issues, not just single issues.

One of the techniques for achieving better governance at the local, national and international level is expanding the concepts of accounting and accountability to incorporate: social, cultural, political, economic and environmental indicators. The fluid structure and process for governance involves working, learning, doing and communicating with:

- People within the community and civil representatives
- Elected representatives
- Corporate structures to link public, private and non government organisations

[16] For a discussion on evolution, making meaning and communicating shared meaning see Bausch, K. 2000, Emerging Consensus in Social Systems Theory and Banathy, B., 2000. Guided Evolution of Society: A Systems View.

It also involves continuous review with all the stakeholders. We need to build social, economic and environmental indicators into planning. The vision, mission, planning, implementation and monitoring need to take many factors into account, in order to do good accounting and accountability for governance.

The goal is to achieve a well-rounded approach to governance that addresses multiple perspectives and voices on human rights. Multiple voices and advocacy for the voiceless are required to developa 'design of inquiring systems' that strives to consider – not on the basis of experts - but on the basis of the ideas, experiences and advocacy for all those who are at the receiving end of a decision. Power and knowledge in the Foucaultian sense is no longer the preserve of the expert in this critical and systemic approach. Thus this volume (and the ones that follow) strives to extend the potential of the 'Enlightenment' and democracy to think beyond citizenship rights and to address the rights of non-citizens and the powerless, including sentient beings.[17]

Human beings need to move from the hubris of certainty that is tied to a culture or to a favoured scientific paradigm to embrace an appreciation of diversity. The will to be compassionate can be addressed through mindfulness, but also through our desire for a sustainable future for those we care about.

Perhaps this openness will help to rescue democracy and the Enlightenment from hegemony. The potential of democracy can be restored through realizing the value of diversity for creativity. Diversity should be valued to the extent that it does not undermine the rights of others.

Knowledge systems are nested and form dimensions, rather than meta hierarchies in a hegemonic sense. The volume of papers will make the case that good governance is about thinking and practice that can lead to a better balance of social, cultural, political, economic and environmental concerns to ensure a sustainable future for ourselves and for future generations. We also need to consider transnational human rights, not only nationalist rights of citizens. The whole system is the context of social and environmental justice. A revised democracy and Enlightenment needs to redress the hegemony of power and exclusion.

[17] This is developed in *Working and Re-Working the Geographical and Conceptual Boundaries of International Relations and Governance*, Volume 3 (also see Volume 2 for the chapter on Molar and Molecular Identity and Politics).

1.8 CONCLUSION

We make a plea for constantly questioning and re-working the conceptual and geographical boundaries of governance and international relations. We also make a plea for re-working democracy and rescuing the Enlightenment from itself.

It was fortuitous that whilst preparing volume 1 of the C. West Churchman Series and Related Works that I experienced Euripides' play the '*Trojan Women*' (November 2004, Adelaide, Dunstan Playhouse). The central theme and quotation from this Greek tragedy is: '*The enemies within*'. The non-combatant women despair the trickery of the Greeks. The Trojan horse quite literally contained their enemy. But at a deeper level Euripides explores the need for men and woman to embrace the serpent and to realize that all of us contain 'the enemy within'. This realization is the basis for systemic thinking and practice.

The women are raped; their husbands, children and city are destroyed. They are left with nothing, other than their ability to make meaning out of the disaster. Compassion for the ravaged Trojan women is balanced by the shock of their violence towards Helen.

Was Helen merely a pawn or the cause of the war? Did her beauty and faithlessness to her Greek homeland cause the war? Were there other more pragmatic reasons, such as envy and materialism? Was Helen merely the excuse for the ongoing rivalry between the Greeks and the Trojans? Euripides suggests through Helen's voice that her husband Menelaus has merely used her betrayal as an excuse for waging war.

In the wreckage of Troy the enslaved women blame Helen for their misery and wish to kill her. They hiss like serpents when she is brought within earshot. Suddenly their misery mutates to violence. Options such as pragmatism and survival are explored by the women; as are idealism and heroic self-sacrifice.

The real tragedy for the members of the audience – who playing gods – are able to see the blindness of the women to the 'enemies within' which leads them to blame the other. It is always easier to recognize faults in others and less easy to see our own blindness to the violence of current events.

It takes the wisdom and courage of the blind seer Cassandra to see that the conquering Greeks are their own worst enemy. But it takes more than Cassandra's blind sightedness to see that the Trojan's are not without fault. Violence must not lead to more violence. It can only be short circuited by embracing the serpent, or the dialectic. West Churchman explores the process of 'unfolding values' and 'sweeping in' the implications of social, cultural, political and economic circumstances. Democracy today needs to be open to the ideas of all participants and inclusive to the extent that it does not lead to decisions that impact in a negative way on others or the environment.

REFERENCES

Banathy, B., 1996, *Designing Social Systems in a Changing World*, Plenum, New York.

Banathy, B., 2000, *Guided Evolution of Society: A systems view*. Kluwer/Plenum, London.

Bateson, G., 1972, *Steps to an Ecology of Mind: A revolutionary approach to man's understanding of himself*, Ballantine, New York.

Beck, U., 1992, *Risk Society Towards a New Modernity*, Sage, London.

Beck, U. 1998, *Democracy without enemies*. Cambridge. Polity.

Bernasconi, R., 2003, Will the real Kant please stand up? The challenge of enlightenment racism to the study of the history of philosophy, *Radical Philosophy*. 117:13-22.

Bevir, M., Rhodes, RAW and Weller, P. 2003, Traditions of governance. History and diversity. *Public Administration. An international quarterly*. Blackwell. Vol.81.1. 1-19, 191-209.

Bhabha, H., 1994, *The Location of Culture*, Routledge, London.

Bogue, R., 1989, *Deleuze and Guattari*, Routledge, London.

Buchanan, I. and Colebrook, C., 2000, *Deleuze and Feminist Theory*, Edinburgh University Press.

Churchman, C. West, 1971, *The Design of Inquiring Systems. Basic concepts of systems and organization.* Basic Books, New York.

Churchman, C. West, 1979a, *The Systems Approach*, Delta, New York.

Churchman, C. West, 1979b, *The Systems Approach and its Enemies*, Basic Books, New York.

Churchman, C. West, 1982, *Thought and wisdom*, Intersystems Publications, California.

Denzin & Lincoln, *Handbook of Qualitative Research*, Sage.

Elkington, J., 1997, *Cannibals with Forks*, Capstone. Oxford.

Etzioni, A., 2004, *From Empire to Community. A New Approach to International Relations*, Palgrave. New York.

Fine, Michelle, 1994, Working the hyphens: Reinventing self and other in qualitative research, in: For whom? Qualitative research, representations and social responsibilities, M. Fine, L. Weiss, S. Weseen, and L. Wong, 2000, in: *Handbook of Qualitative Research*, N. Denzin, and Y. Lincoln, eds., Sage, London.

Florini, Ann, 2003, *The Coming Democracy: New rules for running a new world*, Island Press.

Flood, R. and Carson, E., 1998, *Dealing with Complexity: An introduction to the theory and application of systems science*, 2nd ed., Plenum, London.

Foucault, M. and Gordon, C., (eds.) 1980, *Power / Knowledge: Selected interviews and other writings 1972-1977*, Harvester, Brighton.

Foucault, M., 1967, *Madness and Civilization: A history of insanity in the age of reason*, Routlege, London.

Foucault, M., 1979, *Discipline and punishment: the birth of the prison*, Vintage, New York.

Fukuyama, F., 2004, *State Building: Governance and world order in the 21st century*, Cornell University Press, New York, Ithaca.

Fung, A. and Wright, E.O., 2003, Deepening Democracy. Institutional innovations in empowered participatory governance, *The Real Utopias Project 1V*, Verso, London.

Greenfield, S., 2002, *The Private Life of the Brain: Emotions, consciousness and the secret of the self*, John Wiley and Sons, New York.

Habermas, J., 1984, *The Theory of Communicative Action: Reason and the rationalization of society*, Beacon, Boston.

Habermas, J., 1974, *Theory and Practice*, Heinemann, London.

Habermas, J., Derrida, J. and Giovanna, B., 2003, *Philosophy in a Time of Terror Dialogues with Jurgen Habermas and Jacques Derrida interviewed by Borradori, Giovanna,* University of Chicago Press, Chicago.

Haraway, D., 1991, *Simians, Cyborgs, and Women. The Reinvention of Nature*, Free Association Books, London.

Hirst, P. 2000. in Pierre, J. (ed). *Debating Governance*, Oxford, University Press.

Huntington, S.P., 1996, *The Clash of Civilisations and the Remaking of World Order*, Simon and Schuster, New York.

Long, M., 2002, Beyond traditional boundaries: government in the information age, *Australian Journal of Public Administration*, 61(1): 3-12.

Longo, M., 2004, Hostile receptions: dilemmas of democracy: legitimacy and supranational law, *Australian Journal of Politics and History*, 50(2): 211-228.

McIntyre Mills, J., 2003, *Critical Systemic Praxis for Social and Environmental Justice: Participatory policy design and governance for a global age*, Kluwer, London.

Pierre, J. ed. 2000, *Debating governance. Authority, steering and democracy.* Oxford University Press. New York.

Rawls, J., 1999, *The Law of Peoples with the Idea of Public Reason Revisited,* Harvard University Press, London.

Reason, P. and Bradbury, H., 2001, *Handbook of Action Research: Participative inquiry and practice*, Sage, London.

Rhodes, RAW 1997, Understanding governance. Policy networks, governance, reflexivity and accountability. The Cromwell Press.

Singer, P., 2002, *One World: the ethics of globalization,* The Text Publishers, Melbourne.

CONTEXT, MEANING AND CONVERSATION

As I drove to work I thought about how to summarize my learnings from C. West Churchman. Here is an attempt:

- The closest we can get to truth is through dialogue, exploring many points of view, story telling and reflexive art (in all its forms) and sufficient self knowledge that enables us to laugh at ourselves, not only at others.
- Gentle laughter and playfulness is close to spirituality rooted in glimpsing the links across self, other and the environment, in ongoing recursions – that do not make cuts (or decisions) without being mindful of many ideas and issues.
- A new form of democracy needs to be based on seeing ourselves as others see us - what a prospect!
- Democracy is about the never ending need for dialogue and the pursuit of the dialectic.
- Western democracy does not have the answer, only an answer, because closure leads to the limitation of the ongoing pursuit of truth in an ever-changing universe. Churchman's interest is in the area of working with paradoxes through conversation (not in order to solve the paradoxes) but to understand their meaning within context. Paradoxes are portals for problem appreciation.
- It is the pursuit of truth that is important and the realization that recursive loops link humanity with all life. Western democracy and the Enlightenment needs to be revised to take into account the profound sanctity of the web of life (organic and inorganic).
- Questions are as important as answers. Questions are iterative and ongoing. They are the process.
- Answers are only temporary and are always open for re-reconsideration and co-creation based on a design of inquiring systems that calls for multidimensional mindfulness and dialogue with all those who are to be affected by a decision based on logic, empiricism, idealism, the dialectic and pragmatism that together provide the methodological and theoretical literacy to appreciate the way problems are conceptualized by different stakeholders.
- Structures to support democracy enabling the removal of fear so that people can express ideas openly are needed.
- The will to support democracy and a recursive, revised form of inquiry agenda will only occur if a leap of faith is made.
- Knowledge (as we know) does not necessarily change behaviour. Perhaps resonance can help this leap of faith that is required to move

from single line or single voiced accounting and accountability to multiple voiced interactive accounting and accountability that strives for finding "the liberative potential" (to use Gouldner's 1971 term) in all ideas that support the self, the other and the environment.

- Diversity for the sake of diversity needs to be balanced with the need to preserve the sanctity of all life. Here is the paradox and the central challenge for humanity as caretakers; for those who have hegemonic power and those without power, who are striving for self-determination.
- The challenge needs to be addressed through art/aesthetics, politics, morality and religion. These are 'the enemies within' that make us human. They need to be embraced, not denied and certainly not projected onto others. The role of policy makers and managers is to draw on all these 'enemies' and to build them into a more systemic and recursive design for inquiry.
- Last night I was reading Hofstadter's (1979) *Gödel, Escher, Bach* which explores mostly Western, Greek and Asian Zen philosophy. His theme is the systemic loops of recursion that simplistic western philosophy has tried to simplify. He draws on Mathematics and logic, art and music to show the 'strange loops' of recursion that give a sense of both symmetry and also infinity. His figure (18, page 71) explores the unreachable truths and falsehoods that can only be appreciated by ongoing questioning. Unlike Churchman, he seems to be striving for an answer by exploring strange loops, whilst Churchman accepts their inevitability.
- This morning as I drove to work thinking about this summary, I heard an interview [18]with David Fanshawe who described how in the early 1970's he experienced a sense of awe (perhaps an epiphany) whilst listening to a celebration of Christian mass in Jerusalem and hearing outside the church the sound of the Muslim call to prayer. This harmonized at the time with the Latin liturgy. Harmony exists, if we can only appreciate it and help to create it.
- The challenge is to rework the boundaries (conceptual and geographical) of international relations and governance. Fanshawe became inspired by this experience to develop world music and traveled up the Nile recording the music in the villages. This recording released as 'African Sanctus' was the basis for music that harmonizes the voices of a western choir singing the liturgy with the music of Africa. We can call the harmony recursive or multilayered or perhaps multidimensional, because each voice is part of the overall harmony. This will be the subject of a forthcoming volume in the Churchman series.

[18] Margaret Throsby interviewed David Fanshawe on 17th May 2004 on ABC radio)

Recollections and Thoughts by Colleagues and a Newcomer to his Work

Two personal papers provide an introduction to C.West Churchman and his work. They are by: Janette Young (who was inspired by some of West Churchman's work and won the Vickers award at Asilomar 2004) and Olov Forsgren (a student of West Churchman who applies his thinking to his teaching and his practice and has worked with Russ Ackoff).

Chapter 2

C. WEST CHURCHMAN AND THE NEW WORLD OF CO-DESIGN

Presentation at Memorial Panel

OLOV FORSGREN
University College of Borås, Sweden

2.1 INTRODUCTION

As with many of us, I got the feeling "I have to meet this man!" the first time I read, *The Systems Approach* [1]. I read the book because a colleague of mine, Kristo Ivanov, said to me: "With your ideas you have to read a book by Churchman..." and so my relationship with West Churchman started. Later on I worked with Stefan Persson – a former student of Churchman. He became the supervisor for my doctoral studies. In the late 1970s, I was a rather introverted, inexperienced but reflective youngster from the north of Sweden. I had little knowledge of the wider world. Despite that background I felt that I had to meet West Churchman. So I wrote introducing myself to him and explained that I wanted to visit him. I was really happy and surprised when I got the answer that I was welcome!

The contact developed into a close friendship resulting in a number of overseas travels back and forth for both of us. During this time I felt like an adopted student. But in 1998, I felt that I had graduated, because when West Churchman was invited for a keynote talk at the conference for International Federation for Information Processing in Helsinki, he asked me to take on the role because his doctor advised against the trip. I agreed because I felt ready to do some real work in the spirit of West Churchman.

In this paper it is my intention to give you a brief idea of my view of the spirit of West Churchman and some hints on how I have tried to maintain and nurture that spirit. I will also mention some other researches related to Churchman that has been working in the same spirit.

2.2 THE BASIC IDEA

In a recent conference at the Ackoff Centre in Philadelphia [2] many of Churchman's students were gathered together and in one of the panels the question was: How can we describe the core idea of systemic thinking? From the panel we heard somewhat different ways to describe systemic thinking, but as Peter Checkland also pointed, out many of the ideas could be summarised by Churchman's sentence:

> "When for the first time you can see the world from the view of someone else..." you have started to travel into the land of systemic thinking. [1]

This is more challenging than it sounds! At the same conference many of the speakers seemed to agree that it would take between 100 and 200 years before we really could grasp this change in thinking.

Most of West Churchman's writing is about how to help us overcome this difficulty. The other part of his work is about trying to describe what a difference it makes when we start to act in accordance with systemic ideas. Inspired and supported by West Churchman I will in this paper try to address these two challenges.

2.2.1 My own steps exploring the other side

My own research (together with my research group) has been to work on small examples showing some possible steps to bridge hard and soft systems. This work has been supported and partly made together with West Churchman's students Ian Mitroff [12] and Dick Mason. [13]

As a facilitator of learning I try to enable students to discuss problems beyond the 'box' or limits of a framework or paradigm of knowledge. I ask students to avoid working within a coffee box. But instead of telling them this, they are encouraged to work it out for themselves. I call these working outside the 'coffee box'. Students however start with so-called 'coffee box' examples such as: "What is an apartment?"; "What is a Day care centre?"; "What is a car?"; "What is a mobile phone?"; "What is a bank"; "What is a municipality?"; and so on.

From this perspective, I have then applied a co-design approach that, in the example of the apartment, enables the student to see that an apartment is what someone wants to study about an apartment. If you are interested in golf you may want to design a measurement scale for "Access to golf courses". If you have other interests you may want to use other measurement scales: maybe you want to know about "afternoon sun", cost of the apartment against other kinds of properties, or perhaps the price over the past ten years in one town versus another.

From a co-design perspective there are, as I have discussed earlier, an infinite number of possible measurement scales. So this thinking would be absolutely useless and destructive if we do not set the limit, based on the purpose of the design. From the first example of "What is an apartment?" in the "Live better" system developed together with the local city of Umeå [14] we are now working with companies and organisations all over the world. But it is all about taking the first small steps into a new worldview in which science no longer strives to find the true nature of an apartment or whatever.

West Churchman was the first one to conceptualize this new world. Yet he left a huge amount of work for the rest of us to do. I am sure that he would like us to better co-design this world so it will be a good world for our future generations to live in.

This systemic thinking has been applied and further developed together with companies and cities in the design of their business and services and it has been found to be very successful. At the same time, this thinking opens up for many new questions. As an example, in most cases the ideal co-designed "product" is both a service and a product. Also it often contains service parts both from traditionally public and private business sectors. In all cases the co-design also is a moral and political question – Who's ideal life are we going to support?

The further development of these ideas goes along three lines. **One line** is to summarize the co-design thinking into guidelines for development of business and all kind of services with IT as an important component. **Another line** is to develop software in terms of a co-design platform that supports a co-design approach. Finally, **the third line** (closely connected to the others) is to gather successful examples of co-design thinking in a "co-design platform" that can serve as inspiration for new projects.

2.3 THE EXAMPLE STRATEGY

Some support for crossing the Bridge from the non-systemic to the systemic world.

In one of the final information systems classes at Umeå university I told the students to create a database with all the knowledge they had about 'something' In this case 'something' was a half empty, wrinkled coffee box filled with coloured papers, pictures and text! The students worked in small groups and I told them that if they needed help they could come to my room. Only once did a group came into my room to say that the task was meaningless if they did not know what the knowledge was going to be used for! Some groups started to be very enthusiastic when they realized that they were the ones in the world that knew the most about coffee box bottoms (for example)! It was not difficult to get these enthusiasts to agree on the need for special research programs and professorial chairs for their specializations!

In the end of the class we reflected about what we had been doing and many of the students really felt that they had woken up. They realized that the coffee box meant a paradigm and that the challenge is to work with, rather than within paradigms. In most of their education they had acted like photocopiers for coffee box knowledge without asking: "what is this knowledge good for?" Many years later it happened that students from these classes approached me and told me that class was one of the few things they remembered from their education and that they still used these ideas themselves.

2.4 THE NAMING STRATEGY

"Systemic thinking" may be a better term than "systems approach", but in my view they both are rather badly chosen phrases because they both, to a high degree, refer back to the coffee-box style of knowledge and thinking. As described above when the students decided to just study the bottom of the box it is possible to say that they just took one part of the total system. Again from my point of view it is the normal standard to use the term system when we realize that even a coffee box is too complex to be described at once. The word 'system' is closely connected to coffee box thinking.

Once I spent some time at a research centre that specialized in coffee box system thinking – the International Institute for Applied Systems Analysis (IIASA) in Vienna. In my project we gathered knowledge about environmental problems in Europe. The knowledge was in the form of very complex systems models that were summarized in computer models and could be simulated in different ways. It was really coffee box knowledge. There was absolutely no thinking about the process of how to design the measuring system that defined the environmental data. [8] Looking a bit deeper we can see that designing the measuring systems is to a high degree a political process. What is important to measure for one nation is not at all be important for another nation, sometimes on the contrary. IIASA was set up as a research centre after the Second World War – and it is stated that IIASA should not be engaged in politics – just in pure science. This also makes it really dangerous to start thinking of knowledge from Churchman's systems approach point of view. The result is that IIASA has produced impressive amounts of systems models that were never used (but in most cases connected to successful personal academic careers).

Despite that, IIASA is a great meeting point for reflective people from all over the world, as an example IIASA connected me to Len Troncale and

ISSS. Len Troncale is open for reflection and together we had a number of conversational seminars at ISSS meetings bridging "the hard systems" and "the soft systems thinking" [9]

"Soft systems" is a poorly chosen term mostly connected to Peter Checkland, [10] and West Churchman never liked it. His argument was that soft systems thinking could be regarded as harder than hard systems thinking. For me this naming of the world on the other side of the bridge was a big challenge. Just because West Churchman used the "systems approach" still today many people connect him to a positivistic coffee box research tradition. West Churchman could agree on this mistake but on the other hand, "What is the alternative?" he asked me. "One possibility", he said, "is to call it the Churchman philosophy of knowledge. I was first, but I am too shy to use that name."

I did not have a good alternative name for this philosophy of knowledge other than a "co-design approach". There has to be a design approach to the measuring instruments we use and more important than that the final design has to be a result of negotiations and agreements between stakeholder interests. The co-design process surely is a political process. West Churchman himself used "Design of inquiring systems" as a title of one of his books. This was a dangerous thought at IIASA. It was dangerous because it was so different. The late Donald Schoon (another co-design thinker) once told me: "I can't tell my colleges at MIT about all the implications of my thoughts because I am afraid they would think I am too far out". [11]

A few times I also spoke with West Churchman about this and that he in the book *The Systems Approach* (as Peter Checkland does) separated hard systems like the solar system from soft teleological systems. I understood West Churchman did this separation because he then left a playground for the hard systems thinker rather than go into a conversation with them. He told me: "I did not dare at that time". Soon after this situation West Churchman had a seminar at the natural science faculty at my home university. This time it did not leave space for hard systems thinkers; it was up-front attack with full energy. With a combination of explaining the historical shifts in strategy by painting a picture of the major thinkers, he showed how the shift in worldview had occurred. He also used a number of coffee box examples and he was able to present a convincing argument. One of the poor university professors said: "These ideas are so different, how do we take the first step over the bridge?" West Churchman found it hard to answer that question.

2.5 A FIRST TASTE OF THE SYSTEMS APPROACH

West Churchman's student Russell Ackoff has made great attempts to 'wake up' students and in line with that has also really criticized the school system and the academic 'coffee box' knowledge approach. [3] Bela H Banathy a close friend of West Churchman is the one that took a few steps further and developed the new landscape of learning from a systems approach. He really took seriously a call from a reflective teacher I once met. The teacher said:

> "You philosophers have to come up with a new idea about learning now that you have invented perspectival thinking, because students have started to ask about why they should learn about just one or two perspectives. And we do not have good answers to that question."

Bela H Banathy was a pioneer but still we have oceans of work to do here. [4]

During the time I spent with West Churchman in his classes I saw a number of 'wake up' teaching examples. Of course some of them also went further on exploring the landscape on the other side of the river. The landscapes were explored with new questions: "What perspective of the coffee box is good for whom?" "Who is going to select the perspective?; and so on. Sometimes he called this approach "The systems approach" because the aim was to sweep in all relevant perspectives. Many of West Churchman's writings were also in that style. All the time there was a new perspective challenging the basic assumptions. I know some students missed Churchman's own point of view. Once I put the question to West Churchman: If I could be better than you in terms of the systems approach, how would I be then? West Churchman was unusually quiet for unusual many seconds and then he answered: "You would be more confused than I am." That reflected something of Churchman's point of view.

Trying to accept this challenge of being more confused than Churchman, there is one perspective I think Churchman neglected too much or perhaps it confuses me even more. Churchman was talking about systemic thinking in some discussions when he, in a high voice, said, "Enough is enough!" "We can not go on and on challenging assumptions we also have to act on the basis of the perspectives we have, otherwise nothing is going to happen!" But in his writing it was hard to find this aspect. Instead it was often an ongoing 'unfolding' – 'sweeping in' process.

Another strategy West Churchman used to help us over the bridge into the world of the systems approach was to use **the historic strategy**. In many of his books we could follow the line of thinking about knowledge from the time when the philosopher Immanuel Kant solved the philosophical dispute between the 'idealists' focusing on ideas as the essence of the world, and the 'realists' focusing reality as the essence of the world. [5] Kant's idea was that we need a' priori ideas to interpret the real world. We could follow how this idea today called 'perspective or view', was connected to purpose by the first pragmatist philosophers. Applied into real world problems this was developed in terms of cybernetics and operations research.

In the view of West Churchman, his teacher Singer [6] was important because he told us that we need to design views or measurement scales and calibrate them in order to be able to measure or to gain knowledge. Tom Cowan, [7] who was another of West Churchman's teachers, had many good examples about this. On Churchman's advice I spent some time with Cowan. He had a clear picture painted with many good examples of how we have to design good questions to get good answers. Tom Cowan's mode of thinking was to a high degree adopted by Churchman's student Russell Ackoff. Ackoff with his many good examples is the main force behind making the systems approach as widespread as it is today. One part of this is the change of name from the systems approach to "Systemic thinking".

REFERENCES

1. Churchman, C.W., 1968, *The Systems Approach*, Dell Publishing Co. Inc., New York.
2. The Third International Conference on Systems Thinking in Management, Philadelphia, Pennsylvania on May 19-21 2004.
3. Ackoff, R.L., 1981, *Creating the Corporate Future*, Wiley, New York.
4. Banathy, B.H., 1990, *Systems Design of Education: A journey to create the future*, Educational Technology Publications, Englewood Cliffs.
5. Kant, I., 1987, *Critique of Judgment* (W.S. Pluhar, Trans.), Hackett Publishing Company, Indianapolis.
6. Singer, E.A. Jr., 1959, *Experience and Reflection*, University of Pennsylvania Press, Philadelphia.
7. Cowan, T.A., 1975, Nonrationality in decision theory, in: *Systems and Management Annual 1975*, C.W. Churchman, ed., Petrocelli/Charter, New York.
8. Forsgren, O., 1984, An actor oriented concept for a model inventory and environmental information system, The International Institute for Applied Systems Analysis (IIASA), Vienna.
9. Forsgren, O., 1989, We are many people with the purpose to build a better world: Comments to "We are one people" by Len Troncale, *General Systems Bulletin*, 1-2, 7-8.
10. Checkland, P.B., 1988, Soft systems methodology: An overview, *J. of Applied Systems Analysis*, 15:27-30.
11. Forsgren, O., 1997, Co-constructive reflective practitioners, Plenar for D Schoon, presented at the *Systems for Sustainability, Fifth International Conference*, London, 7-11 July 1997.
12. Mitroff, I., and Churchman, W., 1993, A manifesto for the systems sciences: Outrage over the state of science, in: *Ethical Management of Science as a System - Proceedings of the thirty-seventh annual meeting*, R. Packham, ed., International society for the systems sciences, Sydney, pp. vii-x.

13. Mason, R.O. and Mitroff, I.I., 1993, *Mindshift*, John Wiley, New York.
14. Forsgren, O. and Ivanov, K., 1990, From hypertext to hypersystem, in: *Cybernetics and Systems '90. Proc. of the Tenth European Meeting on Cybernetics and Systems Research, Vienna, April 17-20, 1990,* R. Trappl, ed., World Scientific, Singapore, pp. 275-282.

Chapter 3

SOME NOTES WITH REGARD TO "THOUGHT AND WISDOM"

JANETTE YOUNG
Flinders Institute of Public Policy and Management
Division of Health Sciences, University of South Australia, GPO Box 247, Adelaide, South Australia

---

3.1. INTRODUCTION

I write in response to the passing of C. West Churchman.

I am a relative newcomer to Critical Systems Thinking (CST) and so this paper is written as a somewhat naïve investigation and exploration of a segment of the thinking of one of the more recent great contributors to CST.

I have chosen to focus on C. West Churchman's 1982 contribution called *Thought and Wisdom*. In reading and exploring this book I acknowledge that I am only focusing on a specific point in the writers thinking, and my sense of the person of C. West Churchman suggests strongly that once challenged or queried on points of his argument(s) Churchman would have actively pursued thoughtful responses and changes. This said however, I believe that it is not infrequent in our Western culture of "progress and forward-ho-ness", that we neglect the ideas and thinking of past times, and presume that improvements have of course been unilaterally made. By doing this in fact we lose gems (and even whole truckloads!) of thought and wisdom.

3.2. "THOUGHT AND WISDOM"

The book *Thought and Wisdom* is a compilation of four lectures delivered in 1981 by Churchman (the Gaither Lectures), followed by six papers that closely relate to the subject matter and discussion contained in the Gaither lectures.

- The four papers that formed the Gaither Lectures are entitled:
- On Connectedness: Putting the World Together
- Future Generations: Humanity's Invariances
- Success of Failure: The Joys of Dissatisfaction
- Design of Life: the Aesthetics of Conversation

From the other chapter titles it is clear that Churchman had a passionate interest and concern for wildlife, and in particular endangered species.

I have chosen to use Churchman's initial four chapter structure as the core of this paper, pulling in thoughts and discussions from the later chapters, focusing in on the themes that emerge for me in each and concluding with a summary of my perspective and critique of Churchman ideas and approach in "Thought and Wisdom".

There are six interrelated themes that emerge for me from this book. These are:

- Morality/ethics;
- Action and "Solutionism";
- The nature of reality or facts;
- Failure; and
- Power.

In 1982 when this book was published I was in my early teens. I have found it intriguing to read this book, with its constant references to "future generations" being conscious of being one of the next generations that Churchman identifies as pivotal to his discussion.

3.3 CHAPTER 1 – ON CONNECTEDNESS: PUTTING THE WORLD TOGETHER

Despite my curiosity with regard to chapters 2 and 3 I began reading at the very start of the book. This was not a bad start as I was immediately captured by Churchman's perspective that ethics was integral to a spectrum of arenas including research, city planning, policy analysis and business. He noted that it is frequently only the latter that ethics is linked to directly (p. 3). Having lived through the1980s where Kirk Douglas declares (as the megalomaniac banker of the movie *Wall Street*) that "greed is good", and the fall of high flying businessmen in Australia such as Alan Bond and Christopher Skase, which lead to a consequent focus on business ethics and governance (Burrows, 1993), it was intriguing to reflect on Churchman's criticism. In my experience ethics is still not linked to policy and planning (for towns and cities of any scale), although university based research these days always needs to have ethics clearance, in the form of a committee over viewing the planned process. Perhaps catastrophic events in the business world have continued to attract more attention than less overt corruption in society. Meaning that the impact of unethical planning and social policy with outcomes such as increasing social and fiscal inequities are somehow not viewed as "catastrophes".

In this chapter Churchman gives comment on the title of the book (p. 5). Wisdom he identifies "is thought combined with concern for ethics". His perspective on the definition of ethics is a broad one, "I take ethics to be a conversation with our ancestors and contemporaries on the meaning of right and wrong conduct, policies, management, planning etc" (p. 6). By ancestors Churchman goes on to clarify he is referring to the likes of Plato, Aristotle, St Paul, Hume... endeavoring to understand what the ideas of these (men) "mean in the design of our lives".

On page ten Churchman begins to display one of the characteristics that I find particularly appealing in his writing – his pragmatic, humanity and action focus. He discusses the way in which decision-making affects humanity. Decision-making "cuts off the other possible threads of human life" (p. 10). Our imagination, with its capacity to create alternate films, or dreams of the might-have-beens, highlights the powerful nature of decision-making. Deciding to do A, means that by default we have chosen to *not do* B, and vice versa. To not undertake an action is in fact to choose an alternative line of action (i.e. to not act). This thinking reminds me of the perspective of Dietrich Bonhoeffer (Kuhns, 1967) the Lutheran pastor and theologian who was imprisoned and eventually executed by the Nazi's for his involvement in a plot to kill Adolf Hitler. Bonhoeffer came to believe that to simply ignore and not act against the atrocities of the Nazis was in fact to support them, for him it lead to him moving from pacifism to active resistance.

Churchman discusses the nature of decision making in greater detail in the final chapter of his book as he explores the thinking of Edgar Arthur Singer Jr (p. 125). Churchman acknowledges Singer as the originator of the concept of 'sweeping in' diverse issues, aspects and concepts when seeking to problem solve (p. 125). Where we start a problem, or where we put the boundaries or limits around a topic is significant in terms of where we think the answers may lie. He uses the example of prison overcrowding to illustrate. If the boundary of this problem is seen to start at the prison door it will be responded to with more and /or larger facilities, alternate architecture, for example. Whereas starting at the point of the causes of criminal behavior will lead to very different ideas as to how to address issues such as overcrowding. We would then need to include such issues as poverty and inequality, police discrimination, the legal and judicial system...Churchman notes that Singer identified that "there are no simple facts" (p. 117), in practice seeking to understand does not simplify things but rather complicates them more!

3.4 CHAPTER 2 – FUTURE GENERATIONS: HUMANITY'S INVARIANCES

Churchman begins this discussion with a passionate call for social justice, calling for "moral outrage" at scenarios such as starvation and malnutrition. Calling for his readers to not view such issues with "calm objectivity' (p. 17); yet at the same time cautioning that such passion be integrated with calm intellectual inquiry. Again he displays his pragmatism. He does not believe that moral outrage should just occur for its own sake. It needs to be combined with planning and design in order to rectify and improve (p. 20) situations. Solutions to moral outrages in this world have too often been responded to with activities that at best do little to improve the scenario, and at worst may even feed into problems and make them worse. Churchman uses the term "to secure" as indicating the difference between well thought out, analyzed and change making responses and those that may in fact serve more to assuage guilt, or, to use an Australian colloquialism, provide political and personal "brownie points". This being a scenario whereby there is some sort of reward for the overt helper, and this can in fact be seen to be the main objective, rather than the overtly identified purpose of helping another.

This idea links to a term that I remember a lecturer coining in my undergraduate study days, "solutionism". The drive and desire "to solve" problems, particularly complex social ones is underpinned not necessarily by concerns for shared humanity and addressing human suffering, but rather by political and self interest or the desire to be seen to be caring, competent, able to solve problems, or simply to "tidy up" society at times. As Churchman notes, "thought likes solutions, wisdom abhors them" (p. 20).

Having identified the need to "sit and think" about problems Churchman then proceeds to talk about an action (the Endangered Species Act (p. 22) that had been enacted in order to try to address an issue that was close to his heart. He notes that it is an action that when considered in terms of money can be considered expensive. Costing much to enforce, yet retuning little in terms of dollars. He addresses this using a moral argument, noting that "to propose to a future generation that it accept payment for our immorality is immoral" (p. 24). Here in Australia, and some 20 years on, this same debate and short term approach to wildlife and nature exists as well. I suspect that Churchman would not however be pleased at the counter discussion that has developed, of the potential to cash in on nature and wildlife in the form of eco-tourism. I am sure he would have had some comments to make on this "low road to morality". But it is better than nothing as it is still a road to morality and he was above all things a practical man!

Churchman resorts to judgments of morality much of the time and expresses concern for what could be called cold, utilitarian discussions of morality. I worry a bit at this call to morality. Here in Australia in recent times immigration debate and policy has lead the locking up of children and babies in detention when they arrive on our shores in leaky boats, on the basis of morality- no I would argue this has nothing to do with a moral argument – it is a rationalisation. "They should have joined the queue with other asylum seekers" is seen by many to be a moral stance. If we are to respect all persons perspectives as suggested to by Churchman then we need to take on board others differing perspectives of morality. A response to such contested perspectives of morality could usefully be of pragmatic 'low roads' (to draw on McIntyre's reference to Saul Alinsky, 1972). To present arguments against policies in a way that addresses such things as financial costs and benefits, if this is how those with morality that differs from my own perspective can be engaged and encouraged to give consideration to alternate responses.

Churchman continues his argument stating "morality is not often articulate because the moral sense does not work through words…" (p. 79). He is referring to aesthetic sense of beauty – a sense that some cannot express in words but can express in song or dance or painting- the arts. Surely accountability (a form of morality) requires that arguments, debates and stances need to be articulated. Silence and secrecy are prone to becoming instruments of power and control. The high moral ground, that cannot and should not be questioned, is a commonly appropriated weapon as can be seen by the invasion and continued presence of the US, Britain and Australia in Iraq; the oppression and abuse of the Palestinians by the Israelis.

From this discourse on the nature of morality, Churchman moves into discussion on the nature of facts (p. 29). From his own experiences as a scientist he portrays examples of how people and politics can restrict information gathering and assessment, and hence what becomes defined and accepted as scientific fact. "The models are wrong because they make value assumptions which are wrong …not only are the models wrong but they are precisely wrong" (p. 31). Ackoff was his student. It is the same argument put by Ackoff and Pourdehnad (2001, p. 199) many years later. "The righter we do the wrong thing, the wronger we become. If we made an error doing the wrong thing and correct it, we become wronger". Ackoff and

Pourdehnad. In this paper they do not link such wrongness to politics and interests explicitly.

I include here the quote that Churchman incorporates in his discussion of the development and existence of facts (p. 33). "As one African gentleman once said to me, "we are developed; you are overdeveloped". Here he is highlighting how definitions of society are highly dependent on perspective. It is the old story of the blind men and the elephant. The blind man at the tail is sure that the fact is that the elephant is a fine-featured animal, the one at his leg believes it is strong and "like a tree", the man at the trunk believes it is flexible and perhaps swings through the trees! We need to recognize that the nature of human perception and perspective is such that we can never get a "whole picture or bird's eye view". To use the analogy of the globe in orbit. Seeing clearly on one side will never remove the reality that at any one point in time the opposite side will be in the dark!

3.5 CHAPTER 3 – SUCCESS OF FAILURE: THE JOYS OF DISSATISFACTION

An intriguing chapter title – "The Success of Failure"! Who hasn't been caught by failure? But how many of us wish to openly admit the shadow that seems to hang behind us all. The fear of perhaps not so much just failing, but of somehow *being* a failure, of falling short of some invisible ideal of person and substance. It seems difficult when first considering this chapter title to imagine anything other than a world in which failure is not a negative. Churchman however manages to illustrate (p. 38) very simply and visually how failure could be integrated into our concepts of learning by using failure to highlight and then address difficulties.

In this chapter, Churchman again returns to the topic opened in the last chapter with regard to the nature of facts. In particular, hard scientific, technical facts. For someone like myself with a social science background it is an intriguing discussion in which Churchman utilizes his very different work life experience in technical, scientific laboratory work. Churchman's revelations of the subjective nature of hard scientific facts, and the manner in which politics and human ego fudge the nature of "facts" in this arena is ammunition for social scientists who are often accused of the sin of *subjectivity* and murkiness, and held against so called hard sciences as inferior. Churchman provides a range of examples later on in this book that shows how such things as individual group and work team culture (p. 128) can influence scientific facts and measurements and even things such as (p. 127) the mathematics of astronomy (!).

Having identified that facts and therefore successful identification of facts is not as black and white as generally thought, Churchman uses such discussion to consider failure as not the unsuccessful identification of facts, but rather a source of information, even extending this concept to that of "excellent failures" (p. 41). Churchman uses his own unsuccessful attempts over many years to measure human values as an example of successful failing. He presents his experience of "failure to understand, but success in learning" (p. 45) as having been a significantly enhancing process – a development of wisdom – not a loss.

This discussion of failure has links to solutionism. Churchman alludes to the pull to comfort that often inhibits our ability to "successfully fail". He identifies the uncomfortable nature of such things as "losing battles", "sacrifice" and "complete failure" (pp. 46-47). He suggests in fact that life without discomfort and struggle would be boring and desolate (pp. 47-8). Further on in the book Churchman notes that for Singer (p. 134) struggle was the nature of human life, to not struggle was in fact detrimental and perhaps the cause of suffering. In this regard my mind is drawn to the scenario that has developed here in Australia in recent years. In the late 1990s our Prime Minister included in his political platform a pledge to further the "comfort" of Australians. It seems to me that the living out of this aim in Australia in recent years has lead to the draconian practices of locking up refugees, including babies and children. Our need to be "comfortable" materially is seen to be at odds with (visionary) concepts of compassion, sharing of resources and shared humanity. Churchman was of the opinion that his mentor, Singer would have been shocked at the call for "satisfaction" that existed in the 1980s. I shudder to think what Singer and Churchman's responses would have been to the Australian policy aim of "comfort". Alternately we would do well to adopt one of Churchman's final summary points in regard to failure, "wisdom is a life of endless inspiring dissatisfaction"(p. 49).

3.6 CHAPTER 4 – DESIGN OF LIFE: THE AESTHETICS OF CONVERSATION

Conversation is one of my favorite activities so to have the term enshrined in an academic setting is exciting! In using this term Churchman focuses in on the topic of "other knowledges" (p. 56), in the language of Polanyi (1969), tacit knowledge, or Rowan (2001) experiential knowledge. Knowledge that is acquired by the experience of life and living rather than through academic or other formal pathways, but which is often not considered as important or valuable as the latter form of knowledge. Conversation, as compared to interviewing, assessing, directing, commanding...implies equity, mutual sharing, debate and discussion, and it is these characteristics that Churchman explores in this chapter.

Professional distance and "interviewing" seem to me to be forms of "conversation killers or murders" that Churchman discusses (p. 57). As a social worker working in aged care with, to use Churchman's term the "convicted elderly" (p. 58), much of my job involved "interviewing" people. An unaesthetic conversation as Churchman (p. 59) would have termed it, involving non-symmetry of respect. Interviewing was a one sided process that involved me (the paid, formally well educated professional) asking other people (clients, those with an issue identified as a "problem") pertinent questions about themselves, their relationships, bodies and other aspects that had been identified as significant enough to be listed on the assessment sheet. One day having spent some time with an elderly man working my way through my list of questions to tick and cross, the elderly gentleman said "it is okay to tell you about me, but what about yourself?" This gentleman had dementia and so the social rules which enshrine relationships of power and boundaries had become misty, his mind was operating at the level of human to human. I was a person in his house visiting and quite rightly he thought it unacceptable for the focus to simply be on his needs. A process of mutual sharing and concerns was needed. It was a confrontation of the one-sidedness of this process. As the powerful professional, I held a social mandate to extract information about this other human being with no mutually respectful sense of reciprocation. A process that distorted our shared humanity.

Churchman notes that it is often the process of engagement with other human beings that is crucial to social change and action. That the tools used of themselves are not really as important as we would like to think! (p. 100) "Conversation", it seems to me is a very powerful tool for social change and betterment. It is a term that strips layers of power and authority away. Imagining oneself in conversation with another is perhaps a useful tool for gauging respect. In my experience we are too often outcome focused and neglect processes of mutual respect and engagement with other people. As Churchman points out, (p. 111) unlike machines individual human beings act with agency. You cannot make individuals to behave like machines and for example require participation.

3.7 CONCLUSION

There are many thoughts and ideas that I find stimulating and inspiring in this book. I cannot however, as a female and a member of a next generation help but notice the gender and cultural centricity of it. All of the great thinkers that Churchman refers and defers to are male and western. The voice I hear speaking is of a well educated, comparatively privileged white male.

Writers such as Donna Haraway (1991) and Pierre Bourdieu (1984) with their critical analyses from perspectives of gender and class, also question the reality of facts, and the nature of social power and action. However, to use Churchman's language their analyses also explicitly 'sweep in' social constructions such as gender, class and ethnicity. The example used by Churchman of prisons is a prime example of the need to undertake such analyses. In the United States, the vast majority of prisoners are male, working class, non whites (Gilmore, 1998). Whilst it may be fair to say that all human beings are of equal value, the reality of inequitable distribution of such factors as wealth, power and status will impact on the nature of conversations. This is an aspect that is not addressed by Churchman in this book.

Having identified what seem to be omissions by Churchman, I would like to note that one of the main features of his character that emerged for me from this book was that of humility. Which in my definition is the ability to gently look at one self and the world and seek to engage with both for shared betterment. Humility is one of those characteristics that as soon as you think you have acquired it – you lose it! Thinking one is humble is in fact evidence that one is not. In evidence of this Churchman never makes such a self claim, however in his critique of his previous work (pp. 61, 70, 90) and his ideas as to conversations I am struck by this as a quality he exhibits.

As stated at the beginning of this paper we need to be careful not to neglect the ideas and thinking of past times, presuming we have unilaterally advanced our thinking. Thought and Wisdom has many gems that need to not be discarded. Ideas such as that of actively engaging in learning from our failures; being conscious of the powerfulness of decision making; the need for moral outrage combined with calm analysis, and to set our boundaries of consideration widely so that we can sweep in a diversity of ideas and issues, are all gems that can be taken and applied in trying to create a better world for all human beings.

REFERENCES

Ackoff, Russell and Pourdehnad, John, 2001, On Misdirected Systems, *Systems Research and Behavioural Science*, 18:199-205.

Bourdieu, Pierre, 1984, *Distinction: a social critique of the judgement of taste*, Routledge & Kegan Paul, London.

Burrows, Brian, 1993, The Greening of Business and its Relationship to Business Ethics, *Long Range Planning*, 26(1):130-140.

Churchman, C. West, 1982, *Thought and Wisdom*, Intersystems Publications, California.

Gilmore, Ruth Wilson, 1998, Globalisation and US prison growth: From military Keynesianism to post-Keynesian militarism, *Race & Class*, 40(2/3):171.

Haraway, Donna, 1991, *Simians, Cyborgs and Women: The reinvention of nature*, Free Association Books, London.

Kuhns, William, 1967, *In Pursuit of Dietrich Bonhoeffer*, Burns and Oates, London.

Polanyi, Michael, 1969, *Knowing and Being*, Routledge and Kegan, London.

Reason, P. and Bradbury, H., 2001, *Handbook of Action Research: Participative inquiry and practice*, Sage, London.

Rowan, J., 2001, The Humanistic Approach to Action Research, in: *Handbook of Action Research: Participative Inquiry and Practice*, P. Reason, and H. Bradbury, eds., Sage, London.

The Contribution of C. West Churchman to doing better Governance and International Relations

A meditation[19]

Whilst walking along the boardwalk amongst the sand dunes at Asilomar, just before the C.West Churchman Memorial Panel, a colleague from Israel, Tamar Harel Zohar and I talked about the way in which creativity is sparked through a range of connections with other people and with nature.

On cue another walker pointed out a sea otter floating on his back, using his flippers to knock together two oyster shells so as to get his dinner. He rested balancing the shells on his stomach and repeated the process again and again.

The process was repetitive, but he persisted until satisfied. And then dived again and again for more shells.

The two shells can be seen systemically and metaphorically as both a meal and as tools for problem solving for a particular purpose, namely sustaining life.

We all need conceptual tools for problem solving and we must try as best we can to use the two sides of our brain consciously to address the challenges we face. If we have only flippers then using tools can be frustrating, but we need to do the best we can and we need appreciate the efforts of others.

West Churchman stressed in his works that nature can inspire our design of inquiring systems and that we need to co-create solutions with those who are to be affected by our decisions. A systemic approach is ethical when it does not 'cut off' consideration of others or consideration of the environment.

[19] Memorial Panel for C. West Churchman at the 48th Annual Conference of the International Society for the Systems Sciences: 50 years of Systems Science: Honouring Tradition, Embracing the Future.

Chapter 4

THE SYTEMS APPROACH AND ITS ENEMIES HELPS US FIND THE MORALITY OF A REVISED DEMOCRACY[20]

JOHN P. VAN GIGCH, Professor Emeritus, California State University, USA

Abstract: If democracy constitutes the institution of morality in a society, it can only be introduced when peace, agreement and consent prevail.

My hope is that a review of CWC' *The Systems Approach and Its Enemies* will provided the reader with a broader understanding of this great work. Re-interpreting CWCs' ideas in the context of the search of morality in democracy and the adversarial character of different cultures could further our understanding. I argue that the text is relevant to our present dilemmas and problems.

Key words: Democracy, enemies and adversaries

4.1 A MATTER OF DEFINITIONS: ADVERSARIES VERSUS ENEMIES

I note the similarity/difference between the words 'enemy' and 'adversary.' Other authors use the word adversary (ies) to denote all the forces that impede the progress of his/her own discipline. In the Oxford dictionary (1976), the concepts of *adversaries* and *enemies* are considered

[20] Author's Note: In some passages where CWC is cited, the quotation marks have been omitted. However, unless clearly indicated by names of other references, the ideas and text clearly pertain to the original CWC's source.

synonyms. However other sources show a distinction between these two concepts. An *enemy* is seen as a hated opponent and is usually considered a person who hates another and eagerly seeks his/her defeat. Words used in lieu of 'enemy' include: opponent; hostile army or nation, an alien. An *adversary* is an opponent who is *not* hated; an adversary is someone who is 'in front of', opposed, coming from another direction, averse. Adversary indicates one against the other without intent to harm. As an example, in the conflict between Ronald Reagan and Gorbachev during the Cold War, they called themselves adversaries but not enemies. Reagan acknowledged that he considered Gorbachev a worthy adversary: they faced one another but did not necessarily hate each other.

4.2 THE MORALITY OF DEMOCRACY

We dedicate this volume to the institution of morality in democracy.

CWC's concept of morality is in complete agreement with Hobbesian moral theory according to which

> "the moral actions are rational for a person to perform if and only if such action advances his desires and preferences...Moral actions are rational for almost everyone and 'thus mutually agreeable."

Margaret Mead[21] realized that the process of instituting democracy constitutes a challenge. When facing each other, the two cultures clearly become adversaries. However, no cooperative person can be allowed to become the prey of immoral aggressors (read also: "enemies" and or "adversaries) because moral actions can only be agreed upon cooperatively when peace, agreement and consent prevail. This adversarial position must be assuaged and constrained through cooperation and consent and should not be allowed to end up in violence (Goodin & Pettit, 1993).

4.3 ENEMIES: THOSE WHO CREATE A POLIS AROUND THEIR IDEOLOGY

CWC concentrates in *The Systems Approach and its Enemies* on four main enemies (of the rational mind), namely:

[21] Mead, Margaret, (1965): *And Keep Your Powder Dry,* William Morrow and Co., New York.

a) **Morality**, a concept that usually stirs up countermorality that threatens to destroy morality. More details concerning each of these threats are given below.
b) **Politics** that reveals two versions of evil. The first being the threat on the life of the polis; the second, the need of the polis to threaten another polis in order to survive. One nation against another is the equivalent of one polis against another.
c) **Religion** is a concept that is used widely in CWC' book: It acquires a special meaning which refers to blind 'worship' of an idea, a god, an issue. Thus, in their research, physicists worship 'The Holy Grail Of Truth.' Worship can be the blind adherence to a cause like that pursued by environmentalists and anti-environmentalists on different side of an issue. Those who defend the spotted owl or the salamander in order to save the environment do it in the name of so-called 'religious beliefs' or a 'faith' in the infallibility of their cause. Their opponents are equally convinced of their own righteousness.
d) **Aesthetics** is another enemy that in the name of protecting the quality of life of a polis can result in actions that may be counter-productive because it may act to destroy the quality of life of another polis. "Aesthetics fights the tragedy of those who destroy the quality of life."

4.4 A PESSIMISTIC VIEW OF THE RATIONAL MIND

In the SA&E (abbreviation for *The Systems Approach and its Enemies*) CWC raises the question "How can we be certain that we are making 'good decisions' in view of 'the enemies' which are lurking at every turn to derail our project?"

If the Systems Approach is an attempt not only to be holistic but also to be rational, these enemies are making it impossible not only to be rational but also to be holistic.

What is 'a good decision'? Is a good decision one that is merely rational?

Or does a good decision also needs to be ethical? As we study the world of decisions, we realize that we have no way to prove that a decision is 'good' i.e. that it will result in a success. As an example, in the world of investments, we may choose a stock after careful research. It may or may not be a good investment. Too many variables are at play. To say that 'the future will tell' is admitting our ignorance of where the stock will go. It is admitting that our rationality is flawed or at least limited.

When seeking The Good or The Beautiful, we cannot use the rational mind to prove whether our quest has been successful or not. In every realm, the process of making choices is tied to beliefs (held beliefs) that are not necessarily of the realm of the rational mind.

CWC argues that as we try to reach rationality and truth our minds are hampered by enemies, the first of which is what he calls "Politics," a subject to which we now turn.

4.5 THE FIRST ENEMY: POLITICS

Of the four enemies of the Systems Approach mentioned by CWC - he declares: "Politics is the 'most extroverted enemy' because it basically relies on people's relationships [to other] people around a common cause." The people that surround us do not always act or think rationally (friends, family bosses, politicians etc). They certainly don't want or don't even know how to be 'holists.' (Churchman, 1979: 156).

4.5.1 When 'A Polis' Rules It Becomes An Enemy

CWC uses the expressions "To "make [a] polis around an issue" or "To become [a] polis around an issue" interchangeably. We recall that *'polis'* is the way CWC qualifies or names those who take political action in a partisan way. They want to prove that they are part of the same community and that their actions are centered around a common issue and a common ideology.

How does the polis rule? When polises (CWC's plural for polis) are formed, its adepts or participants "worship" the polis' ideals or ideology and will even act "to defend the polis to death." CWC's states that a "Polis [is a sort of] discipline. Through its cohesiveness and its avowed purpose, the polis works to promote its ideology and becomes an agent with power to effect change. This is a result which is obviously repulsive and shocking to Churchman: "The notion that those who have the power *should* rule is morally shocking: that is, the second enemy (See below) called Morality regards such a policy as evil." "When the polis rules, it is morally shocking." (Churchman, 1979:156-159).

4.5.2 Politics Versus the Systems Approach (SA)

The Systems Approach and the Polis [es] are obviously enemies. They certainly work at cross-purposes. "So we are to picture the SA swallowing its enemy, the political approach, by incorporating into its general scheme the all too human urge to form polis..." Otherwise, it is the Systems Approach that will be overwhelmed by the polis and swallowed up in turn. "Politics co-opts the ideal-system planner or, the other way around, the system planner becomes political i.e. forms polis around science," and subverts the attempts to become more open (Churchman, 1979:162-164). The only way to thwart the polis is to incorporate it in the metalevel inquiring system where its logic can be neutralized or dissolved. This subject will be covered in the conclusion to this essay (van Gigch, 2003).

The urge to form polis [es] works against the Systems Approach. The former tries to become restrictive within the confines of its own constraints, whereas the latter tries to become more comprehensive. As the Systems Approach becomes more comprehensive it combats the efforts of the polis to be more restrictive.

Politics and the Systems Approach do not have the same aim. Whereas Politics tries to maintain the polis so that the smaller the system the better the polis, the Systems Approach works in the knowledge that the larger the system the better the objective is met.

A weak polis leads to its decline. A strong polis is a failure for the Systems Approach. Humans, on behalf of their own causes, end up destroying themselves and their own causes. They plant the seeds of their own destruction.

4.6 THE SECOND ENEMY: MORALITY

CWC considers that Morality can be the reason or the cause for people to break in anger and turmoil. It can cause internal turmoil. Even lead to revolutions. (Churchman, 1979: 165-166). Morality is a process, a tool, an instrument by which we tell people how they should lead they lives. However, people do not like to be bossed around. It is an attempt to impose solutions against the will of people. For that reason CWC states that Morality [by its very nature of being bossy] is basically "antimoral." "The spirit of morality is outraged."

CWC warns us that: "Morality always operates so as to create countermorality, " which is a reaction leading to revolution and strikes. (Churchman, 1979: 167). Here we witness a convergence of morality and politics.

4.6.1 Why Is Morality An Enemy Of The Systems Approach?

Conflict arises when a solution to a problem is proposed. Objections are raised. Reasons and counter-reasons are always argued in the name of reason. The enemy is one who can use reason to raise reasonable objections in the claims of reason." (Churchman, 1979: 169-170).

"Each [proposed] approach creates its own evil." Each enemy creates its own irrationalities. "Thus morality often responds to its perception of evil by fighting, killing incarcerating and destroying everything about." (Churchman, 1979: 171).

4.7 THE THIRD ENEMY: RELIGION

CWC uses the idea of *religion* in a very broad manner. By this word he does not only consider what we would call 'organized religions.'

According to CWC a so-called 'religious approach' can occur in particular circumstances such as the following:

A Superior Something. When we believe in 'a superior something' that 'we worship' and that 'we serve,' It can become a religion, for example: reverence towards Nature, blind service towards a community.

When we decide and compare ourselves to a bigger something (an incarnation of some hero) that we 'worship.', for example technology; automobiles; NASA programs such as the Apollo program or the recently announced future manned mission to Mars.

The Economy. A form of religion takes place when the people who participate in it are "bound by a common belief.", for example CWC names the economy as an example of deity and economists are the priests of that religion.

Art. Art can be sometimes be considered a religion: "[Art] is usually religious, especially when the artist is 'driven' to paint compose, sculpt or write; the creative acts of artists are ways of worshiping the grand or the small."

Science. Scientists and in particular physicists are "bound" by "their religion' and a belief in the supremacy of their science.

Furthermore they regard their quest to find the objective truth as sacred to the point that it becomes a sort of god.

Rationality as a religion. According to Weinberg(2001) : Physics must find a Standard model such as fundamentalism which is the belief in physics that this is the most basic science from which all others are derived: Once we understand the Standard Model of Particle Physics we can explain "everything else." Rationality becomes an enemy of the SA. It does not allow others views to prevail. (Churchman, 1979:176).

I will discuss this subject further in another section of this essay.

A Paradigm has already been covered when discussing the forming (or making) of a polis under the section called the "Second Enemy of Politics". In this context, adherence to a paradigm can constitute a form of religion. As Kuhn pointed out, a scientific revolution is needed to shake lose the worship of one paradigm before it can be replaced by another.

The Fragmentation of Scientific Disciplines was criticized by CWC who thought the division of knowledge along the lines of separate scientific disciplines was problematic. Each trying to develop its own methodology and each claiming to reach its own "truth(s)." Instead of promoting the unity of science, each discipline promotes its own agenda (and ideology), when in reality they are all working the same ground of knowledge.

The Environment can be problematised by environmentalists who take to Worshiping "the small" as a symbol to protect their realm that is Nature. Symbols like the spotted owl or the snail dart do not mean much to the common of mortals but some people will defend them with their lives if need be. Churchman derides these efforts by stating that: "A positive benefit-minus cost for the preservation of the snail dart: the act is anti-systems thinking because in the larger system of nature independent of humanity species die off at a fairly regular rate so that we are intervening unnaturally by protecting all species (or some species)."

The Big and the Small can be considered by a religious frame of mind that worships something big or something small. What are the implications of only worshiping a mustard seed and not the universe? Seeing the connections is the essence of spirituality or mindful religion and of systemic thinking. But worshiping only the big or the small is problematic.

Future Generations is 'the sacred cow' which must be invoked whenever a threat to the inherited past is to be defended. Churchman reminds us that we inherited the habit of Worshiping from our forebears. Thus we defend endangered species such as the buffalo, the egret, the geese and the bear to protect them for the enjoyment of future generations. Worship can extend to an immense and abstract "being" such as invoking the need to preserve Nature, unique paintings, and historical buildings for the sake of 'future generations.' The latter becomes then the center of attention. In summary, to worship is to devote our minds and our efforts to a unique deity at the risk of understating others or negating the comprehensiveness of the Systems Approach. But religion does have the potential to contemplate the connections if mindfulness is pursued, rather than worship. The systems approach in its eternal search for comprehensiveness must seek or 'absorb' all enemies.

Scientific Imperialism. The Adversarial Stance Between Science(s) and Religion(s)

This material can also be applied to consider differences between the Physical Sciences and the Social Sciences and their respective god(s). What does "Scientific Imperialism" mean?

Scientific Imperialism refers to the particular belief held by certain groups of scientists that Science holds the most important position in the ranking of intellectual disciplines, even over and above the Humanities, Religion and the like. This belief is most prevalent among physicists who think that science, the hard sciences that is; hold the key to the truth(s) of the universe and that given enough time they would exhaust all the problems and all the questions concerning the origin and the destination of the universe. It is interesting to compare the attitude of Atomic Physics and atomic physicists who are convinced of the supremacy of their science over all others as well as over other forms of critical thinking with religious conviction. Religion(s) also believe in the supremacy of their doctrine(s) and in this respect resemble Science.

4.7.1 Scientific Progress and Religions. Science Versus Religion

The following remarks are pertinent to this comparison and to provide answers to the question: "Why Hasn't Scientific Progress Caused the Disappearance of Religions?" (Cited in Zarader, 2002, op.cit., pp.897-899). These statements also address CWC' notions according to which Science and Religion are enemies of the S.A. and of each other. That science and religion are enemies have been enemies/adversaries of each other throughout the ages is a fact. In the Middle Ages, Religion and religious institutions were the repository of all knowledge. Beginning with the Age of Enlightenment, in the 18th Century, knowledge (other than religious and sacred knowledge) started to separate itself from Religion. In the last two centuries, Religion and Science have become completely separated and occupy different domains and areas of discourse. However, they oppose each other and still fight to determine which of the two ideologies will influence the most minds. We will not resolve the antagonism except to note that:

We can ask the question of "Whether the rigorous proof of the laws of Nature through the apparatus of Science, removes in any way 'the substance' buttressing 'the unprovable truth(s)' of religion(s)?

As the real-world becomes increasingly secular and loses its sacred aspect, it may exacerbate religious fervor of those who think that to be secular is to become an 'infidel.' (cited in Zarader, 2002, op.cit., pp.897-899). Religious wars are based on this assumption.

Science and Scientific Reasoning by its dogmatism and stubborn ideology, resembles religion(s). It seeks to reach and prove truth (the so-called objective truth) and due to the fervor of its practitioners resembles the adepts of a religion. In that sense it offers the disturbing aspects of a "scientific religion," a term which is to be compared to the worship of a religious ideology. (cited in Zarader, 200), op.cit., pp.897-899).

Scientific Imperialism constitutes a good example of CWC notion of "worship" or deity (The Sacred Deity of Truth).

To counter the claims of scientific imperialism, Zarader (2002, op.cit) notes that scientific knowledge is obviously different than religious 'knowledge,' but nothing in science can leads to think or believe that science uses a form of thinking which is *superior* to that of religious thinking. Furthermore, claims that science by its discoveries will eventually reach a state of development whereby it will supplant religion, are not being corroborated or confirmed. As a matter of fact, on the contrary, the religions of the world are gaining converts at a rate that belies the thought that they will disappear soon. (cited in Zarader, 2002, op.cit., pp.897-899).

It is true that religious statements cannot be refuted 'scientifically.' Neither can they be proven right or wrong. It is a fact that science has a more solid foundation of reasoning and scientific methods are more credible than religious pronouncements. However, religious beliefs are taken as articles of faith and do not need proof or confirmation, whereas scientific theorems are backed by an elaborate and rigorous epistemology. (cited in Zarader, 2002, op.cit., pp.897-899).

Finally, religion should not be taken as a regression with respect to universal knowledge. Religion(s) occupy different domains of discourse than science(s). They may be adversaries/enemies but they don't have to be. In Churchman's language, they constitute and occupy completely different inquiring systems and their epistemologies and methodologies are not competing for the same grounds of knowledge. (cited in Zarader, 2002, op.cit., pp.897-899). For example. The confrontation of Evolution and Creationism is a good example of all the points brought up above.

4.7.2 Science Versus Culture

For our purposes, Culture is taken here as a combination of two meanings:

a) Culture is synonymous with the concept of 'society," (Cooper, 1992,1995: 99) and,

b) Culture is considered "common patterns of social kinship" of ideas, beliefs and values that are shared by all members of society (or culture) (Mead, 1965).

I particular like Mead's expression "common patterns of social kinship," which exactly conveys what we share in common as a nation, as a society and obviously as a culture. Mead shows us how different and diverse these patterns in modern and past social aggregations can be. I address the notion of "culture" at this time, to show that in certain quarters it can also be considered an 'adversary.' In one instance, Weinberg (2001) bemoans the antagonism that he senses in US society towards organized science, where these terms may refer to research universities and large academic institutions that are considered 'ivory towers' away from the mainstream. Weinberg (2001) complains that Science has difficulty obtaining funds for research and that there exists an anti-science sentiment in society that conspires against the value and recognition that his "polis," properly deserves. He cites the example of the Superconducting Super Collider –an accelerator to conduct fundamental sub-atomic particle research—a discussion that took place around 1997. But examples abound in the news as we write this essay. For example, when President Ronald Regan died in 2004, his wife reiterated the need to pursue gene and cloning research more aggressively in the hope of saving future generations from the ordeal of Alzeimer's disease suffered by her husband during the last ten years of his life.

The present US administration is battling in court the state of Oregon to prevent them from implementing the only Right to Die statute in legally in force in the nation. The US Administration is still placing barriers to provide abortion funding and sex-education abroad and has not yet approved an over-the-counter drug to allow easy access to 'the morning-after-pill,'- a drug which ahs been available in Europe for many years.

These few examples show the adversarial stance between Science and Politics.

We could also be citing examples of the adversarial attitude between sociologists, historians and philosophers who see the laws of nature as mere "social constructions." In contrast, physicists like Weinberg laud the objectivity of scientific laws which are 'culture-free' i.e. they are "immutable," they resist cultural influence" (Weinberg, 2001: 136-140). So obviously their is a wide gap between scientists and other intellectuals and between intellectual and the average man in the street. These gaps spell conflicts of purpose, conflicts of culture and conflicts of mindscapes.

4.8 THE FOURTH ENEMY: AESTHETICS

CWC first poses a fundamental question: How is The Beautiful (i.e. aesthetics) interrelated with the other attributes (such as The Good and The True) that make human life? (Churchman, 1979: 190).

CWC considers Aesthetics from the point of view of an inquiring system for planning. He states:

> "In classical philosophy it was common to divide the philosopher's task into considerations of the meaning and significance of the Good, the True and the Beautiful."

These attributes should be present in a good design. However, CWC bemoans the fact that "they are absent from most writings dealing with the whole system." (Churchman, 1979: 188).Whenever one leg of the trilogy (comprising The Good, The True or The Beautiful) is considered, one or both of the others are missing and absent from design/planning consideration. In the inquiring system for planning, logic (searching for The True) prevails. It overlooks the Beautiful (or its opposite the Ugly). That's why CWC calls "Logic unaesthetic." The rational mind does not want Aesthetics to intervene. (Churchman, 1979: 191).When scientists in their research seek the truth, they disregard the beauty of a proof, its elegance. When on the other hand they speak of beauty and elegance of a proof, they dismiss logic. Contrariwise, the positivists dismissed the beauty and elegance of a proof and solely worshipped its truth and logic.

4.8.1 Aesthetics: Why can this be 'An Enemy'?

Why is Aesthetics is considered an enemy, the likes of Politics, Morality and Religion? CWC argues that it is because the planner/the planning system "does not want to allow the rational mind to intervene in the very definition of Aesthetics." Aesthetics is added as an afterthought: "The Beautiful seems to be absent from most writings dealing with the whole system." (CWC, 1979: 188)

4.8.2 What is Art and What is Beautiful?

The art world itself is not clear what constitutes Art (how to define it) and furthermore, what is considered "good" or "bad" art. Witness the addition of ready-mades by Duchamp and more recently (2004), exhibits of 'nothing' (literally *nothing* is shown) and "nothings" (the name given to these exhibits) in art shows. To buttress this point, CWC notes that the kind of art that is at the top of the list one day, changes the next day depending on the whim of buyers, curators and artists alike. On this point, CWC asks whether the admiration of historical structures like the Acropolis may become a passing fad. Today we admire the splendor of the Acropolis: but what will be the next fad be? As a result of these ambiguities, Aesthetics may be its own worse 'enemy': it is not readily recognized and thus summarily overlooked: "The trouble came…when aesthetics was cut off from her sisters, the Good and The True," muses Churchman (1979: 189).

"The part of the human system which produces knowledge is not separable from the rest," by which CWC means that the three parts of the trilogy of the Good, the True and the Beautiful must be kept together and considered in unison to understand and obtain 'real' knowledge of the world and of all the attributes of human life.

4.8.3 Where do we Find the Beautiful?

What is 'aesthetic quality' for CWC? CWC lists some of the qualities of human experience that plausibly fall under the rubric of the aesthetics:

"Peaceful, attractive, exciting, anticipatory, luring, and threatening repulsive, boring, anxious, frightening. At the next level of abstraction: comedy–tragedy; love-hate..." In CWC's mindscape, Aesthetics confers quality. Aesthetics may be any qualitative attribute that cannot be said to be rational. Imagination and intuition can confer aesthetic and beauty. The so-called Theory of the Aesthetic conflates hope, strong faith in the future, even love, joy and peace with the beautiful and the aesthetic. Finally, CWC reminds us of the need to have a guarantor to assess the systems planning process and ensure its integrity and credibility whereby each portion of the trilogy of the True, the Good and the Beautiful are represented.

4.9. IN SUMMARY

Why does CWC consider Politics, Religion, Morality and Aesthetics the enemies of the Systems Approach? I hope to have attempted to address this question in this essay. Basically, CWC sought to explain that efforts to be rational and use rational thinking for planning and decision making is fraught with hindrances and barriers which apart from the inadequacies of our methodologies is due to the way the world is organized. Human beings congregate in tight communities as a result of their adherence to groups which may be political or religious and to which they owe their allegiance. Furthermore, while seeking to be rational or ethical they naturally impose their own brand of rationality and ethics, because that is what they believe in. Consequently they do not only impose their own political and religious beliefs but also in addition impose their own rationality and morality while interrelating with other human beings. While CWC limited himself to four enemies (Politics, Religion, Morality/Ethics and Aesthetics) it is easy to conjure other enemies. I will try to address a new list of enemies in this changing world in another paper. These will include pitting one kind of culture and one kind of science against another, and other obstacles to reach an objective, rational and balanced view on any problem or issue. The world is made up of factions, partisans and enemies. The way CWC *describes the enemies* is original, revealing and certainly not trivial. Having decided that conflicts are the order of the day and that we cannot solve any problem in this world without addressing the notion of enemies, we must suggest at least one approach at resolving them.

4.9.1 On Reaching Consensus: Introducing an Inquiring System at a Higher Level of Logic and Abstraction

Bertrand Russell provided the first hint on how to resolve the conflicting and paradoxical discourses take place at the same levels of logic. He postulated that the conflict would be dissolved if the two conflicting sides of an issue were to be considered at different levels of logic or abstraction, or if the discourses could be discussed from a vantage point of higher logic. CWC used the hierarchy of inquiring systems to show that discourses of divergent logic must of necessity be considered in an inquiring system with a level of logic different from those of the original warring sides. Hampden-Turner (1981) also used Russell's philosophy to confirm that logical paradoxes could be dissolved, absorbed and eliminated by resorting to the metalevel. Later, van Gigch (2003) continued CWC's and Hampden-Turner's account by applying the notion of inquiring systems to any logical conflicts.

REFERENCES

Churchman, C. West, 1979, *The Systems Approach and Its Enemies*, Basic Books, New York.

Cooper, David ed., 1992, 1995, *A Companion to Aesthetics*, Blackwell Companions to Philosophy, Oxford.

Edwards P., ed., 1967, *The Encyclopedia of Philosophy*, MacMillan Publishing Co., New York and The Free Press, London.

Goodin, Robert E. and Pettit, P., 1993, A *Companion to Contemporary Political Philosophy*, Blackwell Publishers, Cambridge MA.

Hampden-Turner, C., 1981, *Maps of the Mind,* Collier Books, MacMillan Pub. Co, New York.

Mead, Margaret, 1965, *And Keep Your Powder Dry: An Anthropologist Looks at America*, William Morrow.

Oxford Concise Dictionary of Current English, 6th ed., 1976, Clarendon Press, Oxford, U.K.

Van Gigch, J.P., 2003, *Metadecisions: Rehabilitating epistemology*, Kluwer/Plenum, New York and London.

Weinberg, Steven, 2001, *Facing Up: Science and its cultural adversaries*, Harvard Univ. Press, Cambridge MA.

Zarader, J.P., ed., 2002, Why hasn't scientific progress caused the disappearance of religions?, in: Chapter on "Religion" , *Les Grandes Notions de Philosophie*, (*The Most Important Concepts of Philosophy)*, Collection Ellipses, Edition Marketing, Paris (in French).

Chapter 5

BE YOUR ENEMY

KEN BAUSCH

Executive Director, Institute for 21st Century Agoras

Abstract: C. West Churchman was a very personal writer who wrote from his own existential concerns in an authentic and original way. We get to know him even if we never meet him personally because of the almost autobiographical nature of his work

In *The Design of Inquiring Systems* (1971), Churchman probed how we seek knowledge either as individuals or as groups. He was trying to determine how we think and know. In particular, he was trying to figure out how he could make himself into a good inquiring system. For these purposes, he investigated the thought of Gottfried Leibniz, John Locke, Immanuel Kant, George Hegel and Edgar Singer.

Key words: religion, morality, ethics and aesthetics

5.1 FIVE EPISTEMOLOGIES

Churchman characterizes the epistemology of **Leibniz** as a system of fact nets that include streams of sentences (or charts), some of which may be true, others false, others irrelevant. The citizen's problem is to converge on one story that seems to hold them together in the best manner (p. 176).

According to **Locke**, we are born with a mind that is a tabula rasa. We receive our original ideas as they are written on our slate by reality. Then we develop more complex ideas by ourselves.

The rightness or wrongness of our ideas is then determined by consensus among thinking people. The careful recording of the steps one has taken to acquire a certain knowledge and the careful examination and/or re-enactment of those steps provides the consensus in a scientific community.

> "The Lockean inquirer displays the 'fundamental' data that all experts agree are accurate and relevant, and then builds a consistent story out of these" (p. 177).

Kant said that we impose reality upon our worlds through the activities of our minds. We construe things in terms of space and time; we organize them in terms of innate categories that exist in our minds. We grasp objects through the representations that we have of them.

> "The Kantian inquirer displays the same story from different points of view, emphasizing thereby that what is put into the story by the internal mode of representation is not given from the outside (p. 177).

Hegel introduces passion, drama, history, and progress into epistemology. He says that societies pull themselves together through coherent worldviews that contain their main action/ideas—their theses. A dominant thesis spawns opposition in the form of its "deadliest enemy." This deadly enemy observes the dominant society from a self-consciously superior position. It produces antitheses that attack dominant society on all possible fronts. Conflict ensues.

> The revolutionary looks down upon the reactionary. The reactionary in his conviction can only think that the revolutionary is crazy or criminal; he must utterly reject him as an unnatural evil or a meaningless mind. But the revolutionary understands the nature of the reactionary full well; for him the reactionary's conviction is based on a natural selfish greed and hypocrisy (p. 173).

Next comes another act of one-upmanship. Another observer focuses not on the conviction of the antagonists but on their very opposition. He or she passes judgment on the conflict and settles it by building

> a new world view in which the nature of the conflict is understandable, but which shows that at a higher level the conflict is merely one aspect of reality and not the critical aspect. The conflict in fact is devoured by the higher-level *worldview* (p. 176).

In the context of the Hegelian dialectic, then, an inquirer interested in discerning truth uses the same data as other inquirers but constructs a dialectical model.

[He or she] tells two stories, one supporting the most prominent policy on one side, the other supporting the most prominent policy on the other side.

> The teleological issue is: Which method of telling the story will produce the optimally informed citizen when each is constrained by the same cost and time resources? (p. 177).

In this way, Hegelian storytelling forces us to go beyond routine rationality. It challenges us to give up explicit knowledge for implicit knowledge. It values passion and conflict because opposition produces ideas. The underlying life of the dialectic

> is its drama, not its accuracy ... Drama is the interplay of the tragic and the comic; its blood is conviction, and its blood pressure is antagonism. It prohibits sterile classification. It is above all implicit; it uses the explicit only to emphasize the implicit (p. 178).

Singer was a philosopher and a student of measurement during the first half of the twentieth century. He developed scientific methods for improving measurement. The everyday process of measuring something with a yardstick, for example, can be improved by making measuring rods more accurate. To improve them, one would "sweep in" other factors like the effects of temperature, humidity, motion, the phases of the moon, and so on. This sweeping-in process increases the accuracy and reliability of the measuring process.

In designing standards for measuring the truth of observations, Churchman works in a similar way. He gradually introduces observations about inquiring systems from Leibniz, Locke, Kant, Hegel, and many others. In this way, he honors a realization "that human knowledge does not come in pieces: to understand an aspect of nature is to see it through 'all' the ways of imagery" (Churchman, 1971, p. 298).

In other words, one refines a solid opinion by 'sweeping in' multiple and divergent viewpoints in order to gain a rounded appreciation of a topic. With this method, organizations open themselves to a variety of viewpoints and encourage diverging views in order to clarify goals, objectives, and procedures through cycles of examining practice, theorizing, applying theory, examining results, re-theorizing, and reapplying, etc.

In the dialectical thinking of Singer and Churchman, systems thinkers are most needed when all is going well in the realm of theory and practice. "Then is the time to rock the boat, upset the apple cart, encourage revolution and dissent" (p. 199). They do not advocate the resolution of the ensuing philosophical dispute, but its intensification.

In the implicit world that is captured between thesis and antithesis, there are no explicit rules that guarantee success and/or esteem. The implicit world of intuition and perilous action is the realm of heroes and heroines. In this world, people stumble, fall, and find themselves. Sometimes they are notably successful for the good of their societies.

Churchman calls his systems approach to inquiry "a general theory of systems" (p. 51). He contrasts this approach with the approach of General Systems Theory, which he defines as "a general theory of systems" (p. 51). His systems approach is a method of inquiry while General Systems Theory attempts to integrate the various existing systems of physics, biology, sociology, etc. One can say that his theory is an epistemological one, which is contrasted to the ontological approach of General Systems Theory. Following Churchman's lead, soft systems theorists are interested in how we know, plan, make decisions, and ensure that those decisions are ethical.

5.2 HISTORY OF THE SYSTEMS APPROACH

Churchman says that the systems approach is "based on the fundamental principle that all aspects of the human world should be tied together in one grand rational scheme" (p. 8). He traces the history of this approach as a learning journey, starting with the *I Ching*, the *Bhagavad-Gita*, and the pre-Socratics.

From Aristotle, who "attempted to put all conceptualization in teleological form," he derives the lesson: "In order to conceptualize social systems, one has to think of conceptualization in general and therefore the ways in which we think about the psychological and social worlds" (p. 39). If we adopt the "Aristotelian notion that the two domains of nature, the physical and the social-psychological, are non-separable . . . [then] we can link the findings of all the disciplines in order to explain what we mean" (pp. 39-40).

The heroic truth-seeker finds an ethical dimension in the Kantian concept of logic as "the basic justification of what we take to be true or false" (p. 54). He concludes that some arguments can and should be justified, while others can and should not. In his systems approach of comprehensiveness, he assumes that the facts of nature comprise a unified whole and, therefore, that those facts are to be "explained by what can be called the 'fact net' of nature's world" (pp. 54-55). The meaning of logic, then, "is roughly displayed by the idea of an explanatory fact net" (p. 55).

5.3 THE SYSTEMS APPROACH HERO

In his continuing quest to define an ideal inquiring system, Churchman adopts a semi-autobiographical approach. He identifies himself with the hero, described by Singer, "who wants the world to act sanely . . . [and he is angry because] his enemies are righteously bent on destroying the peaceful world of humanity" (p. 5).

His enemies are people who attempt to "solve" systems problems head-on by relying upon politics, morality, religion, and aesthetics. They fasten upon on one thing and prescribe simple solutions like "outlaw the use of marijuana," and neglect historic lessons like "prohibition." In Churchman's phrase, these enemies commit the "environmental fallacy," which is committed even by environmentalists and ecologists.

> The fallacy of the environment, says our hero, is far more serious than pollution, depletion of resources, and the like. These so-called ecological phenomena are easy enough to see, but the insidious pervasiveness of the fallacy of the environment seems not to be seen at all. Nothing the prophets of doom predict about population growth can equal the disasters that lie ahead as a result of our repeatedly basing large-scale social policies on the environmental fallacy (p. 6).

Our hero prescribes a corrective to the environmental fallacy, or piece-meal approach. It is:

> "the rather bland prescription to look at matters in as broad a way as possible . . . [This] bland prescription questions the whole existing enterprise of science, which is based on the publication of findings arrived at by 'acceptable' research methods." (p. 6).

5.4 THE GUARANTOR OF DESTINY (GOD)

The systems approach, like every overarching theory of everything, rests upon some ultimate logic. Churchman playfully terms that logic the theory's Guarantor of Destiny (GOD), because in many traditional systems approaches, ultimate logic lies in the mind of God that contains the explanation of the whole system.

Churchman identifies three approaches to this question of God as guarantor that he identifies with Augustine, Spinoza, and today's systems thinkers. For Augustine, "God is the designer of the real system, as well as its decision-maker" (p. 41). For Spinoza, "God *is* the whole system: He is the most general system" (p. 41). For many of today's systems thinkers, the idea of god is only implicit in their ideas of progressive systems designs. For them, "god" is just "the best systems design we humans can create at any period of time" (p. 43).

Such a god is incapable of finding a perfect design (to say the least). And god is not static; if we are hopeful, we can say that he learns or evolves, and that later on in his life he'll know a lot more than he does now (p. 43).

The theory of the "scientific laboratory," formulated by Galileo and others, replaces God's design as the guarantor of truthfulness with methodical testing of hypotheses. This laboratory rests on the metaphysical assumptions that "nature is neither capricious nor secretive" (p. 57). It also rests upon an epistemological assumption that its work can be shielded from social environmental influences. Upon reflection, however, it is obvious that "the planning laboratory cannot be decoupled from the rest of society" (p. 59). In particular, the bias of the scientific researcher, who collects "just those data that appear relevant, and can be obtained objectively . . . even though these data are not 'basic' in terms of human lives," (p. 62) must be confronted.

5.5 THE SYSTEMS APPROACH

For Churchman, "human bias is an essential aspect" (p. 62) of an authentic inquiring system. He asks whether such a methodology is "scientific." He answers, "No, if we doggedly stick to the assumption that the classical laboratory *is* the basis of science. Yes, if 'science' means the creation of relevant knowledge of the human condition" (p. 62). The methodology of an inquiring system must be

> "based on a schema that can be modified—for example, by expansion—as the planner learns more about reality, and especially about the reality of the human being ... The schema ... pictures the world of the decision maker as stages in time. First there is an action: at the next stage are the "outcomes" of the action [some good "ends," some unforeseen "side effects]" (p. 63).

Planners can take different postures in this scenario. They may accept the decision-maker's goals and simply engage in problem solving. They may accept those goals but put some constraints upon their proposed solutions (legality, feasibility, and so on). Or they may refuse to limit themselves to narrow considerations.

They might ask the basic question: "Who should be served by the social system and in what way?" (p. 64). In other words: who are its appropriate clients? The answer "may (probably should) include future and past generations" (p. 65). If that is the case, the questions: 'who should design and how? and, who should plan and how?' follow inevitably" (p. 65).

When such broad questions are involved, the ends of an inquiry are no longer identified "in terms of the intentions of the decision-makers who pay, but rather in terms of a general ethic" (p. 65). In this situation, "the 'inquiring system' becomes a 'systems approach,' and planning is 'ideal planning'" (p. 65).

The ambitions of systems theorists go beyond the ambitions of the classical laboratory. They try to weave all matters of human concern into "one grand imagery of purposive behavior" (p. 67). They seek to tie all our goals together, and try to accomplish that task without resorting "to mysticism or some deep inexplicable human essence" (p. 67).

5.6 ETHICS

If an ideals approach to systems is to come about, it needs to confront issues of ethics. Churchman does this by considering the work on ethics done by Kant and Bentham during the 1780's, and the work of Singer and Jung who dealt with ethical matters during the 1940's and 1950's.

Bentham proposed that actions were good if they were useful in preventing pain and enabling pleasure. He proposed the principle that governments existed to minimize the pain of their citizens and maximize their pleasure. To discern the appropriateness of a piece of legislation, then, one simply totals the pleasures that might derive to people and subtracts the pains.

Kant radically dissents from Bentham. He says that "power, riches, honor, even health, and . . . happiness" can be perverted. Only Good Will "can be called good without qualification" (quoted by Churchman, 1979, p. 122). Good Will does not operate by conditionals; it is categorical. Kant's categorical imperative expresses the big idea that what your action expresses deserves to be considered as a universally valid law. Churchman paraphrases this imperative as follows: "If you can tolerate the burden of having your big idea become a universal law, then your big idea and the action are moral. If you cannot tolerate it, then the big idea and the action are immoral" (p. 123).

Jung elevates "the process of individuation to the pinnacle of the Good" (p. 130). He says, "Individuation is indispensable for certain people, not only as a therapeutic necessity, but as a high ideal, an idea of the best we can do" (quoted by Churchman, p. 132). Jung's version of the categorical imperative, therefore, is "Thou shalt never interfere with the process of individuation of thyself or another" (p. 132). He makes individual morality the keystone of social morality.

Singer espouses a morality based on social ends. For him, "if we are to search in a rational manner for the good, it must be found in our purposes" (p. 135). He proposes a standard of ethics based upon the scientific notion of an "ideal." For him, "an ideal in principle is unattainable but in principle can be approximated within any prescribed limit" (p. 135). According to Singer, "the statements we make in response to questions about the natural world are always approximations" (p. 135). The actual *answer* to any question is beyond precise statement and is always an ideal. For that reason, "the real is always an ideal" (p. 135).

Now, since we humans are goal-seeking creatures, we also seek the power to obtain those goals. On the basis, of this principle, an "enabling" value theory can be constructed by categorizing the essential things "that increase an individual's chances of gaining what he wants" (p.137). The three necessary things are:

1. a richness of means at his disposal—that is, "plenty";
2. an awareness of the appropriate means to select—that is, "knowledge"; and
3. a desire for goals that are consistent with the goals of others—that is, "cooperation" (p. 137).

The need for cooperation is problematical for a person whose only interest is self-interest. "Why should I pursue these ideals to any extent beyond what serves me best? Specifically, why should I seek the betterment of the lives of future generations?" (p. 139).

If one desires the contentment that comes with a progressive life, one understands this need. One experiences dissatisfaction with oneself and one's surroundings, and an urge to renew vital processes. Because of this dissatisfaction, a person may become actively involved in systematically designing a future worthy of our past history and our descendants. That person becomes a hero. In the words of Singer, "to win contentment, one needs all the qualities of the hero" (quoted in Churchman, 1979, p. 139).

Churchman suggests that "the quality of mind that, for Jung, makes a person seek his own individuation is the same quality that, for Singer, makes him a 'hero'" (p. 139). The psychic force that drives the hero to do ideal planning "is the very same psychic force that drives one through the 'stages' of individuation" (p. 139). In summary,

> Jung's basic exhortation is toward the psyche as a reality and toward the individual psyche, while Singer's is toward the collective "mankind" and toward the most general psyche. Jung's hero seeks the completeness of the Self, Singer's the completeness of all Selves (p. 141).

5.7 THE STORY THUS FAR

Our disgruntled hero has organized his life around designing an inquiring system for himself that is as comprehensive as he can make it. He has developed a clear idea of his own ethics, but he has to admit that he does not have an overwhelmingly persuasive case for his concern for future generations.

> He may be overwhelmed by the image of the possible future of humanity going to its living hell, disease-ridden, malnourished, energy-less. But why should such an image disturb him? There may be psychological

causes why it *does* disturb him, either archetypal or in his own past history. But why should it? (p. 143).

In addition, the morality of the heroic mood is not universal. The ideal planner finds himself in a we/they situation. "The hero dearly wishes to be at one with humanity, just one of the boys/girls, but can't help observing that a vast number of people are different from him/her in their moral concerns" (p. 144).

In his strangely alienated position, the hero can still comfort himself or herself with an image of human solidarity.

> He may take some comfort in the reflection that his vision, though not universal, is essential for the survival and the betterment of humanity, and that the vast majority of humans, taken collectively over the ages, agree that if betterment is possible, it should take place. [Still the hero has doubts. He] wants humans to accept some version of holism in their lives. But why? Why should humans obey the prescriptions of rationality? And there is another "Why?" Why are the rational types urging more rationality on the others? What is there about these people that makes them continuously fight battles to overcome perceived stupidity? (p. 155).

A key to an accommodation to these "Whys," if not an answer, is provided by the Socratic "Know thyself." "The rational planner cannot exclude himself from the whole system, else his holism is unholy." (p. 155). Our hero must immerse himself in the reality of his enemies and recognize the limits of his rationality.

5.8 THE ENEMIES

As a first step, he tries to incorporate the insights of groups that he calls "enemies of the systems approach" into that approach. He enters this discussion with the conviction that he has to cast as wide a net as possible. In the midst of the widely diverging information that he gathers with his net he intends to use three closely related strategies: "Bringing different observers more closely together is one strategy. Another is to use the divergences to suggest another area to investigate. A third is to decide that the divergences show the irrelevance of the information" (p. 147). He is convinced that "an observer is not 'objective' if he sticks to but one of these strategies" (p. 147).

The enemies of the systems approach are people who do not use rationality in the manner devised by our hero. For convenience, Churchman gathers them under the categories of politics, morality, religion, and aesthetics. People who think in these categories do not think holistically; they think and feel *issues*. They become emotionally engaged and think principally in a mode of defending preconceived convictions.

Churchman builds his idea of *politics* around the *vita activa* in the Greek *polis* (city-state) as Hannah Arendt described in *The Human Condition*.. "I've taken this idea and called it 'making polis' around an issue" (p. 157). A family becomes *polis*, for example, if it gathers together around a common concern like getting a child through school. Groups and counter groups become *polis* over issues like pollution, anti-development, gay rights, and so on.

There are several aspects of *polis*-making that differ from the rational systems approach. First, *polis*-making takes shape around an issue. If it succeeds, it may go out of existence with a celebration. If it does not succeed, it may go on for a long time. Second, success is measured by feelings of satisfaction. Third, "a definite, often hard attitude toward information occurs in the life of *polis*" (p. 157). Because of their issue-commitment, *polis*-makers take an ideological stance and become incapable of seeing facts that might undermine their sense of rectitude.

Churchman dramatizes the plight of the hero by considering "the tragicomedy of the systems hero as he encounters *polis* at work" (p. 158). He wants to tell conservationists that conservation has to be seen in the context of the larger system.

> But the conservationist's joy comes from saving a redwood forest from destruction, and not from planning the whole future of the state or nation. The issue dissipates when it's put in the larger context ... To a keen conservationist, the real enemy is in the counterpolis, the guy who says, "Once you've seen one redwood, you've seen them all." He cannot even hear the voice that tells him to put conservation in the proper context (p. 158).

Because of their ideological stance, *polis*-makers consider our hero an enemy. They think that he is trying to co-opt them.

Moralists espouse values that our hero holds dear: "fairness, equity, treating humanity as an end withal" (p. 198), but they resist any attempt to balance values by cost-benefit analysis or some other means. They insist that there are no trade-offs when it comes to robbery, suppression, or murder because those acts are sins. Moralists repudiate any attempts to rationalize or make dialectical decisions. For moralists, morality is a universal feeling and rationality is largely irrelevant.

Religion demands that the systems approach recognize that we are not the only designers of systems. The reality of God needs to be recognized. God needs to be worshipped by humans, who are basically non-rational beings who are to obtain their direction from revelation.

The *aesthetic* person demands that our hero look toward the quality and not just the content of his planning. One should take into account the uniqueness of every individual. At this juncture, the hero perceives the trap. "Everyone's uniqueness is a world in itself, incomparable with any other uniqueness" (p. 199). Therefore, "Gone is tradeoff. Gone is adding up values. Gone is any sensible way of assessing change" (p. 199).

5.9 THE RETREAT TO A VISION

In his encounters with these enemies, the hero tries to sustain his vision of comprehensive rationality, but he is done in at every turn by appeals to comprehensiveness that undermine rationality. In these straits, he turns to visions of progress. Kant's vision included the perspective of humanity gradually creating a world in which virtue and happiness begin to coincide—that is, where virtue produces happiness. For this to happen, the vision needed a guarantor, God, the possibility of free choice, Freedom, and an opportunity for the endless struggle, Immortality. But is there any reality in such a vision? Kant calls the three conditions "postulates," as though in effect there was a demand for them; else the whole plot collapses (p. 201).

Singer improves on Kant's vision in two ways. First, he enriches his vision with a "heroic mood," which "stirs us out of our humdrum life of satisfaction to search for new ways to follow the pathway of progress" (p. 201). He finds evidence for this mood in tragedy, comedy, and heroic music. Second, he likens his vision to ideal images that are employed in physical science to approximate measurements.

Just as ever-more-precise measurements reduce the error-factor in a measurement so that they converge on the ideal of zero-difference without ever attaining it, so by using the technique of 'sweeping in' alien perspectives, one converges on an answer. "Hence the real becomes the ideal, because ideals are desirable ends (signposts constantly beckoning us on) which can be approximated but never attained" (p. 202).

Still, can one observe that progress is really taking place? Can systems visionaries integrate their vision with reality? Not necessarily. Some visionaries escape the need for integration by retreating into the realm of abstract models, and proclaiming that their models have no practical consequences. Others cut off a piece of their vision and do something practical by using politics, persuasion, and tricks of the trade—and lose the vision. The visionary finds himself espousing a faith that fails in every instance—but is somehow true, in general.

5.10 BE YOUR ENEMY

As our hero pushes himself toward a greater and greater comprehensiveness, he finds himself in a realm of incomprehension. He is now faced with a paradox: his own reflection on the deep puzzles of systemic reasoning leads him to results that run counter to that reasoning. He grins. He knows that paradoxes are a driving force of rational advances. He feels that "there is much to be said . . . for purposefully designing paradox" (p. 204).

To deal with the quandary, he offers himself the prescription: "Be your enemy" (p. 204). With this prescription, he opens up a "new teaching of rational planning—that is, a new and more general meaning of rationality, which goes beyond dialectical reasoning" (p. 204). This new teaching says,

> Rational humans need to leave the body of rationality and to place the self in another body, the "enemy," so that the reality of the social system can unfold in a radically different manner. From this vantage point he/she can observe the rational spirit and begin to realize not only what has been left out of it, but also what the spirit is like, especially its quality of being human (pp. 204-205).

By releasing the bonds of hard rationality, one can see that rationality is a tool, an expression (among other expressions) of what it is to be human.

In this stepping out of the body of rationality into the bodies of politics, morality, religion, and aesthetics, our hero does not lose his identity; he continues to operate as a deeply involved rational planner. He embraces the kind of "sane schizophrenia" that Otto Rank (1932) offered to visionaries: "at one and the same time . . . [to live] visions and the reality of the collective consciousness" (p. 213).

In stepping out, our hero gains objectivity about his rational self and leavens it with some humor. In looking at himself *as his enemy*, he begins to see how foolishly he pushes "one point of view, of model building, statistical analysis, game theory, ethics, or holism" (p. 214). Churchman expresses the satisfaction that comes with this recognition as follows:

> Once you are your enemy, you at last see yourself as you really are: a human being, wise and foolish, who has a quirk about the destiny and the improvement of the human condition, just as all the rest of humanity has its quirks (p. 214).

REFERENCES

Churchman, C.W., 1971, *The Design of Inquiring Systems*, Basic Books, New York.
Churchman, C.W., 1979, *The Systems Approach and its Enemies*, Basic Books, New York.

Chapter 6

THE SOCIAL SIGNIFICANCE OF CHURCHMAN'S EPISTEMOLOGICAL POSITION:

Implications for Responsible Conduct

NORMA R.A. ROMM

Research Consultant, ABET, University of South Africa

Abstract: Churchman calls on would-be scientists, as well as others involved in inquiry endeavors, to work across "disciplines" and in the process to be more reflective about the possible consequences in society of their very way of framing research questions for investigation. This is because, in his view, our way of "doing science" already delimits options for practically addressing issues of concern. He proposes that people in society (whether professional inquirers or others) express a commitment to ethical thought/practice in their way of thinking/living – which he calls "conducting ourselves properly" (1979, p. 8). This chapter is aimed at casting additional light on the epistemological and ethical proposals raised by Churchman, by considering his argument in relation to realist-oriented critique that might be directed against it

Key words: "Truth", "fact", framing of questions, proper conduct in the world, accountability.

6.1 INTRODUCTION

In considering the contribution that "science" can make as an enterprise in society, Churchman (in an interview with Van Gigch *et al.*) argues that a meaningful science as far as he is concerned:

> "... won't look like today's science. It will come from a collaboration of different disciplines. It will become a co-ordinated science where the experts are not today's experts of rigorous science but of a science whose main concern is the service of humanity. It won't look like today's science." (1997, p. 741).

Churchman pleads for a science which is "controlled ethically" (1994, p. 135), so that it is informed by inquirers' recognition of the non-neutrality of the scientific enterprise in terms of its possible consequences in society. He avers that knowledge-generating processes are led by boundary judgments concerning how to delimit the research problem or questions to be addressed, and these judgments cannot be considered as neutral in their consequences (cf. Midgley, 2000, p. 137).

In putting forward his argument for an ethically controlled science, Churchman does not concede to the view that scientific observation of "facts" is, or could ever be, a process of observing "realities" independently of some point of view (perspective). He argues that the term "objective" when referring to observations refers at best to "the design of a group of investigators who are attempting to keep their observations under control in a language context where it is impossible to obtain exact agreement." (1994, p. 136). Hence, "objectivity" refers to processes for generating some kind of "agreement" on "facts". When discussing the dialectic of modern philosophy (1946a), Churchman notes that when we reflect upon the scientific enterprise we can pose two questions: "Do answers to (some) questions of fact exist?" and "Do answers to (some) questions of law [referring to cause-effect relationships in reality] exist?" (1946a, p. 116). He comments that when considering these questions we realize that:

> "...the basic terms of the problem of philosophy [epistemology], seem most difficult to define; when we shall say a question is answered, we shall say that a true statement has been made which satisfies the demands of the problem, so that the concept of 'answer' entails the concept of truth, and he [or she] who must define the one must take on all the burden history has felt in defining the other." (1946a, p. 116)

Churchman points out that although it may seem common sense to us that it is possible for us to somehow gain access to facts, "we cannot rely on common sense to make entirely unambiguous the concept of 'fact'" (1946a, p. 117). The same goes for other "basic terms" such as laws and theories.

In terms of Churchman's epistemological argument, we cannot ever presume to have created "true" statements about the world; nor indeed can we rely on probabilities that statements can be considered true with reference to some posited object/aspect of reality. Any given statement (no matter with what degree of confidence it is held) relies on us deciding that it constitutes some kind of "answer" to problems that we have set.

Churchman calls upon us to acknowledge the inextricable link between our knowing and our involvement in the world by proposing that as responsible inquirers we need to consider the possible consequences of framing (research) questions in a particular way. Nevertheless, in terms of his epistemology it must be admitted that the manner in which consequences *are* generated in social life (through our inquiries) can of course never be known. Or rather, they can never be known in the objective sense of the term (that is, by using a definition of objectivity as implying some manner of "accessing" reality). How, then, does Churchman justify his suggestion that we take some responsibility for the possible impact in society of our knowledge-generating processes?

One way to approach this question is to refer back to Churchman's proposal that we make provision in the inquiry process for the necessary *social discourse* that can sweep in alternative values, ways of life, ways of seeing, and ways of understanding that form the basis of intersubjectively generated thinking and being (1979, p. 9).[22] This draws attention to his emphasis on helping people, in processes of inquiry, to "understand different points of view" so that alternative considerations can be swept in from the start of the inquiry (cf. Ackoff, 1999, p. 327; Hammond, 2003, p. 257). Churchman's understanding of an ethical existence is that we are able to bring alternative worldviews to bear, as genuine alternatives, in constructing meaningful ways of "improving the human condition" (Churchman, cited in Van Gigch *et al.*, 1997, p. 742).[23] McIntyre explains her interpretation (with a group of people with whom she was working in a specific context) of Churchman's ethical requirement for systems thinkers:

> "To the extent that a system fails in its apperception it is less than God, i.e. it is an imperfect monad (Churchman, 1971, p. 75). This was interpreted as meaning that our best attempts at 'sweeping in' and 'unfolding' all the relevant issues are unlikely to be perfect and that we can only strive toward the ideal; this ideal can be seen as the holistic goal

[22] This proposal of Churchman is clearly not tantamount to encouraging efforts towards creating/inventing "sharedness" as coerced agreement. Churchman indeed criticizes certain (Business School) courses on conflict resolution, which are designed to "teach [managers] how to influence workers so they don't fight with management and what management want. It is about getting the rest of the organization to go along with management. This is not ethics" (1997, p. 742).

[23] Churchman's reference to improving the human condition does not of course exclude a care for so-called non-human existence, as can be gleaned, for instance, from his expressed concern with our "rape of the environment" (1994, p. 139). Romm (1998a) draws on Flood and Romm's (1996) view of triple loop learning to explore various ways in which we can account for a consideration of issues of sustainability in our engagement with our lived worlds.

or spiritual quest. Having a sense of humility could help human knowledge to evolve and to be mindful that knowledge is improved through openness to diversity in thinking and practice." (2003, p. 495).

The openness to diversity that is invoked by McIntyre, is not for her a matter of trying to "find out" the truth (or even to get closer to finding out what really "is" in "the world"); it is rather process of participants in the context under consideration finding a way of working together with the issues considered problematic.

Nevertheless, when adopting some (pragmatically-oriented) view of the systemic link between thought and action, decisions have to be made about "the issues" to be addressed and ways of addressing them – in such a manner that people can account for the decisions made. Van Gigch (1995, p. 87) offers the suggestion that we interpret Churchman's argument to imply that "the ideal inquiring system is one that combines the love of truth with the love of good and that of the aesthetic (or beautiful)". Decisions as to how to proceed in inquiry processes can then be accounted for (and judged) in these terms.

However, in response to these kinds of arguments, authors who commit to a more realist-oriented approach[24] as the basis for all scientific inquiry assert that a pragmatically-oriented insistence that "reality" can be properly approached/explored only through inputting certain practical intentions (such as "service to humanity" as proposed by Churchman), detracts from research efforts toward truth-seeking (cf. Hammersley, 1995; Hammersley and Gomm, 1997 a, b) In this (realist-oriented) view, science – as traditionally understood – enables people to act in terms of more informed opinions than they otherwise would if science did not function as such as an institution of knowing in society, in which scientists strive to generate increased knowledge of extra-linguistic reality. Churchman's plea (1994, p. 134) for a "responsible science" that incorporates some form of practical intention, is thus ill guided according to this view.

My purpose in the chapter is to examine some of the above-mentioned points of contention surrounding processes of "proper" inquiry and the

[24] The term "realism" embraces a variety of positions that might be labeled under this banner. Romm (2001) offers a discussion of four kinds of position that can be considered realist in orientation (albeit in different ways). These are: positivism (a position focusing on the search through scientific inquiry to come to grips with natural and social regularities); non-foundationalism (a position that emphasises the impossibility of obtaining indubitable proof); scientific realism (expressing a focus on excavating structures of reality); and interpretivism (expressing a focus on striving to understand meaning-making in the social world). In this chapter, I concentrate mainly on how a realist-oriented stance implies a critique of pragmatically-oriented epistemological arguments.

"proper" responsibilities of inquirers in the light of a reconsideration of Churchman's arguments.

6.2 RECONSIDERING THE SCIENTIFIC ENTERPRISE: REVISITING NOTIONS OF TRUTH, FACT, AND HUMAN RATIONALITY

Churchman indicates that in terms of the Pearson-Neyman conception of what the proper practice of science involves, scientific hypotheses are considered as being testable with respect to relevant evidence. Evidence (in the form of observations of "sets of data") can be drawn upon to accept or reject specific hypotheses. He notes that according to Pearson and Neyman, one can use statistical tests to determine the probability of "error" in mistakenly considering hypotheses as true when they are in fact false and vice versa (1946b, p. 85). But Churchman asks the question:

> "...what does 'true' mean in this context? Of course it is not difficult to grasp intuitively the meaning that the authors wish to convey: evidently, the suppose that there is some ontological sense in which a hypothesis may be conceived as true (or false), even though the experimenter does not know the truth." (1946b, p. 85)

Churchman identifies that what is implicit in this argument is that:

> "If the number of experiments of this sort were increased indefinitely, then among all the possible methods of making inferences from the various sets of data, that provided by the Pearson-Neyman theory would result in the least number of wrong inferences. The *right* inference would be the limit approached by the total number of observations as the number is increased without limit. That is, the concept 'true', and the concept 'best test' are defined in terms of a construct of nature that pictures an infinite sequence of experiments that are similar, whose cumulative results approach a limit." (1946b, p. 85)

Churchman avers that this way of reasoning still raises the question of why it is taken that "observations" can be regarded as offering information about objects in the world supposedly existing outside of our inquiry process. We can indeed decide to *deny* this assumption by suggesting that "all 'observations' are themselves inferential entities" (1946b, p. 86). Put differently, an apparently observed "fact" does not mean that the "fact" can be taken to exist in the ontological sense of the term. The word "fact" could simply refer to the idea that a statement is less general than other statements (such as "laws" and "theories"). As Ackoff elucidates: "The less general a statement, the more fact-like it is; the more general a statement, the more law-like it is" (1999, p. 302). And Ackoff continues that "theory" is defined as a "still further generalization" (1999, p. 303).

Ackoff notes that it is sometimes maintained that "the confirmation of facts does not require use of theory but that confirmation of theories does entail confirmation of fact" (1999, p. 304). He contends that this is not the case, and he refers to Churchman to support his contention. As Ackoff states it:

> "... it was once assumed that (and still is by some) that, if facts and theory do not agree, then the theory must be false. As the history of science shows, however, the scientist is just as likely (if not more so) to question the validity of the facts as he [or she] is to question the validity of the theory." (1999, p. 305)

Ackoff goes on to indicate that the ideal of "objectivity" has been proposed as a way of encouraging inquirers to be "unemotional, uninvolved and unbiased" in their inquiry endeavors (1999, p. 312). But Ackoff (as Churchman) turns around this (ideal) requirement and argues that:

> "Objectivity does *not* consist of making only value-free judgments in conducting inquiries and making decisions. It consists of making only value-*full* judgments .... Objectivity cannot be approximated by an individual investigator or decision-maker; it can be approached only by groups of individuals with diverse values [that are brought to bear in framing the issues].l" (1999, p. 312).

Thus considered, objectivity for Ackoff (as for Churchman) is tantamount to generating intersubjective discussion that accounts for a variety of views in the addressing of "realities". As Churchman explains:

> "We ... need to design into the inquirer an ability to see the 'same' object from different points of view ... we need to develop the additional idea of an 'object' as a collection of interconnected observations in which each observer can examine how another observer views the world. The 'objectivity' of experience is to be based on some kind of interconnection of observers." (1971, p. 149)

Haynes comments that in terms of this passage from Churchman we can see that he poses the need for some kind of synthesis of *"different points of view"* (2001, p. 30). Haynes suggests that the creation of a genuine synthesis is not necessarily seen as a "logical sense of two opposing forces reconciling themselves together" (2001, p. 38). The process of developing reconciliations may not be one based solely on (narrowly defined) reason distinct from emotion/value-commitment. Haynes mentions that in Churchman's *Thought and Wisdom* (1982), Churchman himself expressed that his earlier proposed designs developed in *The Design of Inquiring Systems* (1971) "turned out to be too rational" (Haynes, 2001, p. 38).

In his *The Systems Approach and Its Enemies* (1979), Churchman grapples with the question of how we can keep in consciousness a dialectic between reason and its supposed "enemies". He points out that

> "those committed to a systems approach would do well to recognize that the apparent enemies of rational thinking (politics, morality, religion, and aesthetics) can be embraced as potential friends (or, at least, need not be seen as enemies)." (1979, p. 197).

The next four sections subsections (below) consider each of these apparent enemies of rationality in turn. But before beginning this discussion, it is worth noting McGregor's comment, in reviewing Churchman's *The Systems Approach and Its Enemies*, that Churchman's style of writing in this book is in the form of:

> "...playing logically opposite points of view against each other. In this way he organizes a dialectic where the 'enemies' of ideal systems planning [can be] cast as 'friends' .... This new-found friendship is itself a product of dialectic in which Churchman helps us imagine how politics, morality, religion and aesthetics could be used to support ideal systems planning." (1980, p. 87)

McGregor proceeds to show that the dialectic journey is unending and involves, in Churchman's words, "*a process of continuously ... and I think eternally - discussing and debating and occasionally fighting over the issues*" (cited in McGregor, 1980, p. 87). Churchman does not presume to offer final "results" on the questions that he poses in regard to our way of addressing the issues. But he does hope to enrich the debates around the topics raised in the book - which he presents as crucial ones for discussion and consideration. As McGregor summarizes: "*Those with an appetite for debate and complexity should profit from the work*" (1980, p. 87).

6.2.1 Politics: Enemy and/or Friend of a Systems Approach?

Churchman highlights that

> "Politics suggests that in order to create change in the social order, one needs to gather together a community (polis)" (1979, p. 197).

He remarks that politics is often associated with the understanding that "politics [differing people's practical interests] decides the rules of the debate [about how to address social issues] and the referees [how the judges will be chosen]" (1979, p. 200). In this understanding, politics can be seen as an enemy that ruins the chances of genuine debate in processes of rational inquiry. In order to explain Churchman's manner of reversing the view of politics as necessarily an enemy of the systems approach, it is worth exploring in some detail the manner in which this view of politics has been posed by others. To this end, I will use Hammersley's argument as an exemplar of this position.

In supporting the view that rational thought should not be tarnished by political interests, Hammersley argues that it is crucial that the scientific enterprise in society is kept strictly separate from the domain of politics. The scientific enterprise, when operating properly, is geared to provide maximum opportunities for rationality to flourish in the sense that the scientific community is involved in correcting any biases that individual researchers might bring to bear in their inquiries. As Hammersley (citing favorably Popper's argument in this respect) puts it:

> "Important as such asceticism [necessary for detachment from political interests] undoubtedly is, one should not underestimate the role of the research community in socializing researchers into it and maintaining their commitment to it, as well as in correcting at least some of the biases which it does not prevent." (1995, p. 115)

Hammersley here refers to Popper's argument (1959, 1969, 1994) in order to affirm the role of the scientific research community in calling its members to account for their way of creating observations and generating inferences in relation to their observations. In line with this argument, Hammersley and Gomm (1997a) emphasize that the admission that we can never have complete certainty in any statements forwarded during the process of scientific inquiry, should not commit communities of inquirers to take the position that there is no point in trying to avoid error. On the contrary, it is still crucial (indeed, more so than ever) to develop a concern to attempt to "avoid the danger of accepting as true what is in fact false" (1997a, paragraph 4.2). Researchers engaged in practicing science, are required to "do their utmost to find and keep to the path which leads toward knowledge rather than error" (1997a, paragraph 4.3).

Hammersley's (and Gomm's) view of the proper practice of science is based on the contention that the immediate goal of scientific inquiry must be the goal of producing knowledge. Any political goals associated with interests in implementing certain actions, runs anathema to demands for rational debate in the scientific community. Scientific debate has to be guided by the principle of seeking interest-free knowledge. This, Hammersley insists, is what is expected by "funders [of research] and lay audiences" in the wider society (1997, paragraph 1.12). As he argues, "if funders and lay audiences believed that research were primarily concerned with serving researchers' (or someone else's) political goals, it would not survive" (1997, paragraph 1.12). For it to survive, scientific activity must be seen to be directed toward the goal of producing knowledge.

In considering the question as to how judgments are made by people within the scientific community concerning what constitutes reasonable behavior, Hammersley and Gomm admit that these judgments are fallible: they are "open to dispute and to subsequent revision" (1997a, paragraph 4.9). The possibility of fallibility of judgment on this score nevertheless does not detract from the fact that people can still judge one another's behavior (though never with complete confidence) in terms of the way in which its reasonableness is presented.

In terms of this view of scientific reason as a mechanism to find out about the world, the scientific community is seen as strictly separable from what Churchman calls "the polis". But Churchman's alternative point is that if the polis (where social discourse on issues of concern is generated) is guided by the (no matter how tentatively put) statements about "facts" and "truths" as reported by scientists, then this *already has pre-framed the social discussion along restricted lines of thought and action.* According to Churchman, scientists can be considered as committing what he calls "the environmental fallacy" (1979, p. 6) when their choices to "cut off the environment of the problem" is not accounted for. Their decision to focus on research questions in a certain way, rather than looking at matters more broadly, is already a political act (with consequences for discussions in "the polis"). Midgley expresses Churchman's position as follows:

> "...there's a tight relationship between boundaries and values, and reflection on different possibilities for making boundary and value judgments becomes imperative if an intervention is to be regarded as systemic." (2001, p. 253)

Midgley, as others (for example, Romm, 1995, 1996, 1997 a, b, 2001, 2002 a, b; Jervis, 1997; Banathy, 1999; Nelson, 2001; McIntyre, 2000, 2003, 2004) argue that all knowing processes at the same time can be considered as interventions that "make a difference" to the subsequent 'unfolding' of outcomes. That is, as Nelson expresses it, "thinking and acting or making" are, in this epistemological and ethical position, "not divided into separate domains" (2001, p. 286). In adopting this position, Nelson proposes that "systems thinkers as leaders [need to] act systemically and responsibly with full accountability for the consequences of their work whether intended or unintended" (2001, p. 286).

Strictly speaking, though, as indicated in the Introduction, we cannot ever know what the "real" consequences (unmediated by human experience/ understanding) are of any inquiries that may be undertaken. In *The Systems*

Approach and its Enemies (1979) Churchman invokes the example of the narrowing of a specific research focus on the physical effects of smoking marijuana without paying due attention to possible consequences of this focus. He argues that by the very offering of "reports" on the effects of the drug (reports supposedly springing from "impeccable" inquiries) scientists already frame the social discourse – which becomes restricted to considering possible means for controlling "the use of the drug" (1979, p. 6). Given this restricted focus, "police search and seizure" then become a justifiable response in "the polis" (1979, p. 4). Meanwhile, this response can have the further consequence that "an underworld drug market" becomes created, leading to the price of the drug going up, potentially "necessitating robbery and even murder on the part of the drug abuser" (1979, p. 4). Now, considering my suggestions offered in my Introduction, although we cannot know in any objective sense about these "consequences" of the initial inquiries, we can be more alert to the way in which our inquiries *might contribute* to the 'unfolding' of people's (in this case policy makers as well as drug users) responses/actions. This would account for Churchman's plea for inquirers to "sweep in" a range of concerns from the start (as well as while inquiries develop), so that broader considerations can become part of the process of defining the issues and ways of addressing them (1979, p. 9).

In short, in terms of these considerations, rational systems thinking can and should be concerned with the question of how issues are likely to become debated in the political arena; that is, inquirers should not deny any responsibility for the 'unfolding' of politics and political discussion in "the polis". But the meaning of rationality hereby of course becomes extended so that the erstwhile dualism of "scientific" rationality and "political" decision-making is transcended.

6.2.2 Morality/Values: Enemy and/or Friend of a Systems Approach?

Churchman defines morality as "the underlying spirit of all action that drives a person to act as he [or she] does" (1979, p. 25). He notes that in many people's views, scientists should not get involved in considering or deliberating upon the quality of the goals or values that people in society (or that they themselves as citizens) may uphold. Morality/values – as driving forces of people's conduct and decision-making – must not interfere in processes of "rational" scientific thought. Churchman summarizes that in this kind of view, the realm of morality seems to amount simply to "blind collective feeling" (1979, p. 198). He asks:

Are we to be left, not only blind, but driven by a blind collective feeling, which is not only whimsical but often downright dangerous as it creates dissent, strikes, wars, and revolutions. Such an enemy is clearly *not* us [as rational systems thinkers] (1979, p. 198).

While Churchman concedes that rationalizing morality would seem to "ruin the spirit of morality" (1979, p. 198), he does not subscribe to the view that morality is a realm that should be excluded (and left untouched) by inquirers. The decision to treat as not amenable to rational inquiry the realm of values, has the consequence that we do not try to direct our efforts at, say, investigating options for nurturing human capabilities for compassion and caring (1994, p. 137).

Churchman is resolute that systems thinking needs to include moral considerations so that it can be directed towards looking at problems "as broadly as possible" (Porra, 2001, p. 21). Churchman explains his position:

> "On the broadest level, the systems approach belongs to a whole class of approaches to managing and planning our human affairs with the intent that we, as a living species, conduct ourselves properly in this world .... The systems approach is, therefore, only one approach to the way in which humans should respond to reality; but it is a 'grand' approach, by which I mean 'large', 'gigantic', or 'comprehensive'." (Porra, 2001, p. 21, citing Churchman, 1979, p. 8).

Churchman expresses his concern that with all the technological developments characterizing the "modern" age, which seemingly allow for control and mastery of the world, we have not yet come to grips with "how to control *desire*" (Churchman, cited in Van Gigch *et al*., 1997, p. 743). He states: "you haven't got anywhere near control if you can't control the desire function". He points out further that:

> "Gita's [the Bhagavad-Gita's] point is that where humans go off the track is because they can't control their desire. For example, just ask Mr. Perot, 'How many more billions of dollars do you need to be satisfied?' If he is honest, his answer will be that he will never reach that point." (1997, p. 743).

Churchman remarks that the message from the Gita is that "we are all in a state of confusion because our ethics are in a state of confusion. We don't know how to control the desire in our lives" (1997, p. 743). Churchman comments that this is a message that "I don't get ... out of Kant". He mentions that he is not a Kantian in the sense of trying to "find *a* moral law" (1997, p. 743). He finds it impossible to define universal moral principles as in Kant's dictum for people to try to "legislate universal law" (Kant, 1785, p. 44). But he suggests (to his interviewers) that it *is* possible to posit an overall ethical purpose to which people should adhere: "Service to humanity. That's an invariant" (1997, p. 743).

In order to serve humanity, Churchman (1994, p. 141) argues that science cannot pose itself as an enterprise detached from a concern with the plight of humanity. He expresses his frustration at the "real confusions of modern science" - a "science" which has thus far been unable/unwilling to investigate questions such as why, say, "political power is being used to keep food from [starving] children, and what can be done to remove the blockages" (1994, p. 137).

Churchman highlights that according to more-or-less accepted figures (1994), "35,000 children die every day in the world from starvation and starvation related diseases" (1994, p. 136). He exclaims: "As a species, we murder a significant number of our young, and keep an even larger number in disease and pain. ... We could say ... that humanity collectively is a child abuser" (1994, pp. 136-137). He wonders what can be done about this, and why science has not directed attention to considering this lacuna in its inquiries. In addition, he points out that we are also clearly not preparing "a world for our progeny". And he pleads: "Can science grow up and begin finding some ways? Why not?" (1994, p. 137).

Churchman believes that one of the reasons why science has not turned attention to these problems, is because it refuses to frame questions in such a way that they can indeed be seriously considered. He would prefer that science directs itself (guided by an ethical intention) to "teach us ... how to bring starvation, the rape of the environment, the tragedy of modern education, into control" (1994, p. 139). This is not a process of "teaching" in the sense of offering "information" that is posed as value-free. It is a process of offering information that at the same time allows people to reflect on their way of seeing "the world" – via a debate between ways of defining experiences and human values.

Of course, from a realist-oriented point of view, the "information" offered by science is of potential use to people only insofar as it is not from the start beset by the infiltration of particular value-commitments that scientists might uphold. For realists, lay people should be able to turn to science to help them to become more informed about the real world (as such). A science guided by an ethical intention cannot perform this information function. As Hammersley argues:

> "... the findings of research are, on average, less likely to be in error than information from other sources. And this stems from the operation of the research community in subjecting research findings to scrutiny and thereby detecting and correcting errors." (1995, p. 57)

According to Hammersley, scientists have a responsibility to direct themselves toward understanding realities existing in the external world, and to "ensur[ing] that, as far as possible, the information provided is valid" (1995, p. 59). The remit of science (to get closer to the truth about reality) cannot be extended to include a concern that people (including scientists) direct their action more compassionately and caringly. Indeed there is reason to believe that the intrusion of this extra-scientific concern will tarnish the truth-seeking efforts of the scientific community.

But despite the seeming appeal of Hammersley's and others' proposal that scientific thinking cannot include as part of its endeavor as science these (moral) concerns, it is still possible to question this proposal on the grounds that it in effect separates out the domains of human thinking (reflection) and human action (cf. Mitroff, 1994; Swanson, 1994; Nelson, 2001).

Midgley offers an account of why the (realist-oriented) insistence that scientists should focus solely on pursuing the ideal of truth, can be regarded as ill founded. He refers to Churchman's suggestion that contrary to common-sense thinking about the status of scientific "observations", agents (including scientists) are involved in a process of interpreting observations, and value judgments guide what is investigated in the first place (and how it is investigated). Midgley (following many others in the critical theoretical tradition[25]) argues that:

> "Once the moral, subjective, linguistic, and other influences on observation are opened to critical reflection, scientific observation has to be seen as a form of intervention [in the sense that it] ... creates change in the knowledge and/or practice of a community of people." (2003, p. 88)

Midgley emphasizes that choices as to "what to observe and how to observe it" must be considered as having "unavoidable moral consequences for action (which may sometimes be anticipated and sometimes not)" (2003, p. 88).

Once we appreciate that our values do influence our own way of seeing "the world" and that this has implications for action (of ourselves and of others in the world), the appropriate moral injunction, as Gregory notes, is for people to engage in what she calls "critical appreciation". This implies a preparedness to "become critically reflexive about issues that others hold dear" (2000, p. 497). Or put in Churchman's language, the requirement is to

[25] See, for instance, Habermas (1976, 1982, 1986, 1993). Habermas contends that the problem with Popper's view of the "facts" that supposedly constitute the empirical basis of science, is that he fails to take full account of the way in which these become accepted through intersubjective encounter (1976, p. 205). Likewise, in considering validity claims of "rightness" in regard to the development of social norms, Habermas argues that these too are rooted in intersubjective encounter. Habermas sees the connection between "facts" and "norms" as types of statements that are linked in that in our viewing of facts we invoke certain underpinning standards (norms), and in the creation of norms we draw on certain "factual" information. (See also Romm, 1991, 2001 for a detailed discussion of Habermas's epistemological argument in relation to more realist-oriented alternatives.)

look at yourself from the vantage point of the enemy and realize "how foolishly you push one point of view" (1979, p. 214). Systemic thinking is then tied to a recognition that what one oneself holds dear (ones way of approaching "the world", including ones values), are choices that have to be accounted for in encounter with others.

Thus, in terms of the above reasoning, the apparent enemy of the systems approach – the blind/irrational appeal to values to ground ones actions in the world – turns out to be a potential friend insofar as the appeal to values is seen as an ethical orientation towards becoming critically reflexive about ones way of approaching "the issues".

Churchman argues that if we are to be "stirred out of our humdrum life" we need to search for new ways to follow the path of progress (1979, p. 201). This is not to say that there is any evidence for the reality of this path – but it is to say that it may be possible to try to realize it (1979, p. 201). A systems thinking that is able somehow to incorporate moral considerations in its way of thinking/being (rather than considering these an "enemy"), is one way of working towards generating a different social existence.

6.2.3 Religion/Spirituality: Enemy and/or Friend of a Systems Approach?

Churchman indicates that religion might well be considered as an enemy of rational systems thinking. The invocation of "religion" seems to imply a mode of blind, irrational worship, or blind adoration, of the object(s) of attention – "the immense and the small" (1979, p. 198). Churchman remarks in this context that even the "reason" underpinning the systems approach to the world, "which our systems approacher worships, is elusive and undefinable" (1979, p. 199). This being the case,

> "... the conclusion is that the hero's persistent urge to improve the human condition is simply the hero's mode of worship; he [or she] is disturbed out of a mood of satisfaction into a mood of dissatisfaction by his nonrational need to worship humanity. Calamity! Certainly here is an enemy which is *not* us." (1979, p. 199)

As with the other "enemies" that he tries to recast as potential friends, Churchman employs irony to show the salutary side of the spiritual urge to improve the human condition. Churchman points out that as far as he is concerned, questions of religion are not necessarily to be couched in terms of whether or not one believes in the existence of God (1979, p. 214). As he notes:

> "...reason tends to believe this [question] is the first, or most basic, issue of religion, whereas Saint Paul doesn't even list it in his trilogy of faith,

> hope, and love. That one can have glorious religious experiences without even considering the issue of a divine being's existence is the revelation of religion." (1979, p. 214)

Churchman thus revisits the view that religion is concerned with the existence of a divinity – and at the same time he asserts that "reason" should incorporate a spiritual concern with "faith, hope and love". Reason thus itself needs to be extended to recognize that these forces can play a part in systems thinking.

Churchman's statement that religion can propel one to be "disturbed out of a mood of satisfaction into a mood of dissatisfaction" (1979, p. 199) is consistent with (and follows up on) his view that the aim of systems thinking is to "maximize dissatisfaction with the goals already being attained" (1952, p. 64). As one moves towards this aim, one "considers ... problems to be more serious and harder, not nicer and easier. In other words, under this thesis, a future generation is better than this one if it considers its problems to be more critical" (1952, p. 64). Churchman defines progress in embracing systems thinking as occurring when:

> "...the next generation finds its problems much tougher to handle and is at the same time much more anxious to study them. Our contribution [as systems thinkers] along this scale consists of posing and urging better problems, not better 'solutions'." (1952, p. 64)

McGregor comments in this vein that the purpose of systems thinking is to "open up the range of planning choice far beyond the options initially considered. The heroic imagery is invoked by Churchman for good reason: ideal planners [inspired by systems thinking] are not uniformly popular" (1980, p. 86). Systems thinking is a process of challenging certain "popular" ways of framing questions (with their attendant ways of developing answers/solutions). To become involved in this challenging task, one needs to have some hope in the human potentiality to respond ethically. As Churchman elucidates: "hope is the spiritual belief in an ethical future" (Churchman, cited in Van Gigch *et al.*, 1997, p. 744).

From this account we can see that Churchman does not concede to the Comtian (positivistic) interpretation of religion/spirituality as functioning as an irrational force in society – to be kept strictly separate from the realm of scientific thinking (cf. Comte, 1974).[26] He recasts the contribution of religion/spirituality so that this "force" (when embraced) can aid people to, as Fuenmayor puts it, "address world problems". This, in turn, entails "thinking and caring about bringing justice all around the world" (2001b, p.

[26] Romm (1991, pp. 11-12) explains Comte's positivist stance regarding the mutual exclusiveness of theological and positive (scientific) thinking.

44). From this systemic perspective, which Fuenmayor attributes to Churchman, addressing world problems means "addressing issues of global justice, e.g., relationships between developed and undeveloped countries, racial discrimination, the so-called 'new international order', etc." (2001b, p. 44). To say that the issues need to be addressed, however, is not to say that their addressal will be a simple process. It is to say that at least the questions relating to such concerns (of global justice) need to be put seriously onto the agenda of our thinking/being.

When it was put to Churchman (by Van Gigch) that "we're [in our society] not kind, because of greed, competition, pressure and the like" (1997, p. 742), his reply was that: "*There are more books written now on this idea of 'care' than anything else .... What this means is that there's a bigger push for the human species*" (1997, p. 742). Nonetheless, he avers that "you have to learn how to care. Learning is what we professors are all about. Why isn't there a course on that. There should be a number of courses on caring. Human care" (1997, p. 742).

Churchman's argument is that the possibility of an "ethical future", where people can activate more "caring", is something that can potentially be made more real – as long as (systems) thinkers choose to concentrate on this (as part of a collective effort in society). The systems hero, though, may need to have courage to utilize his or her spiritual belief in an ethical future in order to turn around the "Creed of Greed", as Koenigsberg calls it (Van Gigch *et al.*, 1997, p. 742). Ignoring/reducing possibilities for studying and deliberating around the "Creed of Greed" on the grounds that religion does not fall under the scope of rational systems thinking, simply serves to further entrench this "creed" (form of worship) in society. These "soft" aspects of human experience should not be excluded from rational systems thinking, for this itself may have (ethically) dangerous consequences in society.

6.2.4 Aesthetics: Enemy and/or Friend of a Systems Approach?

Churchman considers aesthetics as a quality of human experience that makes action "radiant" for us – whether it is experienced as "beautiful, ugly, pleasurable, painful, comic, tragic, whatever" (1979, p. 26). Churchman declares that it seems "extremely reasonable" (to him) to argue that systems thinkers should incorporate a concern with "the quality that people value" (1979, p. 199). He makes the point that planners have tended to ignore aesthetic concerns and yet these concerns should be "incorporated into our reasoning about change" (1979, p. 199). Even if we cannot quantify (measure) the aesthetic quality of experience, this is no ground for simply excluding it from our thinking about change. So he argues that "our [apparently] nonfriendly aesthetic critic has become a helpful friend by calling our attention to an aspect of the human system we had ignored" (1979, p. 199).

But Churchman cautions that this new-found friend can still "turn against us", for it seems impossible to make provision for the "unique quality each of us has" to carry this quality of experience (1979, p. 199). How then, can we "add up values" (embracing what different people may define as important)? And how can we "assess change" once we do not have a way of "measuring" its meaningfulness to different people? The aesthetic enemy seems to have laid a clever trap by asking us to confront these questions (knowing that there is no answer to them). However, instead of treating the encounter with the (seeming) enemy of systems thinking as discrediting the quest for rational systems thinking, Churchman argues that we should "welcome" the questions set by the "enemy" (who could also be considered a critical friend).

One way of incorporating a concern with "quality" in systems thinking is to suggest that systems thinking be directed towards engendering "a world of human equity, in which the quality of each individual's life is of the highest" (1979, p. 201). This is not a matter of suggesting that in terms of this vision (of such a world) "everyone is blissfully happy, because for many bliss may not be the quality that is appropriate to their lives: struggle, frustration, hope, despair may rather be the appropriate life them" (1979, p. 201). The important point is that people's quality of existence should seem meaningful to them, while they also take account the need for what Churchman refers to as "proper conduct in the world" – so as to incorporate as part of their thinking/being, a concern with addressing complex issues confronting "humanity". This is not a matter of knowing rationally in any well-defined sense what it means to embrace the "aesthetic quality of our life". But it is still to embrace it (1979, p. 214).

Fuenmayor offers support to Churchman's views by noting that in any case there is no human requirement to try to separate our aesthetic experience or "feeling" from "the facts" which supposedly can be known independently of our feelings. Fuenmayor points to our attention that in certain traditional cultures, "with a different *constellation* of the realm of beings (where dualism [between thinking and feeling] does not dominate), the bad and the good, the beautiful and the ugly are constitutive of whatever-is-the-case" (2001b, p. 41). Fuenmayor suggests that already to detach "feelings" and "facts" is to adopt a way of living that "wipes out the richness and *wholeness* of the living situation" (2001b, p. 41). For this reason, holistic thinking cannot subscribe to such a dualism. As he puts it:

Obviously *sense* [meaning] can be found neither in the things-in-themselves that are facts' and 'things' nor in those new ghostlike beings (feelings, emotions, etc.) that belong solely to the human mind. This is why holistic (systems) thinking cannot logically take place under a dualistic *constellation* of the realm of beings (2001b, p. 41).

To offer a vision (as Churchman does) of an existence where the aesthetic quality of life is incorporated in our sense of "what is" is thus to (re-)organize our relationship to the world so that we can appreciate that human being-in-the-world includes an aesthetic quality. Churchman laments the fact that thus far:

> "...the history of aesthetics is rarely written, except in histories of art and (occasionally) in biographies, but the true essence of aesthetics is what gives the tapestry [of life] its meaning; what 'really' happened to humans in history is an image of human joys, desperation, love, hate, opulence, and drudgery." (1979, p. 26)

For Churchman, aesthetics – as a quality of human experience that makes action "radiant" – is an essential part of the "reality" of human experience. This conception of aesthetics is linked by Churchman to a way of conceiving the significance of systems thinking. He proposes that because "systems thinking is a kind of action" (in that thinking and intervention in the world are inseparable), we can contend that "its significance lies in its aesthetics, and not in its 'validity'" (1979, p. 25).

By placing in quotation marks the word "validity", Churchman points to the need to put forward an alternative manner of validating our knowing processes to that presented within realist-oriented thinking. The next section turns to a consideration of his views on how his systems approach to/in the world can indeed by justified (1979, p. 200).

6.3 CHURCHMAN'S SYSTEMS APPROACH AS VISIONARY THINKING

As discussed in Section 2. above, Churchman refers to the importance of our creating a vision of a way of living where human reason at one and the same time incorporates the four "enemies" that he has located (politics, religion, morality and aesthetics). He states:

> "The vision contains all the enemies. It contains community (polis), morality, religion, and aesthetics. It is also symbolic [in the sense that] ... it is a symbol of something, something [potentially] real. But where is the reality? The 'real' world contains few of the aspects of the vision." (1979, p. 201)

Churchman accedes that looking at the "real" world through a narrow perspective would lead us to believe that the challenge to realize the vision that he puts forward is "utterly unrealistic" in the sense that such a way of living can never be developed (1979, p. 201).

But he decides to present us nevertheless with his desired vision of a world where "reason" *can* incorporate politics, morality, religion and aesthetics. Now we can ask the following questions:

What is the status of Churchman's "vision"?

How, if at all, can it be "tested"?

6.3.1 The Status of Churchman's Vision

In approaching this question, it is helpful to remember that Churchman presents his vision as offering a specific (systemic) way of approaching perceived realities. His systems approach suggests that:

> "...rational humans need to leave the body of rationality and to place the self in another body, the 'enemy', so that the reality of the social system can unfold in a radically different manner. From this vantage point he/she can observe the rational spirit and begin to realize not only what has been left out of it, but also what the spirit is like, especially its quality of being human." (1979, pp. 204-205)

He maintains that once people can begin to express their quality of being human, the resultant person "is neither a loose dreamer nor a hard thinker" (1979, p. 205). However, the systems thinking "hero" may become disappointed because, while trying to realize a vision of a different form of social existence where "reason" can embrace its "enemies", this way of operating in reality may seem to "fail over and over again". Yet the vision is itself "a part of reality" and for this reason should not simply be "given up" as unworkable in practice (1979, p. 213). What, then, is the status of the vision?

Van Gigch (1995) proposes that one way of considering the vision is to suggest that it offers a way of:

> "...rekindling our faith in the rational spirit [where rationality now embraces its so-called enemies], for giving us renewed strength to improve the human condition, and for showing us the way in which we can redirect our discipline[s] to accomplish the dream." (1995, p. 87)

The vision is a part of reality (to use Churchman's words, 1979, p. 213) insofar as it serves to offer renewed strength to people to indeed think and act more systemically. This is not an injunction for people to try to create in practice a definite design for human living. It is an encouragement for them to appreciate, when looking at problems, that "if a problem is unraveled far enough, we will encounter all other problems" (1979, p. 36). This means, according to Van Gigch, that people are called upon to recognize that:

> "If we push the system's boundaries too far, we have to consider too many systems and the situation becomes too complex and unsolvable. If we do not push the boundary far enough, we face the environmental fallacy which consist of taking too few or not taking sufficient relevant systems into account." (1995, p. 87).

There is no recipe for how people (in Churchman's visionary thinking) might make decisions about appropriate ways of delimiting the relevant area(s) of concern. My suggestion is that what *can* be asked of people is that they act *accountably* – that is, that they show preparedness to take seriously into account different ways of framing "the issues" (and what is "at issue") and their possible addressal. (See Romm, 1998b, 2001, 2002a,b, for a discussion of the concept and practice of "discursive accountability" considered as part of an ongoing process of trust building[27].)

In considering the status of systems thinking in general (and not only that of Churchman's approach) Lazslo puts forward the suggestion that it offers a way of learning and of becoming and thereby *"provides a path by which we may realize the evolutionary opportunities ahead of us"* (2000, p. 11). Seen in this light, Churchman's visionary thinking about how we may approach our living in the world can be considered as an encouragement to us to try to generate new ways of "learning and becoming". But this still leaves unanswered the question as to how, if at all, the arguments that he offers towards adopting a "systems approach" can be tested (see the second question posed above).

6.3.1.1 "Testing" the Systems Approach

Churchman emphasizes that any meaning that we give to the word "test" is itself based on a perspective. He elucidates that: "a test for the narrow approach is one in which qualified observers can agree on 'success' or 'failure'" (1979, p. 200). But a holistic approach does not simply rely on "qualified observers" to make decisions, because it may well be that they are making their decisions by viewing the problem in a too narrow manner in the

27 Romm (2001) develops a way of conceiving human trust building with reference to a trusting constructivist position. The term "constructivism" itself embraces a variety of positions. Romm suggests that critical theory, anti-foundationalist feminism, and discursively-oriented constructivism (for example) can be pitted against realism in terms of the focus on the manner in which ways of knowing as developed in society can be seen as related to our ways of being. The trusting constructivist position explores options for people building trust through processes of discourse while appreciating that we do not have any access to "the world" independently of human mediation.

first place. He offers an example of expert physicists who may judge that the "'problem' of nuclear waste is minimal, or does not exist at all, because, say, they have no interest in the values of generations, a consideration of prime importance in the holistic approach" (1979, p. 200).

Churchman here indicates that there is no device for "testing" the arguments of systems thinkers against arguments that may be presented by others if people are bringing different criteria to bear on defining what seems "reasonable". To a systems thinker it seems *unreasonable* to believe that formulating hypotheses and (supposedly) subjecting them to "empirical tests", will allow the species to advance their responsible action in the world. Churchman refers to his influence by Singer, who "saw that inquiry must be an integrated "whole-system", and not a sequence of isolated and separable hypotheses" (1979, p. 10). Churchman admits that of course, if we are committed to a holistic approach, we can never "sweep in" all the relevancies (as defined by ourselves and others) that might be brought to bear on a problem, and for this reason we can never be fully satisfied with the results of our inquiries. But the strength of the systems approach is precisely that it calls for us to *examine* our cut-off points, so that we are more aware of the possible implications hereof and can defend decisions in an accountable fashion (as part of the process of human being in the world). As Koenigsberg and Van Gigch (1994) comment, Churchman tried to spur people to indeed be prepared to reflect upon "the social consequences of their own work" (1994, p. 2).

Clearly, there is no manner of "testing" through supposedly rational argumentation and/or observation/experience the validity of the appeal to embrace a systemic approach. But Verma summarizes how Churchman's appeal was aimed, as Verma puts it, at:

> "...mak[ing] the enterprise of science more moral, more humane, and more engaged in solving the world's problems. And once we see this, Churchman's recent emphases on the problems of starvation, child-abuse, addiction, and justice acquire a remarkable resonance with his earlier work." (1994, p. 66)

This is not to say that realist-oriented authors will be convinced by Churchman's suggestions for linking "science" with a pragmatic intention towards addressing such identified "problems". Indeed they could easily continue to argue that scientific reason requires that the realm of possible action/implementation is *not* brought to bear on scientific work, so as not to bias unduly the scientific effort to understand the realities in question. For example, Hammersley insists that it is still important that would-be scientists are geared to offering arguments about how they are deriving conclusions based on relevant (cogent) evidence.

Hammersley proposes that rational criticism between scientists about one another's way of developing "conclusions" serves the function of aiding the community to get closer to the truth about the realities under consideration at that point in time. But he adds the caution that, when exercising criticism, they should not try willy nilly to invalidate claims about reality when "in any reasonable judgment" an account is valid. As he states:

> "In the absence of a foundation of empirical givens whose validity is absolutely certain [that is, as is supposed by foundationalism], it is [of course] possible continually to raise doubts about any claim to knowledge: ... there is always further scope for criticism. However, it is not desirable to pursue criticism when, on any reasonable judgment, an account is valid." (1995, p. 43)

According to Hammersley, scientists need to exercise reasonable judgment when deciding whether it is appropriate to operate the process of continuing criticism (of knowledge claims). In Hammersley's terms, Churchman would probably be pushing too far in his continuing challenge of claims presented as springing from supposedly "scientific" inquiry. Yet in Churchman's understanding, the reduction of reason to focus upon what are presumed to be empirically testable hypotheses about certain posited realities, is to be seen as questionable from the start for being indeed too reductionist in its approach.

6.4 CONCLUSION

In this chapter, Churchman's epistemological argument was pitted against more realist-oriented views. I showed how Churchman's conception of the accountability of scientists/inquirers is linked to his understanding that a "science" that strives towards value-freedom in its approach to the world is ill guided. Churchman urges systems thinkers to take some responsibility for their way of setting questions and framing research topics, arguing that these ways of framing are not without consequence in society.

I showed that for realist-oriented authors on the contrary (insofar as they continue to uphold a realist argument) the sole responsibilities of scientists consists in their directing themselves as part of a scientific community in understanding realities to the best of their abilities. Scientists are then able to offer lay people the benefits of the informed accounts of reality that they have created through their adherence to scientific protocol, affected as little as possible by biases that may infiltrate the research process.

I suggested in the chapter that finally we have no way of settling the question as to whether Churchman's appeal to a systemic view of the link between human knowing and acting is more "reasonable" than realist-oriented views. But I emphasized that Churchman's style of arguing for a systemic approach to our being-in-the-world offers a route to (re-)open debate about our responsibilities as inquirers. Churchman argues that a scientific enterprise built on the ideal of striving for value-freedom may well, ironically, become an unethical endeavor. As Verma expresses it: "a value-free science is immoral" (1994, p. 61). I proposed that the failure to take seriously Churchman's epistemological and ethical challenge implies that our accountabilities as inquirers – which requires that we *do* engage in serious encounter with alternative arguments and concerns that may be posed – becomes ipso facto threatened.

My own (constructivist-oriented) position is that inquirers can direct themselves towards earning trust in their processes of inquiry (and in their accompanying decisions regarding options for action in situations) by trying to show that they have indeed engaged seriously with values and concerns brought forth by others in relation to their own (Romm, 2001, 2002a,b,c). This (constructivist) position provides a way of "defending" Churchman's argument that our accountability as inquirers and actors in the social world is linked to our propensity to address opposition between ways of seeing/valuing/living.

REFERENCES

Ackoff, R.L., 1999, *Ackoff's Best*, Wiley, Chichester.

Banathy, B.A.. 1999. The difference that makes a difference: incoming presidential address, delivered at the 42nd Annual Meeting of the International Society for the Systems Sciences, Atlanta, USA, 1998, *General Systems Bulletin*, 28:5-8.

Churchman, C.W., 1946a, Dialectic of modern philosophy, *The Journal of Philosophy*, 43(5):113-124.

Churchman, C.W., 1946b, Philosophical aspects of statistical theory, *The Philosophical Review*, 55(1):81-87.

Churchman, C.W., 1952, Ethics, ideals, and dissatisfaction, *Ethics*, 63(1):64-65.

Churchman, C.W., 1971, *The Design of Inquiring Systems: Basic Concepts of Systems and Organizations*, Basic Books, Inc. Publishers, New York.

Churchman, C.W., 1979, *The Systems Approach and Its Enemies*. Basic Books Inc. Publishers, New York.

Churchman, C.W., 1982, *Thought and Wisdom*, Intersystems Publications, USA.

Churchman, C.W., 1994, What is philosophy of science? (Guest Editorial), *Philosophy of Science*, 61(1):132-141.

Comte, A., 1974, *Positive Philosophy*, translated and condensed by H. Martineau, Ams Press, New York.

Flood, R.L. and Romm, N.R.A., 1996, *Diversity Management: Triple loop learning*, Wiley, Chichester.

Fuenmayor, R., 2001a, The oblivion of Churchman's plea for systems approach to world problems. I. The inseparability of systems thinking and world issues in the modern epoch, *Systems Practice and Action Research*, 14(1):11-28.

Fuenmayor, R., 2001b, The oblivion of Churchman's plea for systems approach to world problems. II. The rise of the modern constellation, *Systems Practice and Action Research*, 14(1):29-45.

Gregory, W.J., 2000, Transforming self and society: a 'critical appreciation' model, *Systemic Practice and Action Research*, 13:475-501.

Habermas, J., 1976, A positivistically bisected rationalism, in: *The Positivist Dispute in German Sociology*, T.W. Adorno, H. Albert, R. Dahrendorf, J.,Habermas, H. Pilot and K.R. Popper, eds., Heinemann, London.

Habermas, J., 1982, A reply to my critics, in: *Habermas: Critical Debates*, J. Thompson and D. Held, eds., Macmillan, London.

Habermas, J., 1986, Hannah Arendt's communicative concept of power, in: *Power*, S. Lukes, ed., Blackwell, Oxford.

Habermas, J., 1993, *Justification and Application: Remarks on Discourse Ethics*, Polity Press, Cambridge.

Hammersley, M., 1995, *The Politics of Social Research*, Sage, London.

Hammersley, M., 1997, A reply to Humphries, *Sociological Research Online*, 2, <http://www.socresonline.org.uk/socresonline/2/4/6.html>

Hammersley, M. and Gomm, R., 1997a, Bias in social research, *Sociological Research Online*, 2, <http://www.socresonline.org.uk/socresonline/2/1/2.html>

Hammersley, M. and Gomm, R., 1997b, A response to Romm, *Sociological Research Online*, 2, <http://www.socresonline.org.uk/socresonline/2/4/7.html>

Hammond, D. 2003. *The Science of Synthesis*. The University Press of Colorado, Boulder, Colorado.

Haynes, J.D., 2001, Churchman's Hegelian inquiring systems and perspectival thinking, *Information Systems Frontiers*, 3(1):29-39.

Jervis, R., 1997, *System Effects: Complexity in Political and Social Life*, Princeton University Press, Princeton.

Kant, I., 1785, *Grounding for the Metaphysics of Morals*, (Ellington, J.W. trans.), Hackett, Indianapolis.

Koenigsberg, E. and Van Gigch, J.P., 1994, in: Celebration of the 80th birthday of C. West Churchman, born Mount Airy, Pennsylvania, August 29, 1913, *Interfaces*, 24:1-4.

Laszlo, A., 2000, The epistemological foundations of evolutionary systems design, in: *Proceedings of the World Congress of the Systems Sciences and ISSS 2000 International Society for the Systems Sciences 44th Annual Meeting*, J.K. Allen and J. Wilby, eds., Toronto, Canada.

McGregor, E.B., 1980, Review of C. West Churchman's the systems approach and its enemies, *Business Horizons*, 1980:85-87.

McIntyre-Mills, J., 2000, *Global Citizenship and Social Movements*, Harwood Academic Publishers, Amsterdam.

McIntyre, J., 2003, Participatory democracy: drawing on C. West Churchman's thinking when making public policy, *Systems Research and Behavioral Science*, 20:489-498.

McIntyre, J., 2004, Facilitating critical systems praxis (CSP) by means of experiential learning and conceptual tools, *Systems Research and Behavioral Science*, 21:37-61.

Midgley, G., 2000, *Systemic Intervention: Philosophy, Methodology and Practice*, Kluwer Academic/Plenum Publishers, New York.

Midgley, G., 2001, Systems thinking for the 21st century, in: *Understanding Complexity*, G. Ragsdell and J. Wilby, eds., Kluwer Academic/Plenum Publishers, New York.

Midgley, G., 2003, Science as systemic intervention: some implications of systems thinking and complexity for the philosophy of science, *Systemic Practice and Action Research*, 16(2):77-97.

Mitroff, I.I., 1994, The cruel science of world mismanagement: an essay in honor of C. West Churchman, *Interfaces*, 24:94-98.

Nelson, H., 2001, Continuing the traditions of ISSS – systems science in the service of humanity, in: *Understanding Complexity*, G. Ragsdell and J. Wilby, eds., Kluwer Academic/Plenum Publishers, New York.

Popper, K.R., 1959, *The Logic of Scientific Discovery*, Hutchinson, London.

Popper, K.R., 1969, *Conjectures and Refutations*, Routledge & Kegan Paul, London.

Popper, K.R., 1994, *The Myth of the Framework: In Defence of Science and Rationality*, Routledge, London.

Porra, J., 2001, A dialogue with C. West Churchman, *Information Systems Frontiers*, 3(1): 19-27.

Romm, N.R.A., 1991, *The Methodologies of Positivism and Marxism: A Sociological Debate*, Macmillan, London.

Romm, N.R.A., 1995, Knowing as intervention, *Systems Practice,* 8:137-167.

Romm, N.R.A., 1996, Inquiry and intervention in systems planning: probing methodological rationalities, *World Futures,* 47:25-36.

Romm, N.R.A., 1997a, Becoming more accountable: a comment on Hammersley and Gomm. *Sociological Research Online,* 2, <http://www.socresonline.org.uk/socresonline/2/3/2.html>

Romm, N.R.A., 1997b, Implications of regarding information as meaningful rather than factual, in: *Philosophical Aspects of Information Systems*, R. Winder, S. Probert and I. Beeson, eds., Taylor and Francis, London.

Romm, N.R.A., 1998a, Nurturing sustainability through triple loop learning: discursive accountability as responsible action, *Systemist* 20(1):40-52.

Romm, N.R.A., 1998b, Interdisciplinary practice as reflexivity. *Systemic Practice and Action Research,* 11:63-77.

Romm, N.R.A., 2001, *Accountability in Social Research: Issues and Debates*, Kluwer Academic/Plenum Publishers, New York.

Romm, N.R.A., 2002a, A trusting constructivist approach to systemic inquiry: exploring accountability, *Systems Research and Behavioral Science*, 19(5):455-467.

Romm, N.R.A., 2002b, Responsible knowing: a better basis for management science, *Reason in Practice* 2(1):59-77.

Romm, N.R.A., 2002c, A trusting constructivist view of systems thinking in the knowledge age, in: *Systems Thinking and Practice in the Knowledge Era*, G. Ragsdell, D. West and J. Wilby, eds., Kluwer Academic/Plenum Publishers, New York.

Swanson, E.B., 1994, Churchman's theory of design integrity, *Interfaces*, 24:54-59.

Van Gigch, J.P., 1995, The paradigm of the science of management: in celebration of C.W. Churchman's 80th birthday. *Interfaces*, 25:81-88.

Van Gigch, J.P., Koenigsberg, E., and Dean, B., 1997, In search of an ethical ccience: an interview with C.West Churchman, an 80th birthday celebration, *Journal of Business Ethics*, 16(7):731-744.

Verma, N., 1994, Organizations and their purposes: a note on Churchman's philosophy of management, *Interfaces*, 24:60-66.

Chapter 7

A RETROSPECTIVE STRUCTURAL INQUIRY OF THE PREDICAMENT OF HUMANKIND

Prospectus of the Club of Rome

Alexander N Christakis
CWA Ltd, USA

Abstract: Churchman dedicated his career to sensitizing systems scientists and other inquirers to the existence of what he referred to as "enormous problems." In his view these are problems that cannot be addressed by a single discipline or a single inquirer but by interdisciplinary teams of inquirers. This chapter tells the story of how a group of researchers, in the context of the Club of Rome (CoR) Prospectus on The Predicament of Humankind, proposed in the early 70s a very forward looking and innovative systems approach. It took over twenty years of research and development for the emergence of a paradigm of Structured Design Science (SDS) that has been tested in the arena of practice for the last fifteen years.

The CoR prospectus introduced the concept of the *Problematique* as the "enormous problem" of the 20th Century. In 1993, twenty-three years after the conceptualization of the *Problematique,* a small team composed of three of the original architects of the CoR proposal employed the SDS paradigm to conduct a retrospective inquiry of the global *Problematique*. The findings from this inquiry demonstrate that no significant progress had been made in terms of resolving the root causes of the *Problematique* in the ensuing twenty-thee years, i.e., 1970-1993. It is conjectured that a similar inquiry conducted today, i.e., ten years later, will reaffirm the lack of progress in resolving this enormous problem of huhumankind. The implications of these findings are significant for all systems scientists, and especially for students of Churchman's work.

Key words: Dialogue, structured dialogue, Club of Rome, *Problematique,* Democracy, systems approach, stakeholders, *CogniScope,* social systems designing.

7.1 INTRODUCTION

In the late 60's an Italian industrialist, named Aurelio Peccei, was traveling around the world meeting with world leaders trying to persuade them that there was an unprecedented global crisis. Peccei had written a book titled *The Chasm Ahead* (Peccei, 1969). In this book Peccei was making the case that there was an ever-growing gap between the technologically developed North America and the rest of the world, including Europe. He was forecasting that this technology gap would materialize in the next twenty years in a major global crisis between the developed and the less developed regions of the world, unless some measures were taken by world leaders to close the gap.

Even though Peccei dedicated all his energy and significant resources to influence the stream of world events, he gradually came to the realization that he was not being very effective in his discussions with world leaders. In 1968, at a conference in Belagio, Italy, Peccei met an American planning theoretician named Hasan Ozbekhan. Ozbekhan had delivered a very thoughtful lecture at the conference titled "A General Theory of Planning." This lecture was published later in a book edited by Erich Jantsch (Jantsch, 1969). Ozbekhan and Peccei became very good friends, and during one of their meetings, Ozbekhan was able to persuade Peccei to adopt the systems approach (Churchman, 1979) for influencing the stream of world events.

Ozbekhan had worked for a number of years in the corporate planning department of the General Electric Corporation in New York City. At the time of meeting Peccei, he was the Director of Corporate Planning for the System Development Corporation (SDC), a think-tank headquartered in Santa Monica, California. Ozbekhan was considered one of the most prominent planning theoreticians in the 70s. This was the golden era of systems thinking, and California with such famous think tanks as the Rand Corporation and the SDC was the Mecca of such thinking. Ozbekhan's proposal to Peccei to employ the systems approach advocated by Churchman and others for studying the global situation was accepted by Peccei in 1969. The intent was to use the findings of such a study in discussions with world leaders in order to maximize his effectiveness.

The two men agreed to a new strategy for the future. A new institution was created called "The Club of Rome." The name Club of Rome (CoR) was decided for two reasons: (a) Peccei was fascinated with the American concept of the "club," and (b) that a new club was to be created, the membership of which would be people of different nationalities, different cultures, and different disciplines. In other words, the major criterion in the formation of the CoR was not homogeneity, as is usually the case in most clubs, but heterogeneity. The mission of the CoR would be to conceptualize the new systems approach and to fund projects by researchers around the world, all of which would be focusing on the impending global crisis. All of these projects would be funded by the Executive Committee of the CoR in the context of a systems approach suitable to meet the challenges of the 20th century.

An Executive Committee of the CoR was formed consisting of a diverse multi-cultural group of members representing such countries as England, the Soviet Union, Germany, Austria, Switzerland, Italy, and the U.S.A. The Executive Committee commissioned Ozbekhan to write the prospectus of the Club of Rome. Ozbekhan produced within six months the first draft of the prospectus titled "the Predicament of Mankind." In 1969 Ozbekhan hired me as a consultant to the Executive Committee.

7.2 THE PREDICAMENT

'The Predicament of Mankind' prospectus was written in 1970 (for a review of the prospectus visit http://www.isss-conference.org/PDFs/ClubRome.pdf). The subtitle of the document was "Quest for Structured Responses to Growing World-wide Complexities and Uncertainties." The prospectus was distributed to the Executive Committee of the CoR, the members of the Club, and a variety of philanthropic foundations and institutes such as The Rockefeller Foundation, The Battelle Memorial Institute, and others. A number of prominent systems scholars of the time, such as Erich Jantsch and West Churchman, reviewed the document. The reviews were mixed. About half of the reviewers acclaimed it as an outstanding prospectus. The other half, especially those trained in traditional analytical disciplines like systems engineering, thought the document was lacking in methodological rigor. Most of the latter reviewers, did not realize that the prospectus was intended as an architectural design concept rather than an engineering blueprint.

In retrospect, the prospectus incorporated the seeds for a paradigm shift in designing social systems. It was sufficiently iconoclastic to be rejected by those in the social science and engineering community who were accustomed to the practice of normal science, and hence were eager to reject any attempt for a shift to a new paradigm. West Churchman was among the systems thinkers and philosophers that expressed great enthusiasm when he read the prospectus. Later on Ozbekhan and Churchman taught together at the Social Systems Sciences program at the Wharton School of the University of Pennsylvania established by Russell Ackoff during the 70s.

As an example of the iconoclastic character of the prospectus I will quote below two paragraphs included in the Introduction of the document:

> "The source of our power lies in the extraordinary technological capital we have succeeded in accumulating and in propagating, and the all-pervasive analytic or positivistic methodologies which by shaping our minds as well as our sensibilities, have enabled us to do what we have done. Yet our achievement has, in some unforeseen (perhaps unforeseeable) manner, failed to satisfy those other requirements that would have permitted us to evolve in ways which, for want of a better word, we shall henceforth call 'balanced.' It has failed to provide us with an ethos, a morality, ideals, institutions, a vision of man and of humankind and a politics which are in consonance with the way of life that has evolved as the expression of our success. Worse, it has failed to give us a global view from which we could begin to conceive the ethos, morality, ideals, institutions, and policies requisite to an interdependent world -- this, despite the fact that the dynamics of our technologies and our positivistic outlooks are global in their impacts, their consequences, their endless profusion and, more importantly, in the promises they proclaim and in the promises they imply.
>
> This failure is often regarded as having created a number of separate and discrete problems capable of being overcome by the kind of analytic solutions our intellectual tradition can so readily generate. However, the experience of the past twenty or thirty years has shown with remarkable clarity that the issues which confront us in the immediate present, as well as their undecipherable consequences over time, may not easily yield to the methods we have employed with such success in the bending of nature to our will. Such apparent resistance could be attributable to many things, none of which must be pre-judged, but about which certain assumptions might be made. It could be due, for instance, to the magnification of the problems we must grapple with -- that is, to the fact that almost all of them are global in scope, whereas the socio-political arrangements we have created are ill-equipped for dealing with issues that fall outside their strictly established jurisdictions. It could be due to heightened yet often obscure interactivity among phenomena, whereas our manner of solving problems owes its strength and efficiency to the identification of rather clear and direct lines of causality. It may be due to rapid rates of change, especially in the technological sector, whereas our institutions, outlooks and minds are geared by long-time habit to beliefs in slow unfolding and permanence -- beliefs which have sustained certain relatively stable concepts of polity, of social order and of

intellectual orderliness. In brief, whatever it is due to, the conjuncture of events that surrounds us is to all evidence world-wide, complex, dynamic, and dangerous".

These statements, written in 1970, appear apocalyptic, and even more applicable and relevant to our contemporary situation, i.e. the beginning of the 21st century.

It is not my intent in this chapter to engage in revisiting the power of the concepts so masterfully described in the Club of Rome prospectus. There are, however, two aspects of the prospectus which provided the intellectual and inspirational foundations for subsequent work by many researchers in the development and validation of the Structured Design Science (SDS) paradigm (Banathy, 1996; Christakis, 1973, 1987, 1988, 1993, 1996, 2004; 2005; Christakis and Brahms, 2003; Christakis and Harris, 2004; Magliocca and Christakis, 2002; Warfield, 1994, 1999; Warfield and Christakis, 1987; Warfield and Cardenas, 1994).

The first aspect is the idea of the *Problematique.* The second is the idea of the "value-base." It is important to appreciate both of these concepts, as well as their relationship to the development of the Structured Design Science paradigm.

7.3 THE *PROBLEMATIQUE*

The idea of the *Problematique* was introduced in the CoR proposal as a way to draw a linguistic, and hence conceptual, distinction between the well-bounded problems that we are accustomed to perceive and articulate and the meta-problem (or **meta-system of problems**) that emerges as a result of the interactivity and interdependency among these problems. In essence, it is the equivalent concept to Churchman's conceptualization of "enormous problems."

It is the nature of our languages, hence our manner of perceiving reality, to see and call the dissonant elements in a situation "problems." Although it is true that there are certain problems (mostly within one discipline and usually in the field of engineering and physical sciences) that can be addressed in their own domains by applying problem-solving techniques, it has become empirically evident over the last sixty years, that critical issues in the social arena are not capable of being solved in their own specific domains.

When we consider the truly critical issues of our times such as "environmental pollution," "poverty," "homelessness," "criminality," "population explosion," "urban deterioration," "racial and cultural discrimination," etc., we recognize that it is virtually meaningless to view these as problems that exist in isolation. The burning of Los Angeles in the summer of 1992, after the verdict on the Rodney King human rights trial, provides evidence that criminal justice cannot be divorced from race, poverty and slum living conditions. Similarly, problems clustered under words such as "hunger" or "malnutrition" in Somalia cannot be separated from social, economic, and geo-political problems on the global scale. Trying to solve any of these problems in isolation exacerbates the intensity of the *Problematique*, whose solution is beyond the scope of the concepts and methods we have traditionally employed during most of the 20th century.

As a means to visualize the interactivity of problems it is useful to look at Figure 1. This Figure provides a graphic portrayal of the dynamic and interactive development of the *Problematique* in the 20th century. It shows problem areas in five frames with approximate dates. In each frame, a particular problem area is symbolized by differently shaped shaded areas. One sees that at the start of the 20th century (Frame A) each problem area is distanced from the others. Because each area is separated from the others it is meaningful to think and tackle these as independent problem areas. It is with reference to this type of situation that most of our problem-solving methods were developed and applied in the 20th century.

By the middle of the 20th century, the problem areas begin to interact because of the escalation of the intensity of each problem area (Frame B). These interactivity trends continued at an accelerated rate during the next twenty years. The individual problem areas began to strongly interact in the 1970s (Frame C). The emergence of large scale metropolitan areas where it is meaningless to separate the transportation problem from the pollution problem from the population explosion problem, and so on, provides an example of this interactivity as shown visually in Frame C.

Figure 1: Dynamics of the Problematique for the 20th Century

Frame A: Year 1900

Frame B: Year 1950

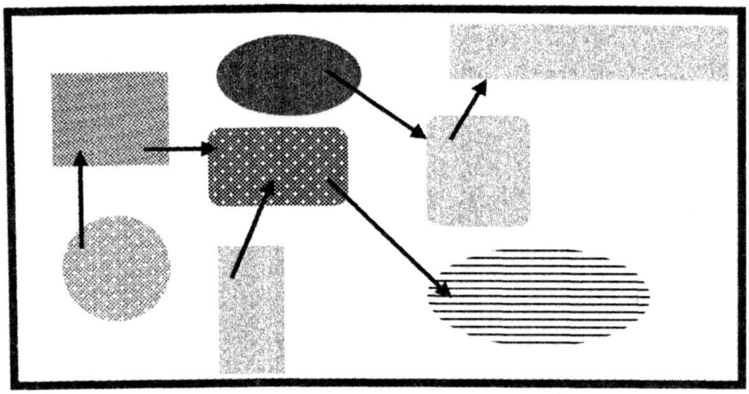

Frame C: Year 1970

Frame D: Year 1980

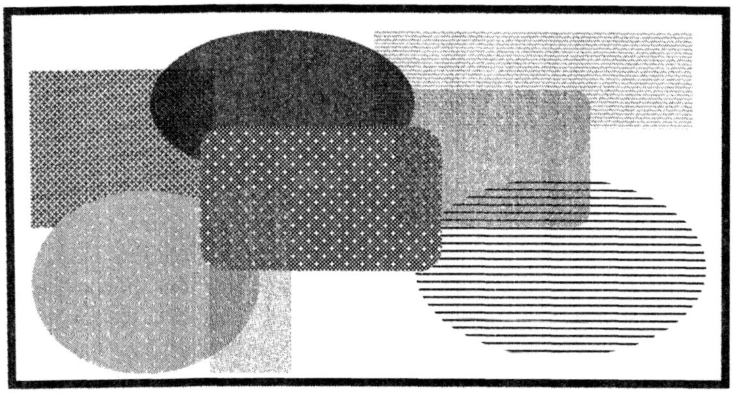

Frame E: Year 1990

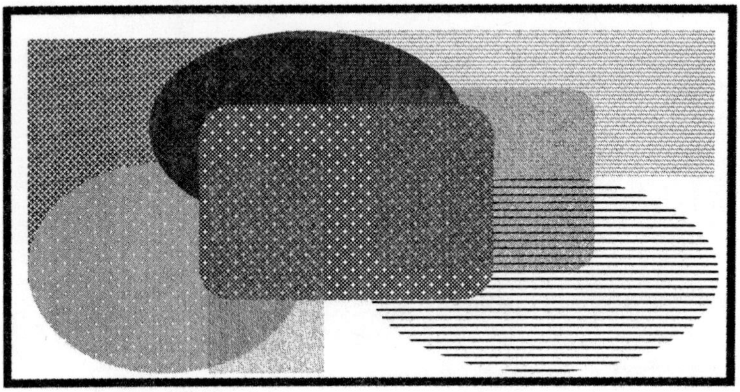

By 1980 the situation becomes to look more as an "enormous" complex problem that cannot any more be dealt with in terms of its individual problem areas but must be addressed in its entirety (Frame D). Attempts at fragmentation of the reality of the *Problematique* over the previous two decades have contributed to serious social and technical disasters, such as the burning of Los Angeles in 1992, the escalation of the nuclear arms race in the 1980s, the mythology promoted by the World Bank of social and economic development of third world countries, and the nuclear disasters at the Three Mile Island and the Chernobyl power plants in the 1980s.

Finally, by 1990 we are experiencing a situation whose morphology is drastically different because of the strength of interdependencies among the problems, giving rise to a composite core that needs to be explicated and understood for its own sake. In other words, the dynamic convergence of all these originally distinct problem areas has given rise to a new entity, namely the *Problematique* of the 20th century. The strength of couplings and overlaps is such that a new approach is required to model the situation as a single complex system, as opposed to the traditional means of breaking it apart in its component parts and assigning responsibility to various individuals or government departments with expertise in the separate areas. Such a systems approach, to be effective and meaningful, should be capable of engaging stakeholders in interdisciplinary dialogue for the purpose of articulating the *Problematique* and integrating the relevant knowledge and wisdom of the community. The approach should also redefine the notion of "the expert" so that all the voices of the people affected by the

Problematique will be heard, independent of their education, social status, or power.

The original conceptualization of the CoR prospectus advocated the position that any attempt to resolving the global *Problematique* founded on traditional elitists, exclusionary, and disciplinary approaches is doomed to failure. As we will show by conducting in 1993 a retrospective inquiry of the *Problematique,* the evidence indicates that no significant progress has been made twenty-three years after its articulation in terms of penetrating its core and resolving it. The inquiry was conducted ten years ago employing the Structured Design Science approach for engaging inquirers in productive and meaningful dialogue on complex issues.

7.4 THE STRUCTURED DESIGN SCIENCE PARADIGM

Even though the complexity of the global *Problematique* demands that we avoid fragmenting reality, the fragmentation of reality is inherent to our conceptual and linguistic make up. It is a limitation that applies to all natural languages. The challenge is to transcend this limitation by enabling stakeholders to use natural language to discover collectively, systemically, and collaboratively the systemic nature of the *Problematique.* Having discovered the essence of the *Problematique,* the same stakeholders should be engaged in designing action plans for making progress toward resolving it. In other words, in a true participative democracy the definition and resolution of the global *Problematique* should not be the responsibility of scientific experts and politicians only, but also of ordinary people if real progress is to be made. It took twenty years of research and testing in the arena of practice for a group of systems scientists to meet this challenge by developing the Structured Design Science approach (Christakis, 2005).

The discovery of the SDS was driven by the appreciation of the critical role of dialogue in the evolutionary guidance of social systems (Banathy, 2001). SDS represents findings from grounded theoretical research starting with the conceptualization of the Club of Rome proposal in 1970. The research led to the discovery of a disciplined dialogue process suitable for interdisciplinary dialogue among stakeholders engaged in designing social systems for the contemporary Age of Information. The SDS is founded on three fundamental axioms, namely: (a) The complexity of social systems designing, (b) The cognitive limitations of the observer, and (c) The relative saliency of observations. Espousal of these three axioms was instrumental in the construction of the terminology, methodology, and architecture of the SDS.

The architecture and the thirty years of evolution of the SDS is shown in

Table 1. The architecture is composed of thirty-five components, six of which are laws discovered by systems scientists during the 20th century. The SDS has been applied for twenty years in the arena of practice for national, international and inter-organizational social systems designing challenges. The results have been extraordinarily satisfactory to the practitioners and the community of stakeholders. For more details the reader is referred to a forthcoming book by Christakis (2005).

TABLE 1: THE ARCHITECTURE AND EVOLUTION OF SDS

The Architecture of SDS is composed of 35 theoretical constructs, which have been grouped into seven modules.

The seven modules, and the approximate time period of their development, are:

a) The 7 Consensus Methods: (1) Nominal Group Technique, (2) Interpretive Structural Modeling, (3) Ideawriting, (4) DELPHI, (5) Options Field, (6) Options Profile, and (7) Trade-off Analysis (Years 1972-1982);
b) The 3 Key Role Distinctions: (1) Context, (2) Content, and (3) Process (Years 1982-1985);
c) The 3 Application Phases: (1) Discovery, (2) Design Dialogue, and (3) Action (Years 1989-2001);
d) The 4 Stages of Inquiry: (1) Definition or Anticipation, (2) Design of Alternatives, (3) Decision, and (4) Action Planning (Years 1989-1995);
e) The (5 Cs) of SDS, namely:(1) Community of Stakeholders, (2) *CogniScope*™ Team, (3) Consensus Methods, (4) CogniSystem Software and (5) Collaborative Facility (Years 1985-1995);
f) The 7 Language Patterns: (1) Elemental observation, (2) Problematique, (3) Influence tree pattern, (4) Options field pattern, (5) Options profile/scenario pattern, (6) Superposition pattern, and (7) Action plan pattern (Years 1970-1989); and
g) The 6 Dialogue Laws: Requisite: (1) Variety (Ashby), (2) Parsimony (Miller), (3) Saliency (Boulding), (4) Meaning and Wisdom (Peirce), (5) Autonomy (Tsivacou), and (6) Evolutionary Learning (Dye) (Years 2001-2003)

7.4.1 Findings of SDS Inquiry

As we mentioned above, recognizing that the fragmentation of reality is inherent to our linguistic make up, the CoR prospectus broke down the global *Problematique* into its major components for the purpose of its tentative identification, and for providing a reference framework. A listing of forty-nine general statements of the most commonly recognized problems in the 1970s was identified and presented in the prospectus. The list of problems was given the general title of **"Continuous Critical Problems (CCPs)."**

In April of 1993, Hasan Ozbekhan, his wife Ann who was a partner in the early phases of the CoR project, and me, decided to use the SDS approach to search retrospectively for the essence of the global *Problematique* as manifested in the time period 1970-1993. The intent of this inquiry was to discern whether humankind has made progress in resolving some of the component problems as conceptualized back in 1970, and especially to investigate if the core of the *Problematique* had been penetrated in the ensuing twenty-three years.

In order to make transparent the systemic, interactive, inter-sensitive, and interdependent nature of the global *Problematique* we used a computer-assisted dialogue methodology -embedded in the SDS approach - called ***CogniScope*** (Christakis, 1996, 2003, 2005). We used this method to build structures, which we shall call problem-clusters and problem-trees.

To build problem-clusters the three of us engaged in a dialogue by focusing on a set of questions such as: **"Does CCP 'X' have SIGNIFICANT common characteristics with CCP 'Y'?"** "X" stands for one of the forty-nine CCPs listed in the prospectus, and "Y" for a second one. Through the help of the computer-assisted methodology we were able to produce and agree on the problem-clusters, shown in Figure 2, in about two hours of focused and open dialogue. Without the use of the *CogniScope* dialogue methodology the derivation of problem-clusters by a small team of three stakeholders might have taken approximately a full day.

For example, one question posed by the computer and displayed on a screen for the three of us to deliberate was:

> **"In the context of the global *Problematique*, does CCP: WIDESPREAD POVERTY THROUGHOUT THE WORLD (CCP-2), Have SIGNIFICANT common characteristics with CCP: PERSISTENCE OF WIDESPREAD ILLITERACY (CCP-6)?"**

The answer we gave to the above question was "No." We agreed that these two CCPs should not be assigned to the same cluster in the context of the global *Problematique*. But when we were asked by the computer if **"Generalized and growing malnutrition,"** should be placed in the same cluster with **"Widespread poverty throughout the world,"** we decided to put these two problems in the same problem cluster. The intangible benefits of this inquiry are that by building up such problem clusters the three inquirers, although they all had made contributions in the original conceptualization of the CCPs, were able because of the dialogue to get a deeper appreciation of the underlying meanings of the CCP statements,

Figure 2: Classification of Continuous Critical Problems (CCP's) into Affinity Clusters

CLUSTER #1: POPULATION GROWTH/DISTRIBUTION

- **(CCP-1)** Explosive Population Growth With Consequent Escalation Of Social. Economic. And Other Problems
- **(CCP-8)** Growing Inequalities In The Distribution Of Wealth Throughout The World
- **(CCP-19)** Inadequate Shelter And Transportation
- **(CCP-20)** Obsolete And Discriminatory Income Distribution System(S)
- **(CCP-27)** Unbalanced Population Distribution
- **(CCP-31)** Widespread Unemployment And Generalized Under-Employment
- **(CCP-32)** Spreading "Discontent" Throughout Most Classes Of Society
- **(CCP-43)** Irrational Distribution Of Industry Supported By Policies That Will Strengthen The Current Patterns
- **(CCP-48)** Irrational Practices In Resource Investment

CLUSTER #2: POVERTY, LAGS & GAPS

- **(CCP-2)** Widespread Poverty Throughout The World
- **(CCP-5)** Generalized And Growing Malnutrition
- **(CCP-9)** Insufficient And Irrationally Organized Medical Care
- **(CCP-39)** Growing Technological Gaps And Lags Between Developed And Developing Areas

CLUSTER #3: WARFARE

- **(CCP-3)** Increase In The Production Destructive Capacity And Accessibility Of All Weapons Of War
- **(CCP-29)** Increasing A-Social And Anti-Social Behavior And Consequent Rise In Criminality
- **(CCP-30)** Inadequate And Obsolete Law Enforcement And Correctional Practices
- **(CCP-33)** Polarization Of Military Power And Psychological Impacts Of The Policy Of Deterrence
- **(CCP-40)** New Modes Of Localized Warfare

CLUSTER #4: URBANIZATION

- **(CCP-4)** Uncontrolled Urban Spread
- **(CCP-17)** Continuing Deterioration Of Inner-Cities Or Slums

CLUSTER #5: EDUCATION

- **(CCP-6)** Persistence Of Widespread Illiteracy
- **(CCP-13)** Anachronistic And Irrelevant Education
- **(CCP-37)** Growing Use Of Distorted Information To Influence And Manipulate People

TIE

LIN

Figure 2: Classification of Continuous Critical Problems (CCP's) into Affinity Clusters (continued)

CLUSTER #6: INSTITUTIONAL ARRANGEMENTS

- (CCP-7) Expanding Mechanization And Bureaucratization Of Almost All Human Activity
- (CCP-25) Generally Inadequate And Obsolete Institutional Arrangements
- (CCP-34) Fast Obsolescing Political Structures And Processes
- (CCP-38) Fragmented International Monetary System
- (CCP-41) Inadequate Participation Of People At Large In Public Decisions
- (CCP-42) Unimaginative Conceptions Of World-Order And Of The Rule Of Law
- (CCP-45) Obsolete System Of World Trade
- (CCP-46) Ill-Conceived Use Of International Agencies For National Or Sectoral Ends
- (CCP-47) Insufficient Authority Of International Agencies

CLUSTER #7: PREJUDICES

- (CCP-10) Hardening Discrimination Against Minorities
- (CCP-11) Hardening Prejudices Against Differing Cultures
- (CCP-28) Ideological Fragmentation And Semantic Barriers To Communication Between Individuals, Groups, And Nations

CLUSTER #8: UNKNOWNS

- (CCP-12) Affluence And Its Unknown Consequences
- (CCP-49) Insufficient Understanding Of Continuous Critical Problems, Of Their Nature, Their Interactions And Of The Future Consequences Both They And Current Solutions To Them Are Generating

CLUSTER #9: ENVIRONMENT

- (CCP-14) Generalized Environmental Deterioration
- (CCP-21) Accelerating Wastage And Exhaustion Of Natural Resources
- (CCP-22) Growing Environmental Pollution
- (CCP-24) Major Disturbance Of The Globe's/World's Physical Ecology
- (CCP-35) Irrational Agriculture Practices
- (CCP-36) Irresponsible Use Of Pesticides, Chemical Additives, Insufficiently Tested Drugs, Fertilizers, Etc

CLUSTER #10: VALUE-BASE

- (CCP-15) Generalized Lack Of Agreed-On Alternatives To Present Trends
- (CCP-16) Widespread Failure To Stimulate Man's Creative Capacity To Confront The Future
- (CCP-18) Growing Irrelevance Of Traditional Values And Continuing Failure To Evolve New Value Systems
- (CCP-23) GENERALIZED ALIENATION OF YOUTH
- (CCP-26) Limited Understanding Of What Is "Feasible" In The Way Of Corrective Measures
- (CCP-44) Growing Tendency To Be Satisfied With Technological Solutions For Every Kind Of Problem

TIE LINE

As shown in Figure 2, all forty-nine CCPs have been categorized in ten clusters. These clusters can be thought of as being equivalent to the geometrical shapes shown in Figure 1. However, in Figure 2 we have given names and substance to each of the shapes shown In Figure 1. For example, we have clusters named: "Population Growth and Distribution," "Urbanization," "Education," "Environment," "Value-Base," and others. These names attempt to capture the essence of the component CCPs categorized under each such problem cluster. By conducting a structural inquiry and constructing problem clusters from the linear listing of the CCPs in the CoR prospectus, we enhanced our understanding of the interdependence and interaction among the original CCP statements expressed in linear natural language.

Similar problem-clusters could have been constructed by engaging a larger team of stakeholders with more diverse perspectives. In this case, most likely the clusters produced by this stakeholder group would be different than those shown in Figure 2, because these clusters would be representative of the perceptions of the global *Problematique* by a different group of stakeholders. This implies that the definition and description of the problem clusters is **observer-dependent**, as it should be. Observer-independent science is a characteristic of First Phase science, such as classical physics. In the realm of social systems designing, the utility of such observer-independent science is marginal. In fact, in the context of the emerging Third Phase science as described by DeZeeuw (1996), the stakeholders are now demanding that their observations be included in the policy-making and allocation of resources for scientific developments at an equal footing with scientific elites. For example these days, AID patients demand that their voices be heard in research and development of antiviral drugs. Similarly, consumers of mental health services are saying "nothing about us without us," thus challenging the expertise of psychologists and psychiatrists in managing their lives.

If a larger group of stakeholders was employed in constructing the problem clusters, it would have taken much longer to produce them because the dialogue would have involved more participants. Given that there were forty-nine CCPs, a reasonable estimate for a group of 12 stakeholders to produce the problem-clusters with computer assistance is about four hours. Without computer assistance the estimate increases by at least a factor of 5, i.e., more than two days. The efficiency gains attributed to the computer support makes it feasible to engage groups of stakeholders in discovering the problem clusters through group work, which is one of the advantages of the SDS.

A cursory study of the problem-cluster presented in Figure 2 would reveal that they are interdependent, and that progress in resolving any problem assigned to one cluster will very likely influence the resolution of other problems in the same cluster, as well as in other distinct clusters. By overlaps among these clusters, I mean something equivalent to the geometric configuration graphically depicted by Frames C, D, and E of Figure 1.

In order to make the interdependencies and interactivities among the ten clusters more transparent, the three inquirers proceeded to build a problem-tree using the *CogniScope* dialogue methodology. Before constructing the problem-tree, the inquirers identified a profile of the *Problematique* by selecting a representative sample of twenty-four CCPs. They subsequently searched for the influence relationships among the members of this profile. The profile of problems that was selected is shown graphically in Figure 2 by those specific problems from each cluster that are connected to the "Tie Line." For example, we selected three CCPs from Cluster # 3, and one CCP from Cluster # 4. At least one CCP was selected from each one of the ten clusters.

The three inquirers focused their dialogue on a different set of questions: **"Supposing that humankind was able in the past two decades to make progress in the resolution of CCP 'X', would this progress help SIGNIFICANTLY the capacity to make progress in resolving CCP 'Y', in the context of the global *Problematique*?"** Again "X" stands for one of the twenty-four CCPs selected for this investigation and "Y" for another.

For example, in constructing the problem-tree one question posed by the computer and displayed on the screen for the three of us to deliberate was:

> **"Suppose humankind was able in the past two decades to make progress in the resolution of CCP:**
> **NEW MODES OF LOCALIZED WARFARE (CCP-40).**
> **Would this progress help SIGNIFICANTLY in resolving CCP:**
> **INCREASE IN THE PRODUCTION, DESTRUCTIVE CAPACITY, AND ACCESSIBILITY OF ALL WEAPONS OF WAR (CCP- 3)?"**
> **In the context of the global** *Problematique*?"

The answer we gave to this question was "No." After engaging in open and focused dialogue on this particular question, we decided that (CCP-40) would not have helped the resolution of (CCP-3) in the context of the global *Problematique*. But when we were asked the reverse question, namely if (CCP -3) would have strongly helped in resolving (CCP-40), the answer we gave was "Yes." The rationale for the "Yes" answer is that if humankind had been able to make progress in decreasing the production and accessibility of weapons we would have also made progress in terms of decreasing localized warfare. As I am writing this chapter in August of 2004, i.e., ten years after conducting this retrospective structural inquiry, I cannot help but reflect on the contemporary disastrous situation with the war in Iraq. If the threat of production and accessibility to weapons of mass destruction had been dealt with, say in the 1980s by the United Nations, there would not be any justification for the invasion of Iraq by the USA and it coalition partners in 2003. It is becoming apparent already, just from the response to the interaction between this pair of CCPs, that no significant progress has been made in resolving the *Problematique* in the last thirty-four years.

With computer assistance the three of us had to answer 120 questions, and provide alternative rationales for our responses. In three hours of structured dialogue, we were able to produce the problem-tree shown graphically in Figure 3. Without computer assistance we would have had to answer 552 pair wise questions, i.e., about 5 times more, and it would have taken at least two days to produce the snapshot of the global *Problematique* shown in Figure 3. It is very unlikely that three busy people (Hasan, Ann, and me) would have allocated two days to search for the core of the *Problematique*. Furthermore, fatigue and cognitive overload would have prevailed, and as a result: (a) we would not have completed the task even with a subset of the original list of forty-nine CCPs, and (b) without the support of the *CogniScope* dialogue we would not have been able to engage in a constructive learning experience by means of focused and open dialogue.

The problem-tree shown in Figure 3 titled "A Retrospective Snapshot of the Global *Problematique* for the 1970's," displays graphically the propagation of problem resolution power among the subset of twenty-four CCPs. The location of a CCP at the seven levels shown in the Figure is an indication of the strength of resolution influence a specific CCP is exerting on the other CCPs included in the snapshot. For example, those CCPs located at Level I, are at the surface of the *Problematique*, and hence they do not help resolve any other CCPs. On the other hand the CCP that is located at the bottom of Figure 3, i.e., Level VII, is exerting an influence on all the others as graphically illustrated by the streams of arrows flowing from the core of the *Problematique* towards its surface (i.e., from Level VII to Level I). One can think of the levels shown in Figure 3 as the layers of an onion, so that the core of the "onion" corresponding to the global *Problematique,* is CCP-18: Growing irrelevance of traditional values and continuing failure to evolve new value systems, which is located at Level VII.

The interpretation of this finding is significant. It tells us that, if in the two decades, i.e. from 1970 to 1993, humankind had made progress in resolving the "value-base problem," the resolution of this problem would have propagated and helped resolve all the other problems in the global *Problematique.* This finding should not be a surprise to environmentalists, enlightened executives, and other global thinkers who have been reflecting on the global *Problematique* for the last thirty years. What is, however, even more interesting, is that the "value-base problem" emerged in the decade of the 1990s, at least in the USA, as a very deep rooted problem in many social

Figure 3: A Retrospective Snapshot of the Global Problematique for the 1970's

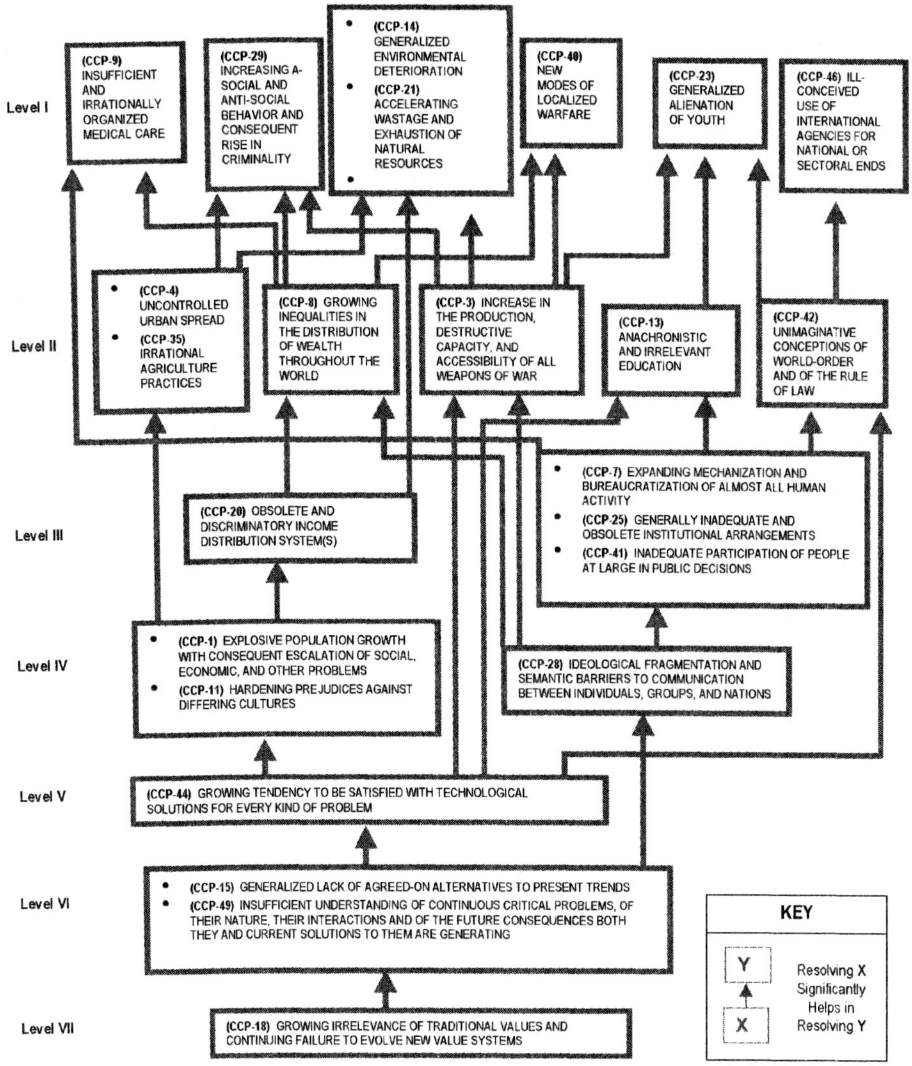

systems, such as: nation states, large multinational corporations, health care provision systems, and even global economic competitiveness for small and large companies. In many instances, the value-base issue was masked under the general term of "organizational culture," which became a dominant theme in the postmodern management paradigm.

With regard to the global scale environmental crisis, Vice President Al Gore in his 1993 book *Earth in the Balance* has converged to a very similar diagnosis. He writes:

> "The more deeply I search for the roots of the global environmental crisis, the more I am convinced that it is an outer manifestation of an inner crisis that is, for lack of a better word, spiritual. As a politician, I know full well the special hazards of using "spiritual" to describe a problem like this one. For many, it is like one of those signs that warns a motorist, Steep Slope -- Truckers Use Brakes. But what other word describes the collection of values and assumptions that determine our basic understanding of how we fit into the universe?"

In the same book he acknowledges the importance of abandoning the conventional wisdom and the status quo that have given us the belief that society can always muddle through. Such an acknowledgment is very similar to the meaning captured by CCP-15: Generalized lack of agreed-on alternatives to present trends, which is located very close to the core of the global *Problematique,* namely Level VI. It is clear to Gore that we cannot afford anymore to assume that the world will somehow solve its problems. He writes: "We must all become partners in a bold effort to change the very foundation of our civilization."

The building of such a partnership, however, can only be founded by enabling the stakeholders of the planet to engage in productive and meaningful dialogue. When a group of stakeholders is actually engaged in making the discovery that the core of the *Problematique* is the "spiritual vacuum", as compared to reading the statement by Mr. Gore in his book, they will be able to internalize the issue and become partners in the resolution of this deep-rooted problem. Without such engagement, the commitment to a common purpose is very unlikely.

Let me elaborate on this point by means of an experience. From 1975 till 1979, I offered a seminar in Washington, DC for Government Executives, under the auspices of the Graduate School of the U.S. Department of Agriculture. The topic of the seminar was "Anticipating the Future." Over the period of four years, approximately 250 executives attended the seminar in small groups of about 15 for each seminar. Each group of executives was asked to focus on the global *Problematique* from their own perspective, and to propose problem statements like the CCPs of the Club of Rome prospectus. On the average, each group produced about forty statements. We then proceeded to use the *CogniScope* dialogue methodology to produce problem-clusters and problem-trees, in a manner similar to the inquiry reported in the chapter.

The most interesting finding from these seminars is that, in the main, the participants did not propose any problem statements that could be categorized under the cluster called "Value-Base" in Figure 2. Only after they studied the initial structure of the *Problematique,* and were prompted by me to go deeper in terms of root problems, some groups generated ideas related to values. In light of the dominant perceptions about the role of the value-base in the context of the *Problematique* in the 1970s, is it surprising that no significant progress has been made in terms of its resolution in the decades of 1970-1990? My assertion is that there is no guarantee that the engagement of stakeholders will contribute to significant progress in resolving the *Problematique.* However, we can be sure that without the participation of the citizens of the planet in discovering the *Problematique* we cannot expect to make progress, as the evidence from this retrospective structural inquiry demonstrates.

Figure 3 has also uncovered another interesting phenomenon, which we are unable to capture succinctly through the use of natural language. The *Problematique* includes four cycles of closely coupled CCPs. A cycle represents tow or more mutually reinforcing problems. These strongly coupled problems in a cycle represent the emergence of new entity, just like the chemical combination of hydrogen and oxygen gives rise to water, which is a new entity. Such cycles require special attention in discovering corrective actions for their resolution.

One of the cycles appearing in Figure 3 warrants some additional discussion because it was instrumental in the discovery of the SDS paradigm. This is the cycle located at Level III and containing the following three CCPs:

- (CCP-7): Expanding mechanization and bureaucratization of almost all-human activity.
- (CCP-25): Generally inadequate and obsolete institutional arrangements.
- (CCP-41): Inadequate participation of people at large in public decisions.

I call this cycle "pseudo-democracy." In other words, this is the cycle that demonstrates the lack of participative democracy of the type practiced in the agora of ancient Athens.

Studying the problem-tree displayed in Figure 3, one sees that four CCPs are located at deeper levels of the tree, namely Levels IV, V and VI, and they are connected to the pseudo-democracy cycle at Level III with arrows. This means that the resolution of the pseudo-democracy cycle depends strongly on making progress in resolving these four CCPs, two of which are also in a cycle. These specific CCPs are:

- (CCP-28) Ideological fragmentation and semantic barriers to communication between individuals, groups, and nations;
- (CCP-15) Generalized lack of agreed-on alternatives to present trends; Which is in a cycle of mutual interdependence with:
- (CCP-49) Insufficient understanding of CCPs, of their nature, their interactions and of the future consequences both they and current solutions to them are generating;

And finally the deepest CCP mentioned earlier, located in Level VII, namely:

- (CCP-18) Growing irrelevance of traditional values and continuing failure to evolve new value systems.

The interpretation emerging from the problem tree displaying the *Problematique* is that without progress in the above four CCPs, it is very hard to make progress in resolving the pseudo-democracy cycle. The phenomenon of pseudo-democracy, as contrasted to participative democracy, is prevalent all over the world, including those advanced industrialized countries of Europe and North America. It is sobering to recognize how difficult it is to re-institute participative democracy without addressing and resolving these deeper problems of the global *Problematique*.

It appears that the *Problematique* has deteriorated in the last three decades, in spite of the end of the cold war. The demise of the Soviet Union has contributed only partially to the amelioration of only one of the forty-nine CCPs, namely: CCP - 3: Increase in the production, destructive capacity, and accessibility of all weapons of war. However, it is hard to find any solace by the amelioration of this particular CCP when we recognize that it has been replaced by another one, namely CCP-40: New modes of localized warfare, resulting in the reemergence and spreading of the phenomenon of Balkanization. This phenomenon was manifested by the war in Bosnia-Herzegovina during the decade of the 1990s, contributing to the death of hundreds of thousand of innocent people.

Following the structural inquiry and the interpretation of the problem-tree of Figure 3, it is now appropriate to discuss the value-base, especially in light of the discovery that the value-base is indeed positioned at the core of the *Problematique*.

7.8 THE VALUE-BASE

The primary aim of any social systems designing activity is to improve a situation. This can only be determined by an objective, which, itself, is external to the situation. Such an objective always entails a value, and the setting of it must therefore create the particular value-base that gives meaning and direction to the whole endeavor.

By constructing a retrospective snapshot of the *Problematique,* we discovered that the "value-base problem" is located at its core. In view of this finding it is essential that the approach to be applied for the resolution of the global *Problematique* must be guided by the value-base. In fact, through our structural inquiry we found evidence to substantiate that the failure to do this in the 70s and 80s is the primary reason for not having made any significant progress in improving the situation.

In light of the complexity of the *Problematique*, it becomes imperative to define a value-base that will govern the development and application of the systems approach from the outset. When we review the problem-clusters (Figure 2) and problem-tree (Figure 3), we find that all of the CCPs are problems in relation to something else -- either other problem-clusters or other problems within a cluster, and always in terms of values we take for granted because they are embedded in our current culture. Thus, for example, uncontrolled population growth is a problem when viewed in relation to a particular state of the environment that we have been experiencing in the last thirty years. It was not a problem in the beginning of the 20th century, when we experience the environment differently. Similarly, the health care problem was not a problem before our technological progress and the increasing tendency to adopt and apply every conceivable medical technology. On the one hand, technology gave us the power to prolong life, and on the other it contributed to the escalation of the health care costs.

Numerous such examples can be generated. In fact these are the type of rationales that stakeholders use in exploring the relationships among the CCPs, and in doing so learning from each other. When we reflect on all those relationships we recognize that in every instance the problematic aspect derives from an imbalance that affects the relationships among the individual CCPs.

Recognizing that the concept of imbalance defines the pathology of an "ecological" system, the CoR prospectus identifies the *Problematique* as being ecosystemic in character. The prospectus includes the statement:

> "The normative statement that describes the value-content of any ecosystem is "ecological balance." Consequently it is the idea of ecological balance that can, and will, be taken as the underlying value-base of the study; for in the terms dictated by our situation the "good" is self-evidently and most generally capable of being defined as the re-establishment of that many-dimensional dynamic balance that seems to have been lost in the modern world."

7.8.1 Revisiting the CoR Prospectus

Revisiting the CoR prospectus thirty years later, and having the benefit of all the research and development effort that has been carried out in the interim by researchers and systems scientists, it is much easier now to make the approach more transparent than it was possible in the original document, primarily because the requisite knowledge was not explicitly available in 1970. However, I can say now that the implicit architecture of the CoR proposal was founded on five fundamental premises. These are:

1. Extending the notion of the "expert" in the social systems designing arena to include interested parties that have a stake (i.e., the stakeholders) in the definition and resolution of the global *Problematique.*
2. Enabling groups of stakeholders to participate in the collective definition of the global *Problematique,* productively and efficiently.
3. Producing high quality observations for resolving the *Problematique* by not allowing the participating stakeholders to push for preconceived, premature choices of options and alternative designs.
4. Using computer-assisted methodologies, such as the *CogniScope* dialogue, to enable participants from all walks of life to structure the variety of ideas and perspectives relevant to the design of alternative solutions without imposing on them a cognitive burden.
5. Ensuring non-threatening and authenticating full participation of all the stakeholders entering the designing arena.

All these requirements have been satisfied by means of the development of the SDS paradigm.

7.9 THE DEMISE OF THE CLUB OF ROME

Unfortunately, the very pragmatic and scientific SDS approach, described briefly above and displayed in Table 1, was not available at the time the architecture of the CoR prospectus was proposed. Because of this, the Executive Committee of the CoR, at a meeting in Boston in the summer of 1970, opted to adopt a much more conventional approach, which in essence invalidated the whole thrust and philosophical foundations of the Club of Rome prospectus. Let me elaborate on this experience briefly.

One of the people who were selected by Ozbekhan and me to become a member of CoR was Professor Jay Forrester from the MIT Sloan School of Management. Forrester was by formal training an electrical engineer and had written a book in 1970 titled *Urban Dynamics*. The book was based on a computer language called DYNAMO, and some notions transferred from electrical engineering regarding state and rate variables introduced in mathematical equations. The purpose of the book was to explain the deterioration of the American cities and to predict their time-development. It was essentially an attempt to apply concepts borrowed from engineering and classical mechanics to predict the dynamics of American cities. The major advantage of this type of modeling is its rigor on account of the use of mathematical equations.

When the Club of Rome members met in Switzerland in the spring of 1970, Forrester made the claim that he could very easily apply the same methodology to model the global *Problematique*. He invited the Executive Committee to visit Boston for two weeks so that he could demonstrate the power of his method.

The Executive Committee accepted the invitation.

When the Committee participated in the meeting in Boston and saw that it was possible to use the computer to predict the dynamics of the world system for the next 250 years they were absolutely amazed. The conventional engineering mentality and the seductive pseudo-rigor of mathematical equations prevailed at the expense of the philosophical and methodological foundations of the original CoR prospectus. They decided to allocate the necessary funds for the development of the "World Model," which eventually led to the publication of two separate books, one by Jay Forrester called *World Dynamics*, and one by Meadows and others, all associates of Forrester, called *Limits to Growth*.

It is my view that, neither one of these publications did justice to the essence of the CoR prospectus, namely the need for a paradigm shift. It is unfortunate that in 1970, even the Executive Committee of the CoR was unable to appreciate the deep meaning of the role of the value-base included in the prospectus. They opted to fund a project that perpetuated the dominant scientific elitism paradigm, and as result the fundamental core of the *Problematique*, namely the value-base, was never discovered or addressed by any of these modeling projects.

When the Executive Committee made this decision, Ozbekhan and I resigned from the Club of Rome. He joined the Social Systems Sciences program at the University of Pennsylvania and continued to work for the conceptualization of the new paradigm. I joined the Academy for Contemporary Problems, where I met John N. Warfield, with whom I collaborated for twenty years, and together with LaDonna Harris, Bob McDonald, Diane Conaway, Larry Magliocca, Kevin Dye, and many others we developed the SDS paradigm.

7.10 DEMOCRATIZING SOCIAL SYSTEMS DESIGNING

The role of social science in governmental policy has been the subject of debate, especially in the United States, ever since President Lyndon Johnson introduced his "Great Society" legislation. The 1960s marked the beginning of an explosive growth in the number of issues that government chose to address by using social science principles. Some of these initiatives were in response to the continuous deterioration of social systems. For example, the bill creating the Department of Housing and Urban Development was signed into law by President Johnson on September 9, 1965, thirty days after the beginning of, and partly in response to, the rioting of the Watts section of Los Angeles. The intent was to ameliorate the deterioration of American cities by using ideas from social science and urban design. Social science was formally introduced in the formulation of government policy at this point in time.

Forty years and hundreds of billions of dollars later, welfare dependency, illegitimacy, violence, sexually transmitted diseases, and other acute social problems are all worse. An investigative report ordered by President Lyndon Johnson and conducted under the chairmanship of former Illinois Governor Otto Kerner after racial and urban riots in the 1960s, concluded: "Our nation its moving toward two societies - one black, one white - separate and unequal." In the spring of 1993, the Senate Banking, Housing and Urban affairs Committee was told by Henry Cisneros, the Housing Secretary of the Clinton Administration, that Kerner's predictions were coming true and the plight of the cities had worsened in the last 25 years.

How is it possible that a society with significant good will and major commitment of resources has not been able to resolve such acute problems over a period spanning at least a generation? The answer is that these acute problems cannot be resolved without using the appropriate social systems designing paradigm.

Even such early proponents of the formal use of social science principles in designing the Great Society programs as Senator Patrick Moynihan, changed their position by the end of the 1960s. The senator was in the late 60s an academic and began to worry about the increasing role of ideas originating in the social sciences being introduced into politics and government. He wrote: "Government has got into the business of promising more than it knows how to deliver; as there is little likelihood of cutting back on the promises, the success of the society turns in its ability to improve its performance (emphasis mine)."

Another academic social scientist, Nathan Glazer, sensing the ineffectiveness of social science in resolving acute social problems, wrote:

"The role of social science lies not in the formulation of social policy, but in the measurements of its results." A "social indicator" movement was initiated in the early 1970s -- modeled after the economic indicators, which have been used extensively by the Federal Government to measure the health of the national economy -- with many social scientists applying indicators to measure the results of social programs. It is apparent, that the social science paradigm espoused by these two prominent social thinkers is the conventional First Phase science corresponding to classical mechanics.

In my view, the role of social science in the Great Society programs was ill conceived from the outset. The shortcomings of the American social science paradigm are to be found in the preoccupation of social scientists with mathematical modeling and techniques, while failing to notice the quality of life their techniques was constructing and destroying. By trying to become scientific through emulating classical Newtonian mechanics, they espoused a paradigm, which was unable to explain and ameliorate the social problems observed in the 1960s, just as the Ptolemaic paradigm was unable to explain the movement of planetary bodies in the Copernican era.

Let me elaborate on this point by using a personal experience. In 1965, after graduating from Yale with a PhD in theoretical physics, I returned to Greece and became an advisor to Dinos Doxiadis, a well-known Greek architect/planner. Doxiadis had a very successful urban planning firm headquartered in Athens with branch offices in various parts of the world including the U.S. He was responsible for designing and building new cities, such as Islamabad, the Capital of Pakistan. Although Doxiadis was very busy with his professional business, his main interest was to invent a new science for the planning and design of cities, or "human settlements" as he preferred to call them (Doxiadis, 1968). He was convinced that unless urban planning becomes a science capable of explaining and predicting urban phenomena, just like physics does for physical phenomena, architects and planners will continue to design cities and towns that are not appropriate for human habitation. He named the science of human settlements "Ekistics," which sounds very similar to Economics. Both words originate from the Greek work "oikos" meaning home. Ekistics was intended to be more comprehensive than the discipline of Economics, which was included as one of the relevant perspectives of the new science.

When Doxiadis heard about my background in theoretical physics he expressed a desire to hire me in an effort to *mathematize* Ekistics. He believed that by introducing the elegance and rigor of the mathematical language used so much in physical sciences, Ekistics will become a science. Adherence by the professionals to the practice of the science will improve the quality of life in the cities. According to Doxiadis "the purpose of a city is to make its inhabitants happy and safe, and to help their human development." In light of the urban crises that became the dominant phenomenon in the 60s, clearly the purpose of those social systems was not being fulfilled.

When I joined the firm in 1967, I found out that urban planners were already using concepts from physics in their city planning and design projects. There was, for example, a mathematical model, called the "gravity model," which was useful in calculating the traffic flow between different neighborhoods in a city by using the number of people residing in a neighborhood as a means to determine its "gravitational attraction." By assigning numbers to the gravitational pulls of the various communities, the model was able to assign different volumes of traffic to the streets and determine areas of traffic congestion.

I worked on building mathematical models for urban planning for a period of three years. With a small group of associates with scientific and engineering backgrounds, and including social scientists and architects, we built a variety of mathematical models, most of which were used in a project dedicated to the design of the Metropolitan Urban Detroit Area for the year 2,000. These models were reasonably good in explaining and occasionally predicting phenomena related to human habitation patterns. Even though some of these deterministic models were elegant and useful, it became apparent to me that they were not appropriate and relevant to the planning and design challenges presented by a social system as complex as a city. Furthermore, the mathematics appropriate for modeling social systems had to be such as to enable ordinary citizens to engage in dialogue. This is in essence the mathematics of stakeholder inclusion, as opposed to the mathematics of exclusion practiced by some social scientists and engineers in the decades of the 60s and the 70s.

While working for Doxiadis I had the unique privilege of being exposed every summer to some of the greatest thinkers of that era. Once a year the firm would conduct a full-month seminar on the science of human settlements and invite distinguished people to participate such as: (a) Arnold Toynbee, the famous British historian, (b) Margaret Mead, the famous American anthropologist, (c) Hasan Ozbekhan, the internationally known planning theoretician, (d) Bucky Fuller, the renaissance humanist, and others. These people will participate in the Ekistics Seminars for one or two weeks, usually spending at least one week on a beautiful ship cruising the Aegean islands, and always visiting the island of Delos on the final day of their cruise. At the ancient theater of Delos, surrounded by the ruins of the ancient human settlement of that island, the participants will read a document produced during the seminar deliberations called the "Declaration of the Delos Symposia."

Most of these documents were making pronouncements regarding the present situation and the future of human settlements. There was a lot of concern about the deterioration of cities everywhere, and the need to take drastic measures to save the planet for human habitation. Because of the expert knowledge of the distinguished participants, and the extraordinary charisma of Doxiadis to orchestrate and to synthesize the deliberations of the seminar, the nine Delos Declarations represent significant contributions to the appreciation by certain publics of the global *Problematique*. The major shortcoming of the Declarations was that they stand out as pontifications by an elite group about the predicament of humankind.

Trying to invent the mathematics of Ekistics, coupled to the extravagant seminars of distinguished scholars, led me to the belief that there was something fundamentally wrong with the dominant social systems designing paradigm. It became clear that unless the social systems designing paradigm espoused the democratic ideal of stakeholder participation in the designing process, it could not become effective in reversing the dismal trends visible today in so many social systems. The dominant paradigm was as ineffectual in terms of resolving the acute problems presented in the *Problematique*, as was Ptolemy's paradigm in explaining and predicting the motion of planetary bodies around the sixteenth century.

I learned three principal lessons from this experience:

1. The theory of social system planning and design should not be the same as that of the traditional scientific approach, i.e., First Phase science, which has largely concerned itself with applying deterministic rationality and principles of observer independent observations to explain the dynamics of physical systems.
2. The methodology for social systems designing must differ drastically from that of the traditional deterministic approach, and as a result it raises severe challenges to some of our most common and well-accepted conceptions underlying the meaning of science, such as validation, replication, and prediction. Accordingly, any activity that relates to the planning and design of social systems should be defined as a "method" if it possesses the following three characteristics:
   - A defined sequence of operations which can be systematically followed in all applications;
   - Ability to communicate the knowledge generated in a form that can effectively lead to improvement of the problem situation;
   - An opportunity by all those (i.e., the stakeholders) who will be affected by the knowledge, when it is translated into action, to participate efficiently in the generation and organization of the knowledge.
3. Documentation of "expert" knowledge, of the type generated through Blue Ribbon Commissions (like the Kerner Commission) or the Delos Symposia Declarations, is of questionable value in terms of actually improving the social systems for the following reasons:
   - The experts do not know enough;

- Their analyses are opaque, even to themselves; and
- Lines of reasoning and consensus are interrupted by opaque analyses, blocking communication, understanding, ordering of thought patterns, and, worst of all, accountability.

These three lessons were instrumental in my thirty years quest to democratize social systems designing. The lessons led me to three fundamental premises:

1. There exists a diversity of values, perceptions, attitudes, beliefs, all of which are superimposed within the context of a given social system. This requisite variety needs to be explicated and appreciated before effective social systems designing can be accomplished;
2. Stakeholders participation is the most meaningful way to obtain representations of the requisite variety of reality perceptions and value structures embedded in social systems. Stakeholders are all those inside or outside of a system, who are directly affected by what the system is doing; and
3. Participation, although necessary, is not sufficient. In order for stakeholder participation to be meaningful and productive it needs to be enhanced by specially designed methods for engaging the stakeholders in a dialogue.

The above three premises motivated me to collaborate with a group of scholars and researchers over a period of thirty-five years to mount an effort, especially after the demise of the Club of Rome ideal, to conceptualize and implement a new paradigm for social systems designing for resolving the global *Problematique*.

7.11 CONCLUSION

I am reasonably confident that the SDS paradigm, described briefly in this chapter, is in consonance with the systems approach espoused and advocated by C. West Churchman in his profound contributions to systems philosophy and science. I also believe that a retrospective analysis of the global *Problematique* conducted today by a group of inquirers employing the SDS approach, will discover that the core is still the value base, as described by **CCP-18: GROWING IRRELEVANCE OF TRADITIONAL VALUES AND CONTINUING FAILURE TO EVOLVE NEW VALUE SYSTEMS.** This is indeed a very painful diagnosis and prognosis regarding the future of humankind. I know that West felt this pain throughout his life. He finally went home in 2004.

REFERENCES

Alexander, G.C., 2003, Interactive Management: An Emancipatory Methodology, *Systems Practice and Action Research,* 15:111-122.

Apel, K., 1981, *Charles S. Peirce: From Pragmatism to Pragmaticism,* University of Massachusetts Press, Amherst.

Ashby, R., 1958, Requisite Variety and Its Implications for the Control of Complex Systems, *Cybernetica,* 1(2):1-17.

Banathy, B. H., 1996, *Designing Social Systems in a Changing World,* Plenum, New York.

Banathy, B. H., 2001, *Guided Evolution of Society: A Systems View,* Plenum, New York.

Bausch, K., 2001, *The Emerging Consensus in Social System Theory,* Plenum, New York.

Bausch, K., 2000, The Practice and Ethics of Design, *Systems Research and Behavioral Science,* 17(1):23-51.

Boulding, K., 1966, *The Impact of Social Sciences,* Rutgers University Press, New Brunswick.

Christakis, A.N., 1973, A new policy science paradigm, *Futures,* 5(6):543-558.

Christakis, A.N., 1987, High technology participative design: the space-based laser, in: General systems, John A. Dillon Jr., ed., *International Society for the Systems Sciences,* XXX:69-75.

Christakis, A.N. 1988, The club of Rome revisited, in: General systems. W. J. Reckmeyer, ed., *International Society for the Systems Sciences,* XXXI:35-38.

Christakis, A.N., 1993, The inevitability of demosophia, in: *A Challenge for Systems Thinking: The Aegean Seminar,* Ioanna Tsivacou, ed., University of the Aegean Press, Athens, Greece, pp. 187-197.

Christakis, A.N., 1996, A people science: the CogniScope system approach, *Systems: Journal of Transdisciplinary Systems Sciences,* 1(1).

Christakis, A. N. (2004). Wisdom of the People, *Systems Research and Behavioral Sciences,* 21, 317-330.

Christakis, A.N., 2005, *Dialogue for the Information Age,* Information Age Publishing Inc., Greenwich, Connecticut.

Christakis, A.N., Warfield, J.N., and Keever, D., 1988, Systems design: generic design theory and methodology, in: *Systems Governance,* Michael Decleris, ed., Publisher Ant. N. Sakkoylas, Athens-Komotini, Greece, pp. 143-210.

Christakis, A.N. and Dye, K.M., 1999. Collaboration through communicative action: resolving the systems dilemma through the CogniScope, *Systems: Journal of Transdisciplinary Systems Sciences,* 4(1).

Christakis, A.N., and Brahms, S., 2003, Boundary-spanning dialogue for 21st-century agoras, *Systems Research and Behavioral Sciences,* 20:371-382

Christakis, A.N. and Harris, L., 2004, Designing a transnational indigenous leaders interaction in the context of globalization: a wisdom of the people forum, *Systems Research and Behavioral Sciences,* 21:2511-261.

Churchman, C. W., 1971, *The Design of Inqiring Systems: Basic Concepts of Systems and Organizations,* Basic Books Inc. Publishers, New York.

Churchman, C.W., 1979, *The Systems Approach and its Enemies,* Basic Books Inc. Publishers, New York.

de Zeeuw, G., 1996, Second order organizational research, *Working Papers in Systems and Information Sciences,* University of Humberside, Hull, England.

Doxiadis, C.A., 1968, *Ekistics: An Introduction to the Science of Human Settlements,* Hutchison, London.

Dye, K.M., 1997, Collaborative design process science, *Working Papers at MIT,* Boston.

Dye, K.M. and Conaway, D.S., 1999, *Lessons Learned from Five Years of Application of the CogniScope™ Approach to the Food and Drug Administration,* CWA Ltd. Interactive Mangagement Consultants Report, Paoli, PA.

Habermas, J., 1984, *The Theory of Communicative Action*, Vols. I and II, Polity Press.
Jackson, M., 1995, Beyond the fads: systems thinking for managers, *Systems Research,* 12:25-42.
Jantsch, E., 1969, *Perspectives of Planning,* OECD, Paris
Magliocca, L.A. and Christakis, A.N., 2001, Creating a framework for sustainable organizational leadership: the CogniScope system approach, *Systems Research and Behavioral Science,* 18:259-279.
McIntyre, J., 2004, Facilitating critical systemic praxis (CSP) by means of experiential learning and conceptual tools, *Systems Research and Behavioral Science,* 21:37-61
Meadows, D.H., Meadows, D. and Randers, J., 1972, *The Limits to Growth,* Universe Books, New York.
Miller, G.A., 1956, The magical number seven, plus or minus two: some limitations on our capacity for processing information, *Psychology Review*, 63:81-97.
Peccei, A., 1969, *The Chasm Ahead,* The Macmillan Company, Toronto.
Roberts, N., 2002, *The Transformative Power of Dialogue,* Elsevier, New York.
Romm, N.R.A., 2002, A trusting constructivist approach to Ssystemic inquiry: exploring accountability, *Systems Research and Behavioral Science,* 19:455-467.
Simon, H.A., 1974, How big is a chunk, *Science*, 183:482-488.
Turrisi, P.A. ed., 1997, *Pragmatism as a Principle and Method of Right Thinking,* State University of New York Press.
Warfield, J.N., 1994, *A Science of Generic Design: Managing Complexity Through Systems Design,* Iowa State University Press, Ames, Iowa.
Warfield, J.N., 1999, The problematique: evolution of an idea. *Systems Research,* 16:221-226.
Warfield, J.N. and Christakis, A.N., 1987, Dimensionality, *Systems Research,* 4:127-137.
Warfield, J.N. and Cardenas, A.R., 1994, *A Handbook of Interactive Management*, Iowa State.

Chapter 8

SYSTEMIC IQ

Systemic Intervention and Problem Appreciation

Ian Mitroff
University of Southern California

Key words: Problem, appreciation, pragmatism

Opening Arguments Given in AAA Wrongful-Death Suit
Tuesday, September 9, 2003
Associated Press

PLYMOUTH—A woman who was murdered in 1999 after her car broke down and she accepted help from a stranger was failed by her Auto Club, an attorney for the victim's family said during opening statements Tuesday.

The family of Melissa Gosul is suing AAA for wrongful death and negligence, claiming that if the auto association had assisted Gosul properly she would not have been raped and stabbed to death.

"This is about a nightmare that should have been prevented," said Michael Paris, the family's attorney. *"All because the defendants failed to do what they were supposed to do."*

But an attorney for AAA of Southern New England said Gosul, whose car broke down on Cape Cod, was in a busy recreation area on a summer evening, near a major highway, a restaurant, and a gas station.

"Melissa was not left in an unsafe location," said Robert Gainor.

On July 11, 1999, Gosul returned from a bike ride in a park to find her car would not start. Michael Gentile, who was later convicted for her murder, let the 27-year-old elementary school teacher use his cell phone.

When the AAA tow truck driver arrived, he told Gosul he was busy and would not be able to take her or her car back to Boston for another three or four hours. Gentile eventually offered to drive Gosul home to her parents' house in Brockton.

Her body was later found in a shallow grave.

The defense said that, according to witnesses, Gosul asked others for a ride back to Boston before the tow truck operator arrived and even described Gentile to another person in the recreation area as "a guy being nice" to her.

"She was already relying on other people, other strangers, for a ride back," said Bobby R. Burchfield, attorney for the National AAA.

The lawsuit names the National AAA; its local affiliates, AAA Southern New England; and the tow truck driver. It asks for unspecified damages. The case is being watched by the legal community to see if a jury will hold AAA liable.

The great poet Rilke wrote: "Be patient towards all that is unsolved in your heart, and learn to love the questions themselves."

Fred Rogers, *You Are Special: Words of Wisdom from America's Most Beloved Neighbor,* New York: Viking, 1994, p. 160.

8.1 THE ARGUMENT

1. The world is inherently fuzzy, complex, and uncertain.
2. The problems of the world are not exercises, i.e., overly simplified, well-structured problems, such as those that are typically found at the end of the chapters in textbooks, e.g., "x + 6 = 11; find x."
3. Exercises have one and only one right answer that everyone is expected to find, and hence, to agree with. This is because exercises have one and only one definition, that which is given to the student. Thus, in "x + 6 = 11; find x;" therefore, from simple subtraction, x = 5.
4. In contrast, problems have as at least as many definitions as there are major stakeholders that affect and are affected by the problem. The more important the problem, then the more that each stakeholder will have a different definition of the problem. Each sees the problem from his or her unique perspective. For instance, why should we expect that every stakeholder should have the same definition of key terms, for example, what is a crisis or what is a terrorist act?
5. The problems of the world are not "given." They do not drop from the sky clearly stated and pre-formulated.
6. The problems of the world are extracted, often with extreme difficulty, from complex messes. They are "taken" from the world by humans, not "given" to us by the gods.
7. A "mess" is a complex system of problems such that no problem or part of the mess exists or can be defined independently of all the other

problems that constitute the mess! In short, messes are highly interactive systems of problems.

8. The problem with conventional education is that for the most part, people are <u>trained</u> to solve canned exercises. They have not been <u>educated</u> to <u>formulate</u> complex problems. As a result, far too many people want checklists, clear-cut guides, and simple procedures for doing Crisis Management (CM). As a result, they want to reduce CM to a series of canned exercises.
9. CM not only deals with problems that are inherently ill-structured and ill-defined, but the field of CM is itself inherently ill-structured.
10. CM also deals with problems that are "messes." In fact, CM is a major form of Mess Management.
11. The failure of many CM scholars, experts, and practitioners to realize and to accept the preceding points is responsible for much of the confusion and the lack of progress of the field.
12. To paraphrase Albert Einstein, "The world's problems cannot even be formulated, let alone solved, by the same mindset(s), i.e., thought systems, that created them."
13. One never gets rid of fuzziness altogether. Indeed, fuzziness is inherent in the world. At best, one copes with and attempts to manage fuzziness, not eliminate it.
14. In this sense, the management of fuzziness is one of the prime challenges facing humankind.
    Every crucial feature of a crisis is unknown in the beginning:
    a. Exactly what is the crisis/crises?
    What were its causes?
    What is the full extent of the damage, i.e., number of injuries, deaths?
    What other crises is it likely to give rise to?
    Were we responsible for it in any way through either our inactions or inactions?
    Did we ignore or block any of the early warning signals?
15. While there are no precise answers to the preceding questions—if there were, then they wouldn't be unknown—there are methods and tools whereby one can at least obtain intelligent responses to them, not necessarily definitive answers.[28]
16. The fuzziness and the uncertainty connected with crises is not an excuse for doing nothing. Instead, it is cause for examining the assumptions behind our actions and the necessity of monitoring the validity of our assumptions as the crisis unfolds. Every crisis validates some assumptions and invalidates others. As a result, the strategies for coping with crises have to change over the course of the crisis as well.

[28] See Ian I. Mitroff, Under Attack, New York: AMACOM, 2005 (in press).

8.2 WILLIAM JAMES: THE QUINTESSENTIAL CRISIS MANAGER

William James, the founder of that distinct brand of American philosophy known as Pragmatism, is arguably this country's greatest philosopher. (James taught at Harvard in the late 1800's and early 1900's.) He is also arguably one of the greatest philosophers that the world has ever produced. James' philosophy is extremely relevant to CM. Indeed, I would argue that James is the philosopher that is most relevant for CM.

James' is especially known for his development of a unique theory of "truth." It is a theory of problem solving for complex problems. It is a theory of critical thinking and a theory of how to formulate complex problems from multiple points of view. For James, a single view or perspective of any problem is automatically wrong. It cannot hope to capture all of the subtleties and the complexities that are characteristic of real problems.

8.2.1 Four Distinct Styles of Thinking

James begins his first lecture on Pragmatism by noting two distinct types of minds, or styles of thinking, that have appeared repeatedly throughout the course of human history: Tough Minded versus Tender Minded. While these two types certainly exist, and are still relevant today, modern psychologists have discovered additional dimensions that need to be added to James' system if we are to capture more completely the full range of differences that underlie human thinking. For want of better terms, I call these two additional types "Earth Bound," "Grounded," or "Bounded" versus "Airy," "Floating in the Clouds," or "Unbounded." Thus, if we take the distinction Tough versus Tender and add the two additional distinctions, Bounded versus Unbounded, then we get the possibility of four, not two, different types of thinkers (see Figure 1).

In order to understand these different ways of thinking, let us take two of them and see how they would respond to the AAA tragedy or crisis that is the frontispiece quote to this chapter.

Tough Minded-Earthbound types tend to frame the AAA tragedy primarily in legal terms. They are also inclined to protect the interests of AAA. As a result, they want AAA to say as little as possible such as to limit the organization's legal liabilities. In slightly different words, those who are inclined to a legal perspective often respond in a language that is perceived as cold as the initial tragedy itself. (To be sure, lawyers representing the plaintiff's side often also use the same dry, cold language and mode of thinking to protect the injured parties. Thus, "lawyering" is not confined merely to protecting an organization's interests.)

Figure 1

FIGURE 1:
Four Distinct Types of Thinking

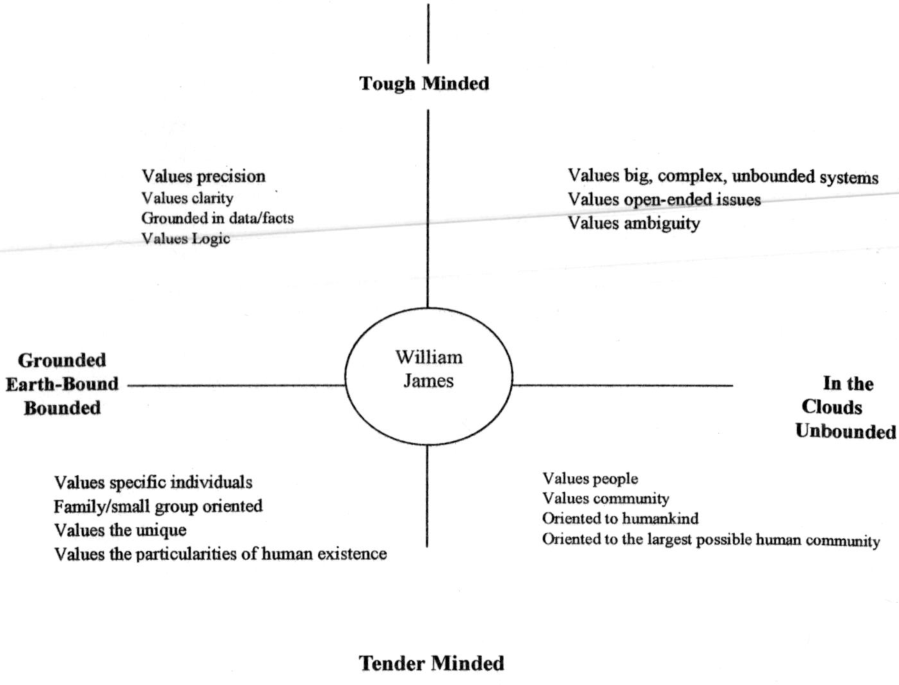

Juxtaposed to this is a completely opposite way of thinking, what I call the Tender Minded- 'In the Clouds type'. This style is the complete opposite of the first. It is the contrast between the upper left-hand quadrant of Figure 1 and the lower right-hand.

This was brought out forcefully in a recent class on critical thinking that I conducted. The students were exposed to all four of the types of characters, or styles of thinking, that are shown in Figure 1. They were then instructed to find a contemporary problem that they could analyze from all four perspectives. Not only were they to analyze the problem from all four perspectives, but even more fundamental, they were to define the problem from each. This was done in order to ensure that they knew how to "speak" each of the four different "languages" represented in Figure 1, and hopefully, by doing so, to learn how to produce richer definitions of any problem.

One of the prime characteristics of complex problems is that they cannot be defined solely from the single perspective of any of the four perspectives in Figure 1. Each of the four naturally emphasizes certain features of complex problems and situations, but neglects others. It is often precisely those features that are neglected that come back to haunt us mercilessly.

One of the students in the class chose the AAA tragedy for the exercise. In the discussion that followed afterwards, which is the whole point of the exercise, an interesting approach emerged that was very different from the legal perspective. To be sure, the entire group had gotten the legal approach down pat because this position is not only completely obvious in today's world, but is especially so in a society as litigious as ours. In the ensuing class discussion, I raised the question, "Why didn't AAA have a van or cab service that they could offer to a stranded motorist such that if he or she felt unsafe in any way then they could be transported to a more secure location?" (Of course, this assumes that the operators of the vans or cabs are themselves not be criminals!)

At this point, one of the students in the class literally jumped out of her seat. And, I mean literally. She not only stood upright, but her eyes flashed as well. She blurted out, "Why didn't AAA think of using their affiliated hotels, which literally span the entire country, as potential safe havens where motorists could be taken if they desired?"

This suggestion opened the discussion up even further. The whole point was that AAA already had an infrastructure in place such that if it were conceived of in slightly different terms then it was already a natural "safe haven infrastructure."[29] In other words, it takes only a slight leap of imagination to convert a system that was designed for one purpose, i.e., to provide discounts to members that were traveling across the country so that they could secure clean and economical lodgings, to serve another human

[29] Walter Kirn, "Birth of a Vacation: Holiday Inn Made it Safe for Americans to Go on the Road," New York Times Magazine, December 28, 2003, p. 12.

purpose, i.e., a nationwide safety system. However, this line of thinking only raises the critical question, "Why is it that most people and most organizations aren't capable of making this leap?"

Most organizations certainly have legal counsel. This is built into their basic operating structure. The job of legal counsel is to protect the legal liabilities of an organization. On the other hand, most organizations also have public relations and human resource specialists, security departments, and so on. Why didn't any of them think about using their affiliated hotel system for another purpose? To be sure, this might have been considered impossible or even ridiculous *before* the tragedy, but it certainly cannot be considered ridiculous *after* it. If one is to learn from such tragedies in order to ensure that they will not happen again, then is not AAA and all other organizations obligated to engage in this kind of thinking so that it will help mitigate future crises?

8.3 THE DEMISE OF WORK AND ORGANIZATIONS AS WE HAVE KNOWN THEM

The jobless recovery is one of the strongest signals to date that the nature of work is undergoing a major transformation, indeed, what some would call a major revolution. It is as profound and as radical a transformation as any that has occurred in human history. To say the least, it is also a crisis as profound as any that has befallen our nation. It certainly promises to alter every aspect of our lives. For another, the transformation has important ramifications for CM.

The nature of this transformation is as follows: Anything that can be defined precisely and unambiguously, i.e. in Tough Minded, Grounded terms, can in principle be outsourced, or exported, to another country where it can be performed cheaper, and in many cases, faster than it can by U.S. workers. The truly radical nature of this transformation is that it no longer applies merely to manual or to physical work, but to higher-level mental work as well. For instance, it applies to the development and the manufacturing of computer software. It applies as well to thousands of other traditionally high skilled jobs such as accounting, financial analysis, engineering design, etc. In short, anything that is well-structured, and relatively independent of context and culture, i.e., once again, Tough Minded, Grounded thinking, can be done by someone that is not a member of the society in which the problem arose initially.

The consequences of this revolution are literally mind-boggling. For one, it threatens to demolish the monopoly and the stranglehold that the traditional academic disciplines have had over knowledge for hundreds of years. The organization of the modern university into separate and autonomous disciplines is as clear cut an example of Tough Minded, Grounded thinking as one is ever likely to find. The boldest prediction one can make is that the modern university will cease to exist, at least in its present form.

For another, this "revolution" also threatens to demolish the monopoly of the traditional disciplines over work. It promises to alter radically traditional corporate functions such as accounting, finance, marketing, law, manufacturing, etc. In other words, it promises to alter the traditional, Tough Minded, Grounded design that has permeated all organizations. Likewise, it will not be alleviated by integrating and even synthesizing diverse disciplines into new forms such as the creation of hybrid disciplines such as bio-engineering (interdisciplinary).

The traditional academic disciplines, professions, and corporate functions are obsolete. If one can no longer expect to work in a single job for a single organization for one's entire life, then one can no longer expect to practice a single discipline, profession, or corporate function for one's entire life. The nature of work has changed dramatically because the nature of knowledge has also changed, and vice versa.

To grasp the truly radical nature of this revolution, it is necessary to understand that multi and interdisciplinary approaches to knowledge will not solve the fundamental problem we are facing. The fundamental problem will not be alleviated merely be combining or bringing more disciplines to bear on major issues (multidisciplinary). Likewise, it will not be alleviated by integrating and even synthesizing diverse disciplines into new forms (interdisciplinary). Instead, it will only be alleviated when we finally realize, and accept, that virtually all of our major problems have significant aspects that lie beyond <u>any</u> of the currently known disciplines, professions, and functions. Furthermore, it will only be alleviated when we finally accept that the solutions to our major problems lie beyond <u>any</u> disciplines, professions, or functions that we can even begin to imagine. As a result, we need to develop both the concepts and the understanding of what it is to know without disciplines, professions, and traditional job descriptions.[30]

The only reliable predictions that one can make regarding the types of jobs that will remain in the U.S. are those that will satisfy two stringent conditions: One, those that require an in-depth, working knowledge of U.S. culture; and, two, those that necessitate the exercise of high-level, critical thinking and creative judgment. In short, we need to learn desperately how to apply and to integrate <u>all</u> of the four types of thinking (see Figure 1) to any and all problems. Anything else is doomed to vanish - forever!

[30] See Ian I. Mitroff and Harold Linstone, The Unbounded Mind: Breaking the Chains of Traditional Business Thinking, New York: Oxford University Press, 1995.

8.4 DIMINISHED EXPECTATIONS, VANISHING DREAMS

By now, the stories are all too familiar and common: the permanent loss of hundreds of thousands of high paying white collar jobs to Asian and Third World Countries; the greatly increased time that it takes to land a similar job at an equivalent rate of pay, if such jobs even exist at all; the substantial lowering of expectations, the fact that often the only jobs available are those considerably below one's level of education, job experience, and certainly, one's previous income; the fact that millions have completely given up all hope of finding any job and have therefore dropped out of the labor force altogether.

The moral of the preceding story is not that individual job seekers are bad or are failures. The moral is that 19th and the 20th Century's notion of knowledge and education is not adequate for the problems and the jobs of the 21st Century. It is the system, not individuals, that has failed.

This paper is necessarily philosophical for the root of our problems is, once again, outmoded assumptions, or more precisely, an undue reliance on outmoded assumptions.

8.5 OUTMODED ASSUMPTIONS

The 19th and the 20th Centuries developed a view of problems that influenced profoundly the nature of education and work. This view is best stated in terms of the key assumptions upon which it was based:

In order for something to be or to count as a problem, it had to be stated (defined) unambiguously and precisely; unless one could state or define a problem in this manner, then one did not know what the problem was, and hence, one would not know what a solution to it was, if one existed; in other words, it had to be stated in Tough Minded, Grounded terms;

The best (superior) language for stating problems was mathematics; the ideal model in this regard was Euclid's geometry where one started with intuitively obvious or self-evident ideas (axioms and postulates) such as the definitions of points, lines, triangles, etc., and from these one derived rigorously (deductively) a potentially infinite set of interesting and important conclusions known as theorems; in the more extreme versions of this philosophy of problems, unless something could be expressed in the rigorous and exacting language of mathematics, then it was not even worthy of the term "problem;"

All complex problems were in principle decomposable into a finite set of separate and simpler problems; the "sum" (synthesis) of the solutions to the separate and simpler problems was then the solution to the complex problem; in fact, for something to even be considered as a problem in the first place, then it had to be decomposable into its simpler problems or "atoms;"

Different disciplines owned different atoms; different disciplines owned different types of problems; as a corollary, the different disciplines were clearly separable from one another; finally, there was a strict hierarchy between disciplines; some disciplines were better than others; "better" meant that one discipline could state its problems more rigorously (e.g., in terms of mathematics) than others; conversely, the more that a discipline could state its problems independently of context, the better it was as well.

Education consisted largely of solving a set of pre-defined exercises (e.g., "x + 6 = 11; find x"); by definition, exercises have one formulation (the one that is given to students in textbooks), and as a result, exercises have only one right answer;

A problem once solved remained solved forever in the same way that a set of facts once established presumably was established forever; for instance, the boiling point of water is a constant, not a variable.

8.6 COUNTER ASSUMPTIONS

This paper is based on a completely different, counter set of assumptions:

A fundamental characteristic of problems is that when they first arise they are highly ambiguous; in other words, problems are not separable from ambiguity; indeed, if anything, problems are extracted from ambiguity; for something to be a problem it has to be infused with ambiguity, for the "nature" of most complex problems is not clear or well known in the very beginning; this certainly applies to crises; rarely is the full nature of a crisis known when it first presents itself;

There is no "one best or superior language" in which to state a problem; the notion of a "best language" already assumes that one knows that the problem is, or at the very least, a great deal about it; certainly, most of our critical problems cannot be stated unambiguously, let alone in the restricted language of mathematics;

In principle, complex problems are not decomposable into a finite set of separate and simpler problems; to the contrary, by definition complex problems must be treated as "wholes;" in other words, complex problems possess properties as a whole that none of the "parts" do;

Different disciplines do not necessarily "own" different parts of complex problems because the disciplines themselves are largely artifacts; if complex problems only exist as "wholes," then the knowledge that is required for formulating and solving problems must be holistic as well; the different disciplines are not clearly separable from one another; there is not a strict hierarchy between disciplines such that some are better than others; mathematical rigor is not necessarily the most desirable property in formulating or in solving problems;

Education first and foremost consists of critical thinking; a critical part of critical thinking is problem formulation; that is, one needs to learn how to formulate problems from at least each of the four different styles of thinking; by definition, complex problems do not have a single formulation;

Problems do not remain solved; indeed, in the process of working on a problem, the nature of the problem, let alone the solution, changes substantially.

8.7 THE MANAGEMENT OF TRUTH

This moral of the story this far is that for William James, epistemology, or that branch of philosophy that deals with the nature of truth—what it is and how we humans can obtain it-- is actually *the management of truth*!

If ever a concept had a dangerous and an ominous ring to it, then surely it is the "management of truth." It implies that humans manipulate the truth solely for evil ends. While to be sure this often happens, the inescapable fact is that truth is a thoroughly human creation. Whether humans manipulate the truth or not, that humans make truth is uncontestable. They create it through their actions in an attempt to achieve desired ends.

For James, Truth is not an abstract concept that is independent of human purposes and ends. Truth is not something that is "just out there" independent of human concerns.

Truth is in fact the management of different ways and styles of knowing. Knowledge can not be decoupled from knowers, i.e., from the processes by which humans know. Truth is thus the management of at least four very different types of knowers.

This is precisely what makes James so powerful and his ideas so relevant to CM.

8.8 THE CRUCIAL DIFFERENCES BETWEEN WELL-STRUCTURED AND ILL-STRUCTURED PROBLEMS

To summarize, real problems are by their very nature ill-structured or ill-defined. At least in the very beginning, when problems first "present themselves," they are not already or automatically well-defined or well-structured. (When we say a problem is "well-structured," we mean it is expressed in an exact language, for instance, algebra or mathematics in general, such that the problem can be expressed precisely so that it yields an exact solution. Another example is that of the field of Corporate Finance where one computes the present value of a future amount of money knowing today's interest rate and how far in the future one is discounting back into the present.)

Consider a simple example, a person who is considering the choice between three available means, or courses of action, in order to get downtown in the shortest time possible (see Figures 2 and 3).

Suppose a person has three available means to get downtown: one, drive his or her car; two, walk; or three, take a bus. Suppose in addition there are three mutually exclusive outcomes: one, get downtown in twenty minutes or less; two, get downtown in exactly twenty minutes; and, three, get downtown in more than twenty minutes. Suppose also that the person places the value of getting downtown in less than twenty minutes at $80; exactly twenty minutes, $15; and, finally, getting downtown in greater than twenty minutes, $5. Suppose also, as shown in Figure 3, that if one takes a car then there is an eighty percent chance of getting downtown in less than twenty minutes, a zero percent chance if one walks, and a forty percent chance if one takes a bus.

The point is that only for well-structured problems, i.e., exercises, do we know all of the available courses of action, their outcomes, i.e., their consequences, and the benefits versus the costs associated with them. In contrast, an ill-structured problem is "precisely" a problem for which the full set of means is not known, or only partially known. In addition, the ends are not fully known as well, let alone their consequences. Thus, Figures 2 and 3 do not represent real world problems at all.

8.9 ALL OF US ARE PROBLEM MANAGERS

The point of the previous discussion is that all of us are not merely problem solvers, which we are, but even more fundamental, we are "problem managers." Even more basic, we are "problem creators" and "problem organizers." Whatever structure or definition problems possess is due to our efforts. Problems become defined and structured through our efforts, i.e., through the process of our working on them.

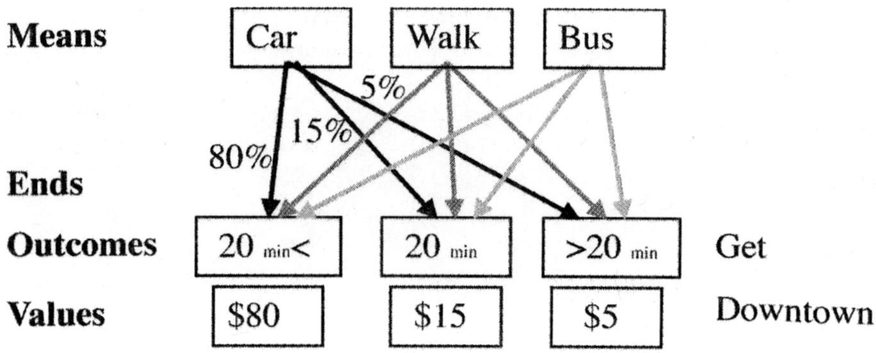

Figure 2: A Well-Structured Problem

Ends

	Values	$80	$15	$5	= $100
	Outcomes	20 <	20 min	> 20	
Means	P(Car) =0.8	80%	15%	5%	Sum % = 100%
	P(Walk)=0.1	0%	1%	99%	Sum % = 100%
	P(Bus) =0.1	40%	40%	20%	Sum % = 100%
	Sum P =1.0				

Figure 3: A Well-Structured Problem

The problems of the world are not normally the captive of a single academic discipline, e.g. Psychology, or a single corporate function such as Finance. For instance, no single discipline "owns" the problem of world hunger. The problem of world hunger can not be defined completely or adequately by any single discipline.

Most "real-world problems" are complex mixtures of different disciplines and functions. Increasingly, the problems of the world are interdisciplinary and cross functional. Even more to the point, the problems of the world are transdisciplinary. That is, they are beyond any of the known disciplines. A good example is once again that of world hunger. The same is also true of CM, i.e., CM is inherently transdisciplinary as well. This means that CM cannot be represented in the form of Figures 4.2 and 4.3. For another, it means that different stakeholders will have different versions of Figures 4.2 and 4.3, if they exist at all. That is, different stakeholders do not see the same courses of action and the same outcomes of the "same" crisis.

For instance, a major crisis is never a single problem or a single crisis. Instead, it is a whole system of problems, and crises, that are highly interdependent and highly interactive. One cannot even begin to define the nature of a major crisis, let alone pretend to manage it, independently of all the other problems and all the other crises to which it is connected, and to which it gives rise.

8.10 THE BASIC PROBLEM OF ILL-STRUCTURED PROBLEMS

The basic problem with ill-structured problems is to define the nature of the problem(s). The problems of (1) problem formulation and (2) problem representation are thus central to ill-structured problems.

In contrast to well-structured problems, different stakeholders generally define ill-structured problems differently-- often radically--depending upon: their differing values, interests, education, personal history, and the organization(s) for which they work, to mention only a few of the many factors that are relevant. For this and many other reasons, ill-structured problems are inherently controversial. As a result, we would not expect two stakeholders to define an ill-structured problem in exactly the same ways. In fact, we would suspect collusion between them if they did.

For the same reason, we would not expect two or more CM scholars, experts, or practitioner to have the same definition of what is a "crisis," "disaster," etc.

The repeated calls for agreement with regard to the basic definitions of what is a "disaster" versus what is a "crisis," etc., ignores the basic character of ill-structured problems. Ill-structured problems are problems for which fundamental and strong differences not only exist, but predominate. Intense disagreement is one of the major features of ill-structured problems.

To insist therefore on agreement as a precondition for studying ill-structured problems is to ignore and to deny their basic nature. It is thus to misrepresent them ontologically, i.e., their reality.

If we could obtain prior agreement on the nature of ill-structured problems, then they would not be ill-structured. They would instead be well-structured. By definition, well-structured problems are the only kinds of problems for which we can obtain wide spread agreement regarding their nature.

When CM scholars, experts, and practitioners call for agreement on the definition of basic terms, in effect, they are committing the same kinds of serious errors that academics accuse practitioners when they ignore the complexity of crises. We have to learn to study and to manage ill-structured problems on their own grounds.

Disagreement regarding the nature of the problem(s) is one of the fundamental characteristics of ill-structured problems. Furthermore, such disagreements will not go away as the result of "better" and "more extensive" or more "precise" analyses. "Better" analyses will only reveal further and deeper disagreements.

8.11 MESSES

In addition, crises possess another essential characteristic that makes them equally difficult, but not necessarily impossible, to study. Crises are not only ill-structured problems, but they are "messes" as well. At the very least they are part of messes.

The eminent systems scientist Russ Ackoff defines a "mess" as a system of problems that is highly interactive, i.e., strongly coupled. In other words, a "mess" is the *product* of the interactions between all of the various problems that constitute the mess. In short, *a crisis is an ill-structured mess.* That is, a crisis is a highly interactive set of problems each of which is ill-structured in itself.

West Churchman put the matter even more forcefully: "Something is a problem if and only if it is a member of the set of all other problems." Thus, something is not a problem, but instead an academic exercise, if it can be defined precisely and independently of at least one other complex problem!

8.12 CONCLUDING REMARKS

This paper has argued that our understanding of CM as a complex process and system has suffered seriously due to the presumption of an outmoded and false philosophy regarding the nature of problems. Indeed, the field of CM will continue to stagnate if we do not revise seriously our underlying concepts.

CM is essentially a theory of complex problems (see the Appendix to this paper). Even more basic, it is a theory of critical thinking. This appreciation is absolutely essential to the management of fuzziness, for fuzziness is an essential and an inherent property of real-world problems.

The end result is that there are no exact checklists for CM (see Figure 4). That is, CM cannot be confined to the upper left-hand quadrant of Figure 4. In the truest sense of the term, CM is open-ended, unbounded problem formulation (see the right-hand quadrants of Figure 4).

Many businesses already practice a version of the right-hand quadrants of Figure 1 if they take a broad view of the conduct and the expansion of their business. In contrast, they often take exceedingly narrow view of the risks, i.e., primarily the upper left-hand quadrant of Figure 4, when it comes to considering the potential crises that could happen to them. In addition, they often compounded this by making the fatal mistake of limiting the particular crises and the risks it could face by restricting them primarily to natural disasters.

FIGURE 4:
Four Distinct Types of Thinking

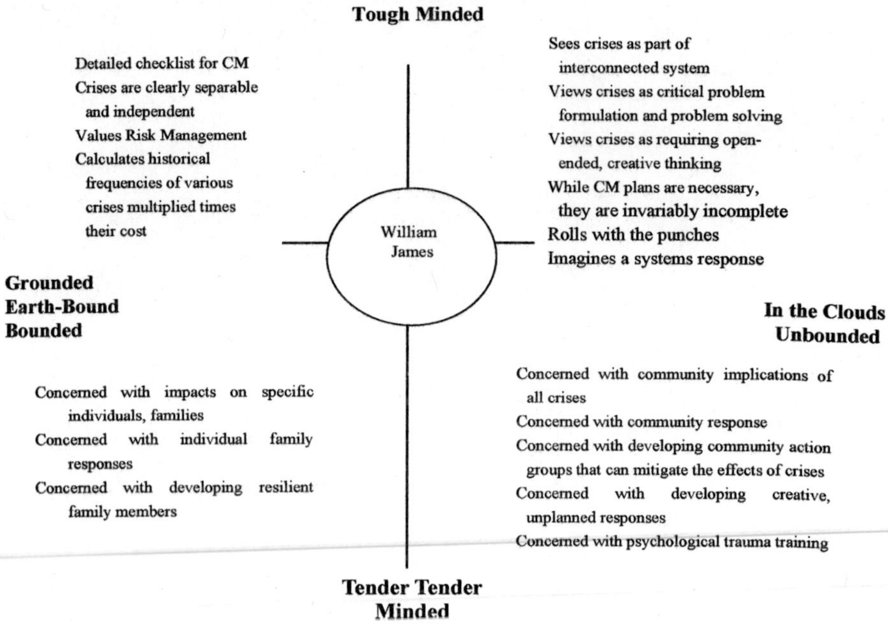

Everything connected with CM is inherently fuzzy. For instance, "What is a crisis?;" "What is the definition of a crisis?;" "What is an early warning signal that a particular crisis is about to happen?" All of these serious issues are fundamentally matters of judgment.

This does not mean that there are no guides in responding to these and other critical issues. Instead, the point is that such guides can only be based on heuristics or rules of thumb. In other words, CM cannot be reduced to the upper left-hand quadrant of Figures 1 and 4. Unfortunately, this is what the majority of the currently existing books on CM would have us believe.

To put it mildly, crises are exceedingly slippery. This is in fact one of their prime properties.

As I have emphasized, every crisis is in effect in every other crisis! If every business is now in every other business, then every business has all of the crises of every business! This is precisely why the right-hand sides of Figures 1 and 4 are so important. Crises demand that we take an expansive, not a narrow, view of the world.

8.13 AFTERTHOUGHTS

By the time everything is fully known about a crisis, it may be too late to act responsibly. One's reputation may already be so damaged by not having acted quickly enough. For instance, witness the flap in April, 2004 with regard to Condoleezza Rice's testimony before the 9/11 Commission. To uphold an abstract principle, namely, that "senior advisors to a president should not testify before a commission" instead of responding to the feelings of the public with regard to an event as heinous as 9/11 comes across as uncaring, to put it mildly.

The supreme challenge of fuzziness is whether to act and how much responsibility to assume <u>before</u> the complete dimensions of a crisis are known. Indeed, a crisis is a crisis precisely because in the beginning, and throughout its entire history, many of its most prominent features remain unknown, fuzzy, uncertain, or full of conflicting opinions-- at best!

In the future, the rewards will go to those who can <u>formulat</u>e novel problems, not those who can solve canned exercises.

APPENDIX: A THEORY OF COMPLEX PROBLEMS

A Theory of Complex Problem Formulation and Problem Solving Inquiry Systems

In a long series of publications, Russ Ackoff, C. West Churchman, and I have explicated the nature of various inquiry systems. Inquiry systems are distinctively different archetypal systems for conducting inquiry. These appear time and again throughout the course of human history. These systems differ radically regarding what they regard as knowledge, i.e., define as knowledge, and the conditions under which once can be said to have attained knowledge about a critical problem or issue.

One of the most important outcomes of the comparative study of various inquiry systems is the recognition that some systems are inherently "better suited" for well-structured problems. In contrast, others are "better suited" for ill-structured problems.

Unfortunately, it is beyond the purpose of this appendix to describe these systems in detail. Suffice it to say that those systems that are better suited for treating ill-structured problems regard the intense conflict between different stakeholders as an important facilitator, and even a necessary precondition, for inquiry itself. They certainly do not regard conflict as an impediment to inquiry. Instead, they use the intense conflict between stakeholders regarding the basic nature and definition of a problem to surface and to expose different underlying assumptions with regard to the proponents' respective views of the world. In effect, these differing assumptions are themselves fundamental parts of the "mess." As such, they can not be ignored or wished away. *The upshot is that the ability to tolerate and to manage conflict is an essential part of the CM process.*

In sum, different inquiry systems differ radically with regard to the recognition of "mess management" as an important area for inquiry and as an important component of inquiry.

Systems Models of Science and CM

Instead of describing inquiry systems more generally and abstractly, I want to present two systems models or frameworks for understanding (1) science and (2) CM. The models reveal the complexity of both activities, and hence, why each are ill-structured messes, or at the very least, has important aspects that are strongly ill-structured.

In 1974, Francisco Sagasti and I published a general systems model of scientific inquiry. We called it simply "The Diamond Model" (see Figure 5.) Its purpose was to show where the major features of scientific inquiry lay in relation to one another, and how and why they were mutually dependent.

The Diamond Model shows that scientific inquiry "typically" starts with the upper top box or activity the "felt recognition or sensing" of a problem. I say "typically" because one of the purposes of the model is to show that an inquiry can start at any point. Also, there are a great number of directions, paths, or ways that one can proceed through the model.

In many ways, where problems come from is as mysterious as where babies do. We think we know the answers, but as soon as we try to describe and to justify our knowledge systematically, then it quickly becomes murky, or in the terms of this appendix, messy and ill-structured.

Figure 5: The Diamond Model of Scientific Inquiry-Transdisciplinary Systems Thinking

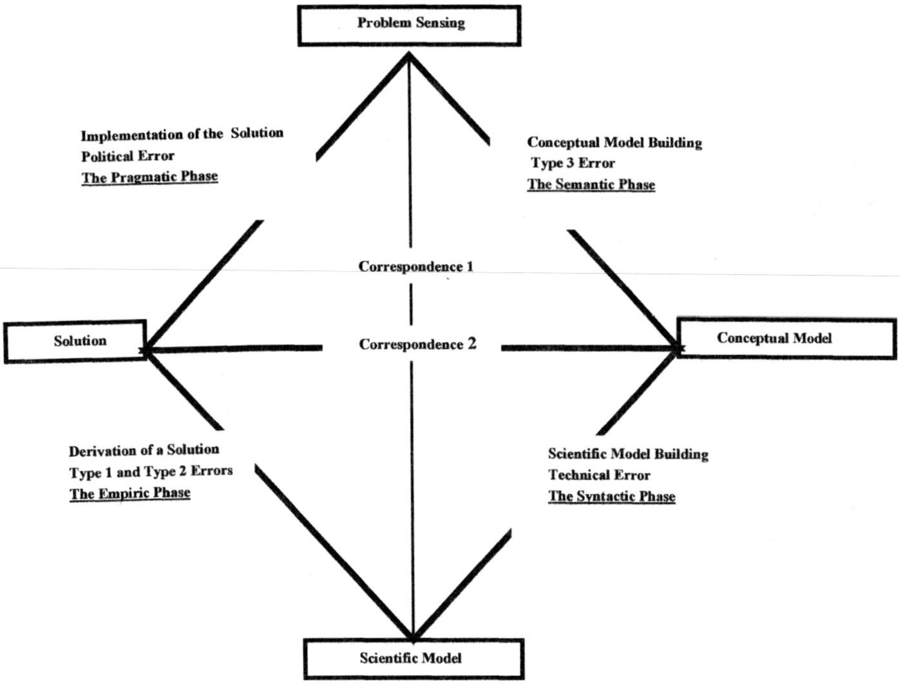

For the purpose of this appendix, *problems are "relatively" structured messes that have been extracted from highly unstructured messes for the purpose of managing better current and future messes.* Every aspect of this "definition" is, and is meant to be, tortuous and problematic. For instance, what do "extraction," "management," and "better" mean? "Extraction" is certainly a key activity because problems are not "given," but instead are "taken" from "reality." That is, problems do not drop preformed from the sky, or already structured. Instead, problems are the result of intensely human activities of which extraction and shaping are merely two.

To be sure, many problems are "given" in the sense that they are merely extensions of previously predefined problems of one's discipline, field, profession, etc. That is, one's discipline supplies one with problems, etc. However, even in such cases, the situation is not completely clear, for one has to explain how "extensions" are arrived at. With out a doubt, problem sensing is one of the more creative human activities. The most creative is certainly problem invention as in the cases of Freud, Gödel, and Einstein.

John Dewey regarded problem sensing as an inherently ethical activity in the sense that many problems originated in "moral outrage." For instance, many of our most important problems begin with "moral outrage" such as "How could they--The Catholic Church, Enron, the Governor of the State of California, etc.—do or failed to have done such and such?" Certainly, many crises arise in moral outrage, and in this sense, possess features that are undeniably ethical.

The important point is that problems are extracted from messes, and that the process of extraction is itself one of life's most important problems. Notice that in saying this—i.e., uttering the immediately proceeding sentence--I am thereby being self-reflective, one of the hallmarks of philosophic thinking. I am turning the model or thinking about problems back on themselves.

The upper diagonal line that proceeds from problem sensing, the felt existence or "fuzzy recognition" of a problem, to the existence of a conceptual model of the problem denotes the *process*_of building a conceptual model or representation of the initial feeling or felt sense. It represents the *semantic* or the *problem formulation* phase of scientific problem solving because at this point we are concerned primarily with the general and broad meaning of a problem. We are concerned with the choice of an initial and even tentative discipline, disciplines, or languages in terms of which to represent the "basic meaning or definition" of the problem.

Notice carefully that in terms of the diamond model, defining is part of the *total process* of inquiry. Defining is not a "thing" per se that exists apart from inquiry. Instead, defining is a tool that either enables or restrains the total process of inquiry. As such, it can not be ripped artificially from its moorings. Defining neither exists nor functions apart from the whole system of which it is a part. Thus, we can not understand it in isolation.

In contrast to the type I and the type II errors of statistics[31], the error that pertains to this phase of the scientific problem solving process is known as the type III error, or E3 for short. E3 is defined as "the error of solving the 'wrong' problem precisely." It is also defined as "the probability of solving the 'wrong' problem when one should have solved the 'right' problem."

Now, "right" and "wrong" are obviously relative. If we knew for sure that we were solving the "wrong" problem, then we would not be committing an E3. More importantly, the initial problem would already be well-structured so that there would be no need for problem formulation or problem structuring! Even in this case, there could still be enormous dispute as to whether we were indeed "solving the 'right' problem."The point is that this phase of scientific problem solving is meant to call attention to the fact that the determination of an E3 can only be made if we produce at least two very different representations of a problem. Thus, the process of "rational" scientific problem solving calls for at a minimum the use of those inquiry methods that are founded upon conflict, i.e., the sharp challenging of the different assumptions that different stakeholders make about problems. This is merely one of the ways that scientific inquiry calls for other inquiry systems that are currently in conventional or widespread use.

Although there is always the possibility of revisiting for the purpose of revising one's initial conceptualization—the diamond model is not meant to imply that the process is linear and proceeds only in one way—the process of problem formulation "terminates" with the production of a conceptual model of the problem. The conceptual model lays out the main variables and the broad, general features of the problem. It is not yet an exact scientific model of the problem, although often many features of the diamond model go on simultaneously such that they can not be easily differentiated from one another. The diamond model exists after all to lay out the various features and processes that are inherent in scientific inquiry and not to imply that they are totally separate. Indeed, one of the basic definitions and properties of systems is that they are strongly inseparable.The purpose of the next phase is to construct using the tools of mathematics, flow charts, and the knowledge of physical, social, historical processes etc., a precise representation or exact model" of the problem. This phase is known as the syntactic phase of

[31] The Type I and the Type II errors are as follows: consider the case of a drug manufacturer who wants to determine whether a new drug that it has just produced is better than an old one. In general, there are two types of errors that one can make. One can say that the new drug is better than the old one when it is not. In contrast, one can say that it is not better when it is. Both of these errors are examples of the Type I and the Type II errors of statistics.

scientific problem solving. It necessitates detailed knowledge of the rules for building exact models from word or conceptual models, and the ability to execute the rules so as to produce a scientific model.

Science is of course replete with numerous examples of outstanding models. One of the best is surely Einstein's transformation of his highly intuitive thought experiments and notions regarding space, time, mass and gravity into complex mathematical equations.

The first vertical line of correspondence between the initial felt sense of a problem and its "exact" scientific representation should not be interpretted as the correspondence between an "objective, external reality" and a "subjective, internal idea." This is wrong, plain wrong. Whatever "reality" is, it is a s much a "representation" by and of our minds and culture as anything is. And surely, if human experience and perception mediate anything, it is the "felt sense or existence" of a problem.

This is not to deny the existence of external objects or crises. It is to deny that whatever "they are," they are not known and experienced independently of our minds and culture.

In this sense, the common distinction between "objective/subjective" ought to be completely purged from common usage. It is a carry over from an earlier, primitive philosophy. It hinders knowledge and inquiry more than it helps. Modern philosophers have rejected completely the notion that our minds are blank tablets that do not participate actively in the experience in and of the world. John Dewey for one rejected this notion by referring to it dismissively as the "spectator theory of knowledge."

The next phase consists of deriving a solution to the scientific model, or more generally, testing a scientific hypothesis empirically. For this reason, it is the empiric phase of scientific problem solving. And, it also concerns the traditional type I and type II errors of statistics.

Notice carefully that the solution is not necessarily the solution to the initial problem, but it is the solution to the scientific model of the problem, which is in turn dependent on the conceptual model of the problem. Thus, the solution is at least twice removed from the initial problem. The point is once again how heavily the entire process is on human experience and perception. Science is an intensely human enterprise through and through.

The so-called last phase- implementation-- is the pragmatic phase of scientific problem solving. It is concerned with the social/political of getting a solution accepted and adopted in a complex organization to remove the initial problem. Thus, implementation tests whether a problem can be removed or managed in practice, not merely in theory.

Implementation also examines whether putting the solution into practice creates new problems or necessitates new formulations. Thus, the process is iterative.

Notice also that there is no requirement that the model is to be run in any particular direction or that it is to start with any particular point. Indeed, different distinct forms of inquiry can be identified by the special paths through the model with they are associated.

The essential point is that scientific inquiry is a tightly coupled and highly interactive system. Its various parts neither exist nor function independently of one another.

Finally, also notice that the process of defining –anything-- is equivalent to the process of inquiry outlined in the diamond model. That is, the initial felt sense of a problem is equivalent to the working or starting definition of a problem. The definition of a problem is thus the outcome of inquiry, not the start.

We are now in a better position to give a better sense of the differences between well-structured and ill-structured problems. Well-structured problems are problems for which a relatively few iterations of the diamond model are sufficient for the definition and the solution to the problem to converge. The implementation also tends to remove the initial problem. In other words, the mess generally lessens over time.

In contrast, ill-structured problems misbehave over time. They generally get worse in the sense that the mess or the messes of which they are associated with grow, sometimes uncontrollably.

This difference is often, but not always, the difference between disasters and crises. It is not always true that disasters are more well-structured and that crises are more ill-structured because in today's complex world both are increasingly intertwined as inseparable parts of messes.

A Diamond Model of CM

Figure 6 is a parallel or complimentary diamond model for CM. As such, it is similar to Figure 5. To my knowledge, it is the first such model of its type for CM. Its purpose is to show the full range of interrelated activities for CM.

Figure 6 shows that in a similar fashion to Figure 5, the—or more accurately, <u>a</u>-- diamond model for CM "begins" with the "felt sense or recognition" of a "potential threat" internal or external to an organization. At this point, the full nature of the threat or threats are not known, just the "sense" that something is potentially remiss.

The process of building a conceptual, or initial, model of the threat consists of the use of the model presented in Figure 5 or something similar to it. That is, the model serves as a basis for conducting a crisis audit or a threat assessment of an organization.

The point is not that the model in Figure 5 is <u>the model</u> for accomplishing this. The point is that something has to be used to perform an initial or a preliminary crisis audit of an organization in order to begin a systematic assessment of the threats, risks, etc., facing the organization.

The initial assessment sets the tone for all future efforts. As before, it can and needs to be revised as one learns more. It constitutes a crucial part of the Before phase of CM.

The next phase consists of using the initial assessment to begin developing the necessary organizational capabilities to manage an actual crisis. This generally in crisis plans, a trained CM Team or CMT. It also includes crisis simulations, etc.

Even with the best of plans, preparations, and capabilities, virtually all crisis experts and scholars agree that in today's world, a major crisis is virtually guaranteed to happen to every organization. Thus, the next phase consists of the enactment of an organization's crisis plans, preparations, and capabilities. This is the During or reactive phase of CM. Damage containment is in effect the parallel to the solution phase of scientific problem solving.

Finally, continual learning is the real test of an organization ability and willingness to implement CM.

Notice that as with the case of scientific problem solving, the definition of what is a crisis changes, often dramatically, as we move around the diamond model. In other words, the definition of what is and what is not a crisis can not be fully decoupled from a systems model of crises. In the language of systems, the definition of critical terms and events is not separable from the entire process of diagnosing and treating them.

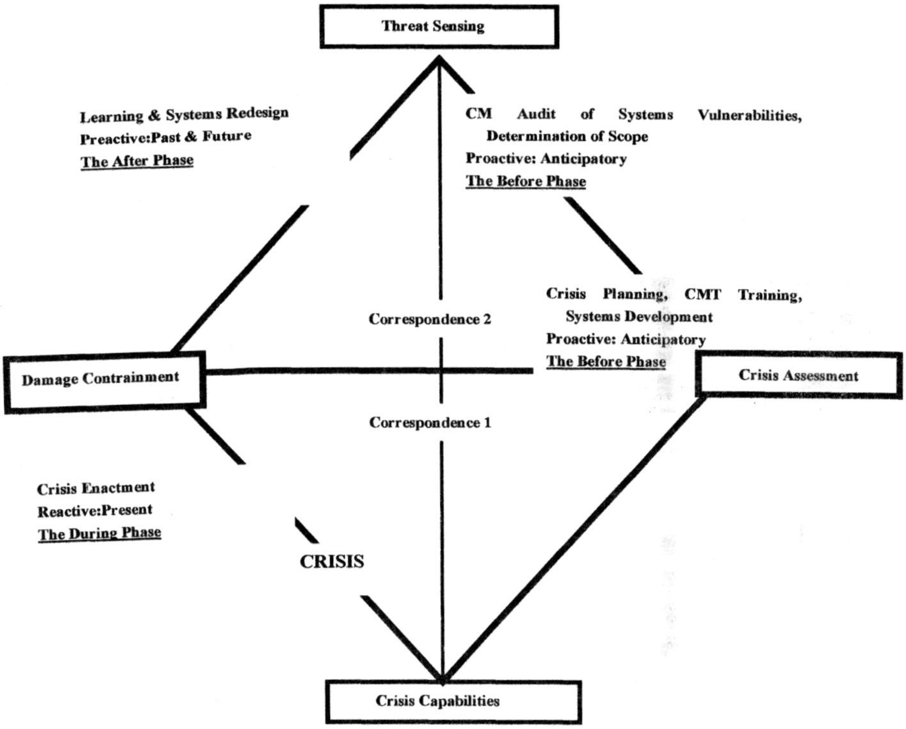

Fig 6--The Diamond Model of CM: <u>Interactive</u> Systems Thinking

Chapter 9

INFORMATION SYSTEMS DEVELOPMENT AS INQUIRING SYSTEMS

Theoretical Discussion

ANDREA CARUGATI
IESEG School of Management
France

Abstract: The paper presents a framework to interpret information systems development (ISD) as composed of activities for the creation and exchange of knowledge. The framework is based on the five inquiring systems presented by Churchman (1971).

The main contribution of the framework is to show that the inquiring systems, when applied in practice, are complementary. They include rationalism, empiricism, idealism, dialectic, and pragmatism are used by participants in ISD projects in accordance to the scenarios they face in an opportunistic way. For each ISD activity there is a predominant inquiring system. Since the framework focuses on the knowledge created during the interaction of individuals and groups (or the lack there of) rather than on the specific development process the framework is robust enough to be adaptable to different ISD methodologies. While methodologies are largely centered on specific paradigms for knowledge creation, the presented framework focuses on micro-level activities and suggests specific management practices to support the different knowledge creation situations

Key words: Information systems development, knowledge creation, inquiring systems

9.1 INTRODUCTION

Information systems development (ISD) is usually intended as a process that emerges from two interconnected activities: software producing activities and managerial functions (Abdel-Hamid and Madnick, 1991). While this dichotomy is highly intuitive, current research tends to pair these two dimensions with a third - the dimension of knowledge creation activities. This branch of research focuses on knowledge issues in ISD either explicitly (Richardson and Courtney 2004; Markus, Majchrzak, and Gasser, 2002) or implicitly (Beck, 1999; Avison, Wood-Harper, Vidgen, and Wood, 1998; Checkland and Holwell, 1998). In both cases they fundamentally agree on a social constructivist approach to ISD where systems and their requirements emerge from the interaction of multiple stakeholders.

This paper continues this research tradition presenting a theoretical framework that sets the spotlight on the knowledge creation and exchange facets of ISD projects. In particular, the goal of the paper is to provide a framework to interpret information systems development in terms of the inquiring systems[32] proposed by Churchman (1971). The intentions are threefold. First, it is desired to show that while methodologies are based on overreaching ontology and epistemologies, focusing on micro-level ISD activities shows that multiple way of creating knowledge are used depending on the task and physical situation of the stakeholders. Second, it is necessary to understand which inquiring system to use in the specific situation to learn the most from it. The third purpose is to highlight the specific knowledge creation situations in ISD activities in order to provide managers with the appropriate means for their identification and management. The contribution of this article is therefore a framework to help managers to see ISD activities as knowledge creation processes and to provide them with the principles for their management to improve the chances of success of ISD projects.

A criticism can be raised that Churchman's inquiring systems is outdated. However using Churchman's work has several advantages. First, the Churchman approach to inquiring systems is based on long standing philosophical traditions that have been refined over time and therefore, albeit biased by Churchman's interpretation of these philosophers' work, these inquiry systems are robust and when used in combination are a useful means to appreciate problems. Second, inquiring systems are integral to the information systems literature and as such their use eliminates the need to use ideas from reference disciplines (Richardson and Courtney, 2004; Richardson et al., 2001). Finally, this paper does not focus on philosophy in

[32] An inquiring system is a set of components that interact with the goal of understanding a situation, learning from it and creating knowledge for the individuals in the system.

information systems and Churchman's work is not used as a way of perceiving reality. Churchman's work is used to highlight, in a way familiar to the IS community, the way in which knowledge is created and exchanged during the different ISD activities.

This paper starts with an introduction to Churchman's inquiring systems followed by a discussion focused on the creation of the framework. Since the interest of the paper is to provide a robust framework that can be used in conjunction with different methodologies, the framework will map Churchman's inquiring systems on generic ISD activities. Such activities are not considered as part of a structured process but rather as modules whose specific sequence and content varies according to the methodology used or not used (Hughes and Wood-Harper, 2000). Then an ISD case will be presented to show the use of the different inquiring systems in a real life case. The paper will be concluded by presenting a primer on the managerial implication of using inquiring systems and knowledge creation activities as a guiding light in ISD.

9.2 INQUIRING SYSTEMS

According to Churchman (1971) there are five basic modes of inquiry to create knowledge. These five inquiring systems find their basis in the philosophies of Leibniz, Locke, Kant, Hegel, and Singer. In summary, these five modes are:

9.2.1 Rationalism or the Leibnizian Inquiring System

In this mode the inquirer discovers within himself, through formal logic, the truth governing the world. The system under study is considered as closed and governed by logical connections. The inquirer creates networks of hypotheses and theses and proceeds to build on these hypotheses until logic reveals a counterhypothesis that invalidates a part of the network. These networks are called *fact nets*. In ISD, the Leibnizian inquirer is the isolated individual, the monad (Churchman, 1971, p. 30), reasoning on the causal relationships governing the system that he wants to improve. The individual's logic becomes the guarantor of the truthfulness of the fact net. Pure structured methodologies (Boehm, 1988) are built on the ideas of the Leibnizian enquiring system: once requirements are specified the programmer proceeds on his own to develop the information system based on his internal logical reasoning. The information system will therefore reflect the fact nets developed by the inquirer (e.g. the programmer).

9.2.2 Empiricism or Lockean Inquiring System

This mode of inquiry differs from the Lockean giving more weight to data than theory. The inquirer is open to the environment and finds the truth in the world surrounding him. The inquirer is not isolated like the monad but participates in creating knowledge with a community of inquirers, the Lockean community. The inquirer uses data and expert opinion, especially the opinion of his community, to explain the world. The community acts guarantor of the truth through agreement and consensus. In Lockean inquiring systems there is only space for one truth but there is no guarantee that two communities will agree on a common truth. In ISD, the groups of users and developers can be considered as two Lockean communities, where agreement can be found within a community but it is difficult to find across communities.

9.2.3 Idealism or Kantian Inquiring System:

This mode of inquiry gives equal weight to data and theory. It is recognized that there are multiple ways in which a problem may be analyzed using multiple models that can be applied to the observations. However the inquirer does not accept multiple truths; the objective is to find the ideal, and only, truth. In ISD it can be imagined that the Kantian inquiring system is used when developers have to find a model to translate something from the client system to the information system. Many possibilities are tried and recursively eliminated until only one, the one with the best fit, the truth, remains.

9.2.4 Dialectic or Hegelian Inquiring System

Sees the truth emerging from opposing views. Debate between different worldviews (Churchman, 1968; Linstone, 1984; Orlikowski, 1992; Davidson, 2002) is seen as the only way to develop theses and antitheses to arrive at a synthesis that accommodates both worldviews. Hegel introduces the idea of multiple interpretations of reality even though he believed in a final "grand synthesis". He envisioned two groups of people engaged in an ardent debate respectively defending their thesis and trying to destroy the other group's antithesis. The issue is than resolved by an impartial observer that synthesizes the arguments to create a synthesis. The guarantor of the truth is the belief in the efficacy of debate between extreme points of view. Possibly one of the best examples of this inquiring system is in the use of the scenario approach to strategy used at Royal Dutch/Shell (Senge 1990).

9.2.5 Pragmatism or the Singerian Inquiring System

This inquirer is based on the ardent debate of the Hegelian enquirer to create progress but accepts multiple sources of data and multiple interpretations of reality and consequently multiple truths. It is called pragmatic inquiring system because the truth/s produced is/are relative to the overall objectives of the inquiry. The multiple truths are found via an approach that continuously attacks currently held beliefs from multiple points of view. The world inquired is interpreted as an open system where all components interact with each other. The truth is temporary and contextually dependent because the more elements are "swept in" the system the more the truth is likely to change. Progress is measured quantitatively whenever possible otherwise the groups' intuition that progress is made becomes the guarantor for qualitative measures (Richardson et al. 2001). In ISD the use of the Singerian inquirer can be found in the more recent methodologies like the ones coming from Soft Systems Methodology (Atkinson, 2000; Winter et al., 1995), Multiview2 (Avison et al., 1998) and in agile methodologies (Beck 1999, Fowler, 2002). These methodologies let the information system emerge from the debate between developers and users letting multiple truths be created pragmatically by the participants and not imposed top down.

The traditional assumption is that the Singerian inquirer suits ill-structured problems (Richardson et al. 2001). This view however attributes the characteristic of the situation as existing outside the individual. Taking an interpretive approach (Orlikowski and Baroudi, 1991), ill-structured and well-structured is not a characteristic of the problem but of the observer: it is the result of individual and social construction. This is reflected in the fact that the way in which a problem is perceived by the individual changes according to the context in which the individual is. From the individual perspective it is therefore acceptable to use different inquiring systems according to the context rather than according to an externally defined characteristic of the problem. Harrison and Bramson (cited in Kienholz, 1999) suggest that individuals display favourite modes of inquiry but can use different modes according to the situation. People working in isolation might tend to rely mostly on rationalism and idealism to move forward. Members of the same group will rely mostly on rationalism, idealism, and empiricism since members of one group possess a shared worldview (Boland and Tenkasi, 1995). Dialectic and pragmatism are instead inquiring modes used when two groups with different worldviews meet with the intent of learning from each other.

9.3 FRAMEWORK DEVELOPMENT

The ISD process can vary radically depending on which development methodology has been chosen. However, there are five basic activities that are shared by all methodologies:

1. Some members of an organization voice a desire for a new system; from this desire concepts for a new information system are created.
2. Groups of developers and users meet and define the requirement for the information system.
3. Developers discuss among themselves how to address the requirements.
4. Each developer develops the part of the information system of his competence.
5. Developers and the users meet to test the system and eventually implement it.

These activities can be performed in multiple sequences, recursions, revisions, focusing more or less on the customer organization. According to the specific sequence adopted different ISD methodologies can be recognised.

By interpreting these five activities in terms of systems for creating knowledge, it is possible to allege a connection between them and the use of one or multiple inquiring systems. The following part maps the inquiring systems on the micro level ISD activities.

9.3.1 Identification and Concept

As a norm this activity is carried out within a company that experiences problems or has identified some opportunities that wants to seize or solve with the use of information systems. This group will be referred to as the "users". The users discuss their situation and how they could expect an IS to help them improve it. The scenario dictates that the users are not expert developers and therefore they can conceive some functionalities for the IS but neither the technical details nor the additional functionalities that developers can create for them through technology.

In this phase the group cannot create knowledge through debate and they are likely to consider their world as an open system. The most likely inquiring mode in which the group will operate is empiricism. In this case members will try to seek consensus within the group that acts as a guarantor for the created knowledge.

9.3.2 Requirements Definition

This activity is carried out in cooperation between users and systems developers with the scope of specifying what the information system has to do in the users' organization. Functional requirements emerge from a debate between users and developers with the purpose of reaching an agreement on the goals of the project.

In this phase the groups behave as Singerian inquirers. The two groups debate ideas about problems and possibilities. They will tend to include into the discussion elements from all interconnected systems and they will come to an idea for what the information system has to do. When using this inquiring system ideas are not necessarily shared by every individual. This in known and therefore contingency plans to do rework can be made in anticipation. The phase is over when enough progress is made. Progress can be measured by changes in the requirement document, in the use cases, or in the user stories (depending on the ISD methodology chosen).

9.3.3 Design

During the design phase the developers work together as a group to plan how to bring the requirements into life. A variety of technical decisions is taken e.g. about the programming language, object orientation, modularisation, interfaces etc. These decisions usually allow developers to work independently during the implementation activities. During this phase the developers adds to the knowledge "obtained" from the users creating their own hypotheses about the users' system and the solution requested: in this step knowledge is created through interpretation of the information gathered from the users filtered trough previous programming experiences.

The development group behaves as a Lockean inquirer. It is open to the environment but creates the truth, the knowledge of how the new system has to operate, within the group. Like before, the members seek consensus within the group that acts as a guarantor for the created knowledge.

9.3.4 Implementation

In this activity the developer, now isolated, begins to elaborate on his knowledge, on what he learned from the users and from his colleagues. In the process of writing code the developer interprets the information in his possession and creates, consciously or unconsciously, a network of hypotheses that bring him from ideas to actual code.

The system developer acts as a monad in a Leibnizian enquiring system. During this phase new knowledge is created through a process of deduction. Fact nets are created based on some hypothesis on the behaviour of the users' system. The fact nets grow as more and more elements are added into the program.

Occasionally the developer faces uncertainty regarding how well a particular algorithm replicates the behaviour observed of the users' system. If multiple algorithms have to be compared the developer enters a Kantian mode of inquiry trying to find which algorithm fits best the behaviour of the users' system. In this mode of inquiry the developer might follow his own judgment or search for answers outside. Even though the answers come from the outside the goal of the search is only to justify the use of a certain algorithm or solution, not to put under discussion the work done up to the point. The search finishes when an algorithm is found that fits the particular situation.

The result of the knowledge work in the design and implementation activities is a continuous evolution in the knowledge of the developers. These activities are not simply an operationalisation of the requirements agreed with the users but are the arena for criticisms, suggestions, logical development, and experimentation. These are all processes that contribute to the creation of software artefacts that can be different from what the users expect.

9.3.5 Testing and Operation

In this phase, the code, in form of prototype or release, is shown to the users for testing. Both users and developers are present during the operation phase. The users operate the software and compare the experience with their expectations. Expectations are debated to achieve new agreements: the Singerian inquiring system is used again. If the software is implemented, the users have the opportunity to use it for regular work and can request additional features or modifications at the following meetings. New knowledge created by the users during operation is outside the modus operandi of the debate and is created using the Lockean inquiring mode.

The framework is presented in table 1. Each activity (vertical axes) is mapped on the inquiring systems (horizontal axes) that are most likely to be active during the phase.

ISD Activities ↓	**Inquiring Systems active during ISD activities**			
Identification and Concept	Activity carried out by the information systems stakeholders in the user's organization.			
Requirements Definition				Activity carried out by users and developers. Achievement of agreement cannot be postulated. Groups and individuals might bring home different views of the truth.
Design	Activity carried out by the development team.			
Implementation		Activity carried out by the individual developer. Fact nets are created based on internal logic. New knowledge is created outside the control of other stakeholders.	Activity carried out by the individual developer. When in doubt the developer tries multiple models until the best fit is found.	
Operation	If the information system is put in operation, testing can occur in real working situations. New ideas are generated among the users to be included as part of the requirements.			The phase of testing is carried out by users with the developers' supervision. New ideas generated from debate are included in the requirements. Progress in this phase is the guarantor for the ISD project.
Inquiring Systems →	**Empiricism**	**Rationalism**	**Idealism**	**Pragmatism (includes Dialectic)**

Table 1. Theoretical Framework

9.4 AN EXAMPLE

This article is intended to present a theoretical framework nonetheless it is appropriate to present observations from an ISD project that highlight the use of the different modes of inquiry as well as highlight some of the pitfalls of not seeing ISD activities as inquiring systems. The author has been participant observer in an ISD project that ran from 2000 to 2003 where a team was developing software for manufacturing scheduling and control for a shipyard. The project had two characteristics that made it interesting for being a good example for the argumentations presented in this paper. The first characteristic was that the IS under development was based on a mathematical technique whose properties were new to the users. The second characteristic was that the development team was external to the shipyard, and the developers were working from distributed locations. Both characteristics point at knowledge creation and exchange as key factors for the success of the project. The first characteristic made it important for the users to learn about the mathematical technique in order to be able to specify the requirements, the second made it important for the developers to learn from each other about their work. The following is a simplified view[33] of the activities that underpinned the approach:

9.4.1 Identification and Concept

This activity was carried out by the client. The client had some experience with the use of the mathematical technique to solve complex planning problems and decided to start a larger project aimed at developing a new integrated scheduling and control software to manage the activities of the inventory and the production lines. This phase was carried out by expert project managers of the company without consulting developers. The results were included in a project description brief that was later used as starting point of the requirement definition phase. The project managers arrived at the decision to start the project working as Lockean enquirers: as a group they decided that the idea was favorable.

9.4.2 Requirement Definition

This activity was mostly done by the developers alone. The developers were taken around the factory one afternoon and than were stationed for 2

[33] Projects of this size and length are much more complex than the simple view provided here. However an exhaustive report of the case is not appropriate here. This section aims to present few instances of the activities discussed above to show that also in real cases multiple inquiring systems are active.

weeks at the users workstations to observe their work. Based on these two observations the developers were required to create a requirement document. The requirement document was presented to the users in a meeting but the debate was very weak and the developers were left freedom to do basically whatever they pleased. The creation of the document was based on the use of empiricism with the developers as guarantors of the knowledge created rather than based on the Singerian inquirer where multiple perspectives should have collaborated to make the requirements emerge.

9.4.3 Design

The design activity was exclusively carried out by the development team. In this phase the developers decided that the software had to be divided in modules; they decided about processes for material handling; they decided about use of overtime. All these decisions were made thinking that they were either "logical" or "possible" and therefore good. These decisions became the hypothesis at the base of the fact nets, the networks of causal relationships that characterized the software. The inquiring system used here was the Lockean one. Certainty in the truthfulness of the knowledge created was guaranteed by the group consensus. The few voices that spoke against taking decisions without consulting the client were ignored. The customer was never consulted about these decisions because it was thought that eventual comments would emerge during the testing of the systems modules.

9.4.4 Implementation

The implementation activity was very problematic. The complexity of the mathematical technique made the code very difficult to implement. During this phase the developers worked mostly alone like Leibnizian inquirers developing their own modules on the hypothesis obtained in the previous phases and further developing these hypotheses on individual basis. For example, one developer arrived at the decision that for testing purposes it would not be necessary to consider machine breakdowns. This decision was logic in the moment: details were less important than speed. The developers thought that the validity of this simplification would have been discussed later during the testing phase.

The developers also used idealism as mode of inquiry to a large extent. Modeling of the arrival of material, modeling of the distribution of delays, modeling of the arrival of material requests emerged from a process of data collection and research for the best fitting model (usually a statistical distribution). As for the other simplifications, the developers thought that the users would validate the use of these models during testing.

9.4.5 Testing and Operation:

During the development project the development team held three demonstrations of the software at the 16th, 19th, and 21st month of the project. The idea behind testing was that the users could control the correctness of the software and could eventually specify new requirements. The three sessions functioned in similar way. The testing was organized like a presentation where the developers used the system and showed it to the users. Not knowing exactly what they were observing, the users were not able to sustain the "ardent debate" required to create knowledge in a pragmatic inquiry. The developers maintained firmly the steer of the discussion negating the users the possibility to give a definitive answer on the validity of the simplifications in the software. These testing phases were not done accordingly to the Singerian inquiring system seeking ardent debate. There was very limited knowledge creation that would bring the software closer to being successful.

This project joined the other 84% of ISD projects that fail each year[34]. The software was never implemented and when, approaching the end of the project, interviewing the company project managers and asking them what they learned about the use of this mathematic technology they replied that their knowledge was still superficial. This answer shows that not only did the project members failed to create the software but they also failed to create knowledge for approaching similar problems in the future.

9.5 MANAGERIAL IMPLICATIONS

Considering ISD as a knowledge process changes the idea of what information system success is. Informal processes like conversations, brainstorming and pursuing ideas become the primary source of value creation while formal processes contribute to the value creation to the extent that they are in the service of knowledge processes (Wenger 2000. p. 244). The framework makes it possible to see the micro level ISD activities in terms of creation of knowledge rather than creation of information systems. In this section some managerial principles are provided deriving from this way of interpreting the ISD process.

1. *Disagreement is inevitable.* Despite the fact that on the surface people seem to agree there is likely to be a difference between what people say and what people do (Senge 1999). This derives from the fact that in the

[34] Number provided by the Standish Group. Site: www.standishgroup.com/ (visited in August 2003). Report at: www.scs.carleton.ca/~beau/PM/Standish-Report.html (visited in September 2003)

activities of identification, requirements, and design consensus is reached in different groups. Even when using the Singerian inquiring mode consensus is, a priori, not obtained. However the more the Singerian inquiring system is used the more likely it is that the resulting information system would serve all involved interests. In ISD this means going through the testing and operation activity many times putting developers and users in the condition of debating the work done.

2. *Requirements are not the information system.* Developers tend to develop software mixing users' input with own experiences and reasoning. In the implementation activity it is always present an element of creativity. If this creativity results in small fact nets than these can be easily analyzed and invalidated during testing but if the nets are very large than discovering all the causal relationships that have been embedded is almost impossible. In the example above the first prototype was shown to the users 16 months after the start of the project, in this period the fact nets of the developers grew to such an extent that is was impossible for the users to comment on them. Time intervals between testing phases should therefore be kept short to limit the uncontrolled growth of fact nets. Consequently also the size of the development task should be controlled so that small prototypes can be developed in the short time allocated.
3. *Debate does not come automatically.* Entering in the Singerian inquiring mode is not easy. A presentation, like in the example case, does not lead to debate. The Singerian inquirer requires cooperation and no locus of authority within the systems: instead authority should be diffused in every participant (Richardson, Courtney and Paradice, 2001). Presentations are a communicative genre that implies a well specified locus of authority (Yates and Orlikowski, 2002), in this case the presenter, and therefore it does not fit the requirement of diffused authority. Project managers on both sides should agree to mediate the discussion instead of leading it.
4. *Testing should be done by users.* In ISD projects most of the work on the code is done by developers. They know the code inside out and assumptions become easily blackboxed (Latour, 1987), in other words the developers cannot attack with critical spirit their own fact nets because these have emerged from the developers own logic. Therefore users should be able to personally test the information system in order to use the knowledge gained during testing as basic for debate (Carugati, 2004). An information system prototype built for this purpose should be al least visual (Carlile, 2002, Brooks, 1987) and functional (Brown and Duguid, 2001). Only by building the software like this and by letting the users operate it we can expect "ardent" debate to emerge.

These guidelines partially confirm the recent interest in emergent methodologies like eXtreme Programming (Beck, 1999) that in fact include many of the principles include here. However these guidelines can be applied also to structured methodologies where, at the level of the single activity, the knowledge dynamics presented in the paper are active too.

9.6 CONCLUSIONS

This paper presents a view of ISD where the software creating activities and managerial activities are joined by knowledge producing activities. In doing so, this article continues and extends the tradition of research in prototyping and participatory design further specifying the knowledge aspects of these traditions.

The paper shows that it is possible to map the basic ISD activities on the inquiring systems presented by Churchman. The article shows that, despite the recent focus of researchers on the Singerian inquiring systems as the superior system for knowledge creation, human nature and physical constraints prevent the application of this inquiring mode in all ISD activities. As a result, the different inquiring systems are active in a scattered fashion and provide therefore diversified ways of creating knowledge which require multiple practices to be managed.

An example of the application of the inquiring systems has been presented not with the aim of providing background for generalization but to show the soundness of the argument and also to show how easy it is for learning not to take place. The extent to which ISD projects show the use of the inquiring systems has to be further investigated. The other aspect that has to be further investigated is the effectiveness of different managerial practices in accordance to the inquiring modes. The hope is that the issues raised in this article will prompt other researchers to pursue this effort in their future endeavors.

REFERENCES

Abdel-Hamid, T. and Madnick, S., 1991, *Software Project Dynamics: An Integrated Approach*, Prentice-Hall, Englewood Cliffs, New Jersey.

Atkinson, C.J., 2000, The 'soft information systems and technologies methodology' (SISTeM): an actor network contingency approach to integrated development, *European Journal of Information Systems*, 9.

Avison, D.E., Wood-Harper, A.T., Vidgen, R.T., and Wood, J.R.G., 1998, A further exploration into information systems development: the evolution of multiview 2, *Information Technology & People*, 11(2).

Boehm, B., 1988, A spiral model of software development and enhancement, *IEEE Computer*.

Boland, R.J. and Tenkasi, R.V., 1995, Perspective making and perspective taking in communities of knowing, *Organization Science*, 7(4).

Brooks, F.P, 1987, No silver bullet: essence and accidents of software engineering, *IEEE Computer*, 20(4).

Brown, J.S. and Duguid, P., 2001, Knowledge and organization: a social-practice perspective, *Organization Science*, 12(2).

Carlile, P.R., 2002, A pragmatic view of knowledge and boundaries: boundary objects in new product Development, *Organization Science*. July/August.

Carugati, A., 2004, *Multiple Perspective Approach for the Development of Information Systems Based on Advanced Mathematical Models*, Unpublished Doctoral Dissertation, Technical University of Denmark, Lyngby, Denmark.

Checkland, P. and Holwell, S., 1998, *Information, Systems and Information Systems*, Wiley, Chichester.

Churchman, C.W., 1971, *The Design of Inquiring Systems*, Basic Books, New York.

Fowler, M., 2002, The New Methodology, (visited in June 2003), www.martinfowler.com

Hughes, J. and Wood-Harper, T., 2000, *An Empirical Model of the Information Systems Development Process: A case study of an automotive manufacturer*, Blackwell Publishers Ltd., Oxford.

Kellogg, K.C., Orlikowsky, W.J. and Yates, J., 2002, Enacting new ways of organizing: exploring the activities and consequences of post-industrial work, *Academy of Management Conference Proceedings*, Denver.

Kienholz A., 1999, Systems re-thinking: an inquiring systems approach to the art and practices of the learning organizations, *Foundations of Information Systems*, (visited in May 2003) www.cba.uk.edu/~parks/fis/ inqre2a1.htm

Latour, B., 1987, *Science in Action*, Harvard University Press, Cambridge.

Linstone, H.A., 1984, *Multiple Perspectives for Decision Making*, North-Holland, New York.

Markus, M.L., Majchrzak, A. and Gasser, L., 2002, A design theory for systems that support emergent knowledge processes, *MIS Quarterly*, 26(3).

Orlikowski, W.J. and Baroudi, J.J., 1991, Studying information technology in organizations: research approaches and assumptions, *Information Systems Research,* 2(1).

Richardson, S.M., Courtney, J.F. and Paradice, D.B., 2001, An assessment of the Singerian inquiring organizational model: cases from academia and the utility industry, *Information Systems Frontiers*, 3(1):49-62.

Richardson, S.M. and Courtney, J.F., 2004, A Churchmanian theory of knowledge management system design, *Proceedings of the 37th Hawaii International Conference on System Sciences*, Hawaii.

Senge, P.M., 1990, *The Fifth Discipline*, Random House, London.

Senge, P.M., 1999, *The Dance of Change*, Nicholas Brealey, London.

Wenger, E., 2000, Communities of practice and social learning systems, *Organization,* 7(2).

Winter, M.C., Brown, D.H. and Checkland, P.B., 1995, A role for soft systems methodology in information systems development, *European Journal of Information Systems*, 4.

Yates, J. and Orlikowski, W.J., 2002, Genre systems: structuring interaction through communicative norms, *Journal of Business Communication*, 39(1).

Chapter 10

PRAGMATISM MEETS SYSTEMS THINKING: THE LEGACY OF C. WEST CHURCHMAN

DAVID MATTHEWS
Department of Applied Mathematics, University of Adelaide

Abstract: The aim of this essay is to survey the scope of Churchman's thought, place it in the context of the American pragmatic tradition and highlight the enormous influence Churchman has had on some of the more prominent systems thinkers of the past two decades.

Key Words: Pragmatism, Neo-pragmatism, philosophy, systems thinking, critique, critical thinking, boundary judgements.

10.1 CHURCHMAN'S PHILOSOPHY

Historical accounts of the career of C. West Churchman suggest that much of his thought can be traced to his background in philosophy (Britton & McCallion, 1994). If so, then it is conceivable to trace his thought to the pragmatism of William James. Apart from the intellectual influence of James over much of American philosophy at the time, a direct line of supervisor-student descendancy can be traced from James through Edgar Singer to Churchman. Accordingly, this essay attempts to embed Churchman's school of thought within the broader American pragmatic tradition championed by James and pursued by such followers as Dewey and Rorty.

10.1.1 The American Pragmatic Tradition

So what is pragmatism? As with most schools of thought, pragmatism is no one thing. At best it can be thought of as a tradition, primarily an American tradition. Indeed, it could be argued that pragmatism is to America what post-structuralism is to Europe. Not only were the founders of the movement American, but most of its subsequent followers were as well. Whereas, Nietzsche, Heidegger, Derrida and Foucault spearheaded the rise of European post-structuralism, the rise of pragmatism on the other side of the Atlantic is associated with such thinkers as Charles Peirce (1839-1914), William James (1842-1910) and John Dewey (1859-1952).

At its heart, pragmatism is a belief about the nature of all beliefs. According to Menand (2001), the rise of pragmatism in the late 19th century is intimately linked to the US civil war. America had just witnessed the most spectacular carnage of its short history. The civil war was a particularly dangerous war for all involved. It was fought with modern weapons and pre-modern tactics. Moreover, the war itself was fought in the name ideology – a sort of crusade for both sides. The lesson that the pragmatists took from the war is perhaps best expressed by Oliver Wendell Holmes, who claimed that *"certitude leads to violence"*. According to Holmes, *"when you know that you know [i.e. are certain], persecution comes easy. It is as well that some of us don't know that we know anything"*. The generation of Bostonians who fought in the war in their youth, and began to influence intellectual life at Harvard upon their return, picked up on this notion and gradually transformed it into a movement.

The early pragmatists claimed that the desire for certainty had led European epistemology (and philosophy of science) into a series of 'dead-ends'. Seeking to find a way around these, the pragmatists attempted to find 'resolution' in 'dissolution'. That is, they attempted to circumvent the pitfalls of epistemology by seeing things in a different light, a light in which traditional problems (such as the quest for certainty) cease to be problems at all and new, hitherto marginal, issues become central[35].

[35] Actually, the method of 'dissolution' is not the sole property of the pragmatists and can be seen in the writings of such predecessors as Kant, Hegel, Kierkegaard and Nietzsche as

The first great pragmatic writer was Charles Sanders Peirce (1932), whose **pragmatic maxim** stated:

> *"if one can define accurately all the conceivable experimental phenomena which the affirmation or denial of a concept could imply, one will have therein a complete definition of the concept, and there is absolutely nothing more in it".*

'Meaning', is therefore a matter of the conceivable experimental consequences of a concept's application. This sounds very much like positivism, and in many ways it is, as Peirce claimed that the role of pragmatism was to highlight that *"almost every proposition of ontological metaphysics is gibberish"*[36]. According to Peirce, *all "meaningful thoughts were signs"* (representations of objects)[37].

In the *Fixation of Belief*, Peirce (1877) argued that the aim of all inquiry was to eliminate doubt (i.e. 'fix' belief). Among the possible methods for fixing beliefs, Peirce claimed that: *"the scientific method was the most successful"*. As unremarkable as this statement may at first sound, within it lies the seeds of a remarkably different train of thought from that of European science on its road to the positivism that characterised it in the first half of the twentieth century. By choosing to use pragmatic terms such as 'successful' and 'appropriate', Peirce distanced himself from the pre-Nietzschean European obsession with finding a methodological path to certainty. Indeed, Peirce even pre-supposed Popper's famed critique of positivism, claiming that scientific inquiry is characterised by *"contrite fallibilism"* and that the scientist is distinguished by *"his readiness to dump the whole cartload of his beliefs the moment experience is against them"*. Thus, whilst Peirce begins mapping out his position in positivist/empiricist terms, he finds the traditional scientific quest for 'truth' and 'certainty' unsolvable, and the human condition as 'fallible'. Accordingly, he claims that the combination of a strict empiricism with the scientific quest for transcendental truth is unworkable. **Pragmatism, according to Peirce,**

---

well as contemporaries such as Wittgenstein. Indeed, Wittgenstein (1953) even gave a label to the technique: **perspicuity**, claiming that perspicuity was needed to see "that the most thorny problems with which philosophers were confronted were, in fact, pseudo-problems" and that once these problems could be seen correctly the problems were "not seen to be problems at all". It was not that they were 'solved', but were 'dissolved'.

[36] Positivism is an empiricist (or anti-metaphysical) philosophy of science that rose to prominence in the early to mid twentieth century.

[37] For positivism to succeed in its 'anti-metaphysical' position it required that there be no facts beyond sense experience and therefore all meaningful statements must be reducible to sense experience (i.e. language is a system of sense datum).

chooses to retain empiricism, thus necessitating the re-formulation of such traditional Occidental concepts as 'truth'.

According to Peirce, the scientific method is characterised by three different types of reasoning: abduction (the postulating of hypotheses to explain observed phenomena); deduction (the discovery of the logical consequences of these hypotheses); and induction (the verification of these hypotheses through repeated testing). Against Humean scepticism of induction, Peirce claimed that all induction resembles statistical sampling and that such reasoning has a self-correcting character. Although there is no logical reason for relying on induction in the short term, repeated use will lead the community of inquirers towards ***"fixation of belief"*** (Peirce, 1877).

Peirce's understanding of induction introduced a subtle variation on the European understanding of scientific truth prior to Nietzsche. Whereas the European position had always been transcendental, in that truth embodied 'correspondence to mind-independent reality', Peirce argued that truth is *"that concordance of a statement with the ideal limit towards which endless investigation would tend to bring scientific beliefs"*. This subtle variation leads Peirce to propose an entirely new theory of truth and to view science not as the transcendental method of securing transcendental truth, but as the best method we have conceived thus far for achieving consensus on where to fix our beliefs.

Peirce's theory of truth begins with his observation that different people using different methods to explore the same question, more often than not, will arrive at the same conclusion. Indeed, Peirce argues that, given enough time and information, all minds will tend to agree.

> *"Let any human have enough information and the result will be that he will arrive at the same conclusion that any other mind will reach ... human opinion universally tends, in the long run, towards truth"*

By itself, this statement may seem at best naively optimistic and at worst downright absurd. However, Peirce argues that it is not just that different minds tend to agree (which could imply that all minds reach the same false conclusion) but that **truth is consensus** - by definition, the consensual conclusion is true. Accordingly, Peirce defines a proposition as true if and only if it is agreed to by everyone who investigates the matter.

> *"The opinion which is fated to be ultimately agreed to by all who investigate is what we mean by truth"* (Peirce, 1932).

It does not matter how this consensus is achieved, whether by the scientific method, mass hypnosis or the teachings of a guru, **it is the**

consensus that is the truth-determinant and not the method of arriving at it.[38] Thus Peirce states: *"If a general belief can in any way be produced, though it be by the faggot and the rack, to talk of error in such belief is utterly absurd"*.[39]

The question that immediately comes to mind is *what is the source of Peirce's optimism?* Why is Peirce so confident that all investigators of a particular problem will come to agreement? At first glance it may seem that behind Peirce's pragmatic definition of truth lies a realist ontology. That is, the only propositions that everyone would agree on are those that accurately reflect a single objective (mind-independent) reality. However, Peirce (1932) argues that it is a neo-Kantian idealism, rather than a realist ontology, that drives his theory:

> *"My social theory of reality, namely, that the real is the ideal in which the community ultimately settles down ... involves phenomenalism, but it is the phenomenalism of Kant, and not that of Hume ... It was the essence of [Kant's] philosophy to regard the real object as determined by the mind ... it was to regard the reality as the normal product of mental action".*

According to Peirce, it is the 'collective mind' that constructs what is taken to be reality (by way of agreement) and not 'reality' that constructs the agreement.

The most significant difficulty facing Peirce's pragmatism may now seem obvious. Peirce's optimism regarding the ubiquity of truth does not follow from his social idealism. If there is no underlying 'real world' forcing itself upon us in a 'mind-independent' manner, then how is it that *"human opinion universally tends, in the long run, towards truth [consensus]"*. In response to this inconsistency, Kirkham (1997) argues that *"Pierce's theory of truth is plausible only because it is parasitic on another, hidden theory of truth – the correspondence theory"*. For example, consider a community that have had access to sufficient information and time to achieve consensus (i.e. 'created' truth). Peirce would have that the observations, which caused this group to reach the final conclusion were forced on them in a reverse chronological order by the final consensus, which at the time had not yet been achieved (Kirkham, 1997).

[38] It is interesting to note that the reason why Peirce thought science such a valuable enterprise was not because its results corresponded with some objective reality, but because it was such a powerful method for achieving consensus.

[39] This is similar to the position Strawson (1964) takes in his 'performative theory'. According to Strawson, truth ascriptions are performative utterances, similar to 'I agree'. Therefore, to utter the sentence "it is true that snow is white" is not to **say** anything at all. Rather, it is to **agree** with the proposition that "snow is white", signalling the performance of the act of agreeing.

Kirkham's argument is a compelling one, however he is incorrect to suppose that he has shown an inconsistency between Peirce's theory of truth and his social idealism. What he has shown is that Peirce's optimism does not necessarily follow from either – there is no reason why Peirce's social idealist ontology and pragmatic theory of truth cannot coexist with a thoroughgoing scepticism about the attainment of truth. Indeed, it seems reasonable to assume that the presence of optimism is related to the belief in either a Kantian-style transcendental apperception or a single unified world-view governing society (thereby causing people to construct essentially the same 'picture' of 'reality'). In the relative monoculture that characterised late 19th century, post-civil-war, Harvard, Peirce may justifiably claim a single world-view, and therefore regular consensus. However, in the multicultures of the early 21st century postmodern West, it is less likely that consensus will be arrived at by all who investigate. If consensus equals truth, then today, pragmatic-truth would tend to be the exception rather than the norm.

Whilst Peirce was no revolutionary, the subtle reforms he made to the entrenched European philosophy of science at the time were to become revolutionary in the hands of his friend and contemporary, William James.

Whereas much of Peirce's work remained unpublished during his lifetime, James became known as the champion of pragmatism, publishing several books and establishing a large student following from his teaching chair at Harvard. According to Menand (2001), James invented pragmatism as a favour to Peirce in a speech at the University of California, Berkeley, in 1898. Introducing the term to the world for the first time, James told the audience that its source was:

> *"A philosopher whose published works ... are no fit expression of his powers. I refer to Mr Charles S. Peirce, with whose very existence as a philosopher I dare say many of you are unacquainted".*

James' lecture made pragmatism a subject of intellectual debate across the US (and indeed the world at large) and helped rescue Peirce's thought from the obscurity it seemed destined to. However, whilst James and Peirce shared much in common, there were also significant differences in their world-views (including their conceptions of truth, science and inquiry). Indeed, if it were not for the fact that James also called his position pragmatism and continually acknowledged his debt to Peirce, it is debatable whether the two would have been thought to be unfolding related strands of thought.

Of the differences between James' version of pragmatism and Peirce's, one of the most telling was their respective position on logic. Whereas Peirce was a pioneer of modern mathematical logic and conceived of philosophy and science as coming fully under its scope, James harboured a deep suspicion of its formalities, claiming that it failed to represent the **process** of scientific (and philosophic) inquiry, or the **structure** of scientific (and philosophic) theories[40]. James saw his work as somehow following on from Kant's *Critique of Pure Reason* (or, critique of formal logic, as he saw it). Whereas Kant tried to set strict limits on formal reasoning, and by so doing make room for such things as values and faith (which he associated with 'practical reasoning'), James (1897; 1907) suggested that formal reasoning had very little to do with how we fix beliefs.

According to James, **when we think we rarely consult fixed principles. Rather, the act of thinking is often what 'fixes' these principles for us.** When we reach a belief (or principle) that we are happy with, it is usually because we have worked out that it 'fits' well with the whole inchoate set of assumptions constituting our world-view. These assumptions act as a set of criteria for fixing beliefs, yet they can never be proven to be true by any standard outside the process of fixing beliefs itself. Thus, a strange kind of circularity takes place in fixing beliefs of all kinds. **Truth, according to James, happens to an idea.** It is, in effect, the compliment we give to the outcome of our process of fixing belief. James associated this process with what he termed the 'faith ladder' and argued that all forms of inquiry involve 'steps of faith', including what are typically labelled as scientific forms of inquiry.

According to James, pragmatism was firstly a method and secondly a theory of truth[41]. The 'method' of pragmatism was derived from Peirce's pragmatic maxim, first articulated in his paper *How to Make our Ideas Clear*. In the hands of James, Peirce's pragmatic maxim attained a radical empiricism. Indeed, James attempted to analyse every statement in terms of statements that explicitly referred to sense experience. Because of this, commentators have suggested that James' commitment to radical empiricism is the core of his entire program. However, it is argued here that behind this

[40] Indeed, James was highly sceptical of logic's ability to represent the structure of reality, claiming that the formalities of mathematical logic could never adequately represent the chaos and randomness of the natural world.

[41] Like his successor Peirce, James maintained that truth is something that happens to an idea and not something inherent within it. However, unlike Peirce, James understood truth in teleological terms. That is, **'truth'** was connected to **'purpose'**. James' radical empiricism led him to suggest that ideas were merely rules for action and that to understand the meaning of an idea was to understand what conduct it was fitted to produce. 'Ideas', for James, became **'instruments'** and not **'answers'**. James' pronouncements on the nature of truth in Pragmatism evoked howls of indignation from Enlightenment scientists and philosophers, most notably Bertrand Russell, who dismissed the entire pragmatist position as "irrational".

commitment lay a deeper motivation for James' thought. That is, to defend **pluralism** (James, 1909b). Indeed, it is James' conflation of empiricism with pluralism that is perhaps the most significant factor in his defence of the former.

On the face of it, James' radical commitment to empiricism would seem to suggest that he was a positivist, however like his mentor Peirce his empiricism led him in the opposite direction to what he saw as the pretensions of positivism. According to James, empiricist forms of inquiry were characterised by *"explaining wholes by parts"*, whilst rationalist forms were characterised by *"explaining parts by wholes"*. Furthermore, James (1909b) argued that whereas the empiricist inclination to explain wholes in relation to parts led to pluralism, the rationalist inclination to explain parts in relation to wholes led to monism. This conflation of seemingly unrelated binary oppositions (empiricism/rationalism, pluralism/monism, reductionism/holism) is highly idiosyncratic and by no means self-evident. Consequently, it requires further explanation, not the least for which, it provides insight into James' empiricism (which is sometimes said to be incompatible with several of his other doctrines, thus giving credence to the criticism that James' writings were incoherent).

According to James (1909b), the question of *"the one and the many"* is central to all philosophic problems. James claims that:

> *"If you know whether a man is a decided monist or a decided pluralist, you perhaps know more about the rest of his opinions than if you give him any other name ending in 'ist' ... to believe in the one or in the many, that is the classification with the maximum number of consequences"*.

James relates monism, holism and rationalism (and hence pluralism, reductionism and empiricism) by supposing that the rationalist sees the world like a grammatical sentence (where the whole is more important than the parts). Accordingly, the rationalist is concerned with apprehending this 'whole' (alternatively described as a 'unity', 'order' or 'fundamental law'). Once apprehended, the rationalist explains all observations in the light of this fundamental law. On the other hand, the empiricist conceives order as imposed upon the disorder of sensory observation by the observer selecting objects and tracing relations. According to James *"we carve out order and the world is conceived thus"*. Thus, the empiricist is concerned with different parts of the world and, as such, rejects totalising rationalist impositions (i.e. laws of nature).

Whilst agreeing with James' central concern of defending pluralism and rejecting totalising impositions, this paper suggests that his assumption that empiricism leads to pluralism is mistaken on both logical and historic grounds. It is beyond the scope of this present work to delve too deeply into this, but for now it should suffice to say that where empiricism has been most dominant (e.g. the positivist episode) monism and reductionism have followed (e.g. the positivist program for the unification of the sciences). Furthermore, reductionism has no privileged relationship with empiricism and can be found in rationalism (Cartesian foundationalism) and empiricism (positivist unificationism) alike. Indeed, if there is a privileged relationship between any of these concepts it is the relationship between reductionism and monism (Matthews, 2004). Unfortunately, James' central task of defending pluralism can be lost in his unique characterisations of the reductionism/holism and empiricism/rationalism dualities. Accordingly, it is this aspect of James' thought that will be the focus of the rest of this overview.

At the heart of James' pluralism was his love for the variety he saw in the world and resentment of its dismissal as 'mere appearance'. Accordingly, he defied the logic that sewed the world up into what he saw as a spurious unity. In his *A Pluralistic Universe* (originally a series of lectures delivered at Oxford), James (1909b) carried out a critique of monism and a subsequent defence of radical pluralism. According to James, 'history' is the key concept separating the monist from the pluralist. Whereas the monist emphasises the 'timeless character' ('nature', or 'order') of things, the pluralist suggests that nothing in the universe is static or eternal or timeless. **All disciplinary knowledge is 'historic', in the sense that it is based on the conceptual schemes developed within the disciplinary tradition and not on the 'timeless character' or 'true nature' of the objects of its inquiry.** James turned to Hegel to understand this 'historic' (or 'contingent') aspect of disciplinary knowledge. Directly quoting Hegel, he argued that there was a dialectical movement at work within the history of disciplinary knowledge. Rather than the disciplines progressing through the application of some transcendental (or statistical in the case of Peirce) method toward 'truth', James (1909a) claimed that the history of knowledge is characterised by discontinuities as new and 'more satisfactory' arguments replaced old and increasingly 'unsatisfactory' ones.

According to James, Hegel's admission that the subject is part of the universe s/he is attempting to account for implies that the subject will always be situated in a historical context[42]. Reality, according to James, exists in the

[42] However, James did not agree with Hegel that history would itself provide the key to transcending this historical context. According to Hegel, dialectics work to transcend the contingent perspective over time in the following manner: Suppose you affirm A and

"each-form" rather than the *"all-form"*. Thus, defining a *"multi-verse"* as opposed to a *"uni-verse"*. According to James (1902), all disciplinary knowledge is historically situated and, as such, what is deemed to be 'true' attains its truth-value with respect to the historical and social context in which the inquirer is operating. James' pluralism is, therefore, supported by his conception of the faith ladder, which he argued was characteristic of the rise of all knowledge. The faith ladder is described as follows:

1. A conception of the world arises in you (no matter how). **Is it true**? You ask yourself.
2. It **might** be true if it is not self-contradictory you think.
3. It **may** be true here-and-now if it passes some test.
4. It is **fit** to be true for all time if it would be well to be true[43].
5. It **ought** to be true.
6. It **must** be true (something inside you whispers)[44].
7. It **shall be held** to be true (it shall be **'as if'** true).

James argued that not one step in this entire process is logical, yet it is the way in which we fix beliefs of all kinds. Accordingly, he rejected the pretensions of the European positivists and claimed that an understanding of the process of fixing scientific belief exposed an *"epistemological naïveté"* in the positivist position (i.e. it was impossible to ascribe to the sciences a *"God's eye view"*). Furthermore, James rejected the modern claim that science had replaced pre-modern superstitions with indubitable facts. According to James, scientists are both enabled and confined by their

---

simply leave it thus, by so doing, you leave it open to supervenience by one who claims ~A, but B. The only way of making your affirmation self secure is by getting your retaliation in first and stating A in a form that will negate all possible negations in advance. Therefore, what is posited as A must already have cancelled the alternatives. Hegel maintained that the **"final truth"** at the end of historical dialectics would be something to which there was no imaginable alternative, because it contains all possible alternatives inside itself as "moments" already taken account of and overcome. Whilst none could argue that Hegel's vision of a single self-securing "truthful whole" was not indeed sublime, James and others have argued (and argued well) that historical dialectics do not work in anything like a Hegelian fashion and that rather than progress toward a single timeless picture, disciplinary knowledge is characterised by discontinuities. James, in particular, thought Hegel's historical vision was sheer 'wishful thinking'.

[43] Here James suggests a teleological element to 'truth'. Truth, according to James is a property that happens to an idea when it allows us to get into satisfactory relations with other ideas.

[44] Here is the classic 'leap of faith' that James suggests is behind every claim to truth and therefore undermines the monist position.

historical, cultural, social and psychological contexts. Thus, James celebrated what he termed *"openness of mind"* over *"dogmatic arrogance"*.

More than anything else, James' pragmatism argued for a radical mutability to what had been hitherto described as 'scientific truth'. According to Schiller (1903; 1907): *"facts are not simply discovered, but selected, even made by us"*.

The criticisms of the pretensions of modern science carried out by James have led some to argue that James articulated a parallel critique of Enlightenment rationality to that of Nietzsche (Rorty, 1991b). Indeed, Rorty (1991b) sees Nietzsche *"as the figure who did most to convince European intellectuals of the doctrines which were purveyed to Americans by James"*. Whereas post-Nietzschean thought is traditionally said to include figures such as Heidegger, Foucault, Derrida, Lyotard, Baudrillard and others, its counter-part in the US (perhaps post-Jamesian thought?) can be said to include such self-proclaimed pragmatists as Dewey, Mead, Quine, Lewis, Putnam and Rorty.

Whilst it would be simplifying in the extreme to suggest that contemporary pragmatism is synonymous with contemporary post-structuralism, it is reasonable to suggest that both traditions contain complementary ideas[45]. Both traditions critique the rationalist (Cartesian-subjective) and empiricist (Newtonian-objective) controversy over the foundations of knowledge. Both traditions see this controversy as being perpetuated by the discourse of subject-object dualism and its bedfellows, **absolutism, monism and certainty**. In its place, both traditions emphasise that subjectivity is an irreducible component of knowledge of objects (an epistemic position that has been variously described as constructivism, neo-Kantianism etc), leading to a commitment to **contingency, pluralism and uncertainty**. Thus, both the pragmatists and the post-structuralists share a similar distrust of the totalising narratives of legitimation characteristic of the modern era, which is what Lyotard (1979) tells us is the defining characteristic of postmodernism.

10.1.2 Neo-Pragmatism, Anti-Foundationalism and Postmodernism

The line of argument presented above suggests that associating pragmatism with post-structuralism and thence postmodernism is, at least on

[45] Indeed, even labelling the works of Peirce, James, Dewey, Mead, Schiller, Rorty and others under the general heading of pragmatism is dubious enough. As Schiller (1903; 1907) himself testifies: "there are as many pragmatisms as there are pragmatists".

the face of it, coherent. However, it has been largely thanks to the efforts of Richard Rorty that this association has begun to gain serious currency.

Richard Rorty is often considered the heir of American **pragmatism.** Whilst schooled in the Anglo-American analytic tradition, upon commencing his professional life Rorty quickly established himself as one of its most articulate critics. Rorty's first major work, an edited book entitled *The Linguistic Turn* (published in 1967) describes the attempt by analytic philosophy to turn away from 'the mind' as the source of knowledge towards 'language'. However, even at this early stage, Rorty suggested (in the editorial) that the analytic tradition shares a fundamental premise with the tradition that it has devalued. That is, the idea that *"there are philosophical truths still waiting to be discovered"*. Following the publication of *The Linguistic Turn*, Rorty regularly contributed to the open literature, but did not complete another book for over a decade until the publication of his much-heralded *Philosophy and the Mirror of Nature* (Rorty, 1979). In this book, Rorty reviewed the various theories of mind and knowledge that had been put forward over the course of philosophic history and argued that Western philosophy's obsession with the mind as a 'mirror' of nature was the source of the modern problematic of finding the 'foundations of knowledge'. **Rejecting the 'mirror' metaphor (i.e. rejecting representationalism) led to rejecting foundationalism in regard to knowledge.**

By 1982 Rorty had come to associate his peculiar style of anti-foundationalism and anti-representationalism with the American pragmatic tradition and argued as much in *The Consequences of Pragmatism.* According to Rorty (1991a) the core of the pragmatism of James and Dewey was the attempt to replace the notion of true beliefs as accurate representations of reality with the notion that they are successful rules for action. Thus, knowledge is not a matter of *"getting reality right"* but of *"acquiring successful habits of action for coping with reality"*. Knowledge is not about representing reality and not something that can be said to have solid foundations. This position followed from Dewey (1909), who claimed that knowledge was: *"an instrument, or organ, of successful action"* and that *"the chief service of pragmatism, as regards epistemology, will be to give the 'coup de grace' to representationalism"*.

After the publications of *Philosophy and the Mirror of Nature* and *The Consequences of Pragmatism,* Rorty's reputation as a postmodern thinker was firmly established (albeit even though he rarely used the term). Within these two books, Rorty articulated many of the 'postmodern' critiques of epistemology that the post-structuralists were expounding in Europe at the same time. Thus, the 1970's and 1980's saw the beginnings of a parallel set of critiques of 'modernity' from both the pragmatic (Rorty) and structuralist (Foucault, Derrida etc) traditions.

In his early works, Rorty acknowledged three thinkers who shaped his positions (and cemented his rebellion against his analytic upbringing) more than any others. These were Dewey, Wittgenstein and Heidegger[46]. What impressed Rorty about these thinkers was the way in which they either dissolved or radically changed the shape of traditional philosophic problems as a result of employing new assumptions or vocabularies. According to Rorty, Dewey, Wittgenstein and Heidegger each in their early years tried to find a new way of making philosophy 'foundational'. Dewey attempted to construct a naturalised version of Hegel's vision of objective-knowledge-through-historical-dialectics. Wittgenstein's (1922) *Tractatus* provided a new theory of representation that was based on the structures of a sense-datum language (and therefore had nothing to do with the traditional object of epistemology: the mind). And Heidegger tried to construct a new set of philosophic categories that had nothing to do with science, epistemology or the Cartesian quest for certainty but found its foundations in 'Being'.

What Rorty respected about each of these thinkers, however, was not the novel theories they articulated in their youth but that each of them eventually came to see these efforts as self-deceptive and in their later work broke free of the conception (inherited from Descartes and Kant) of philosophical thinking as 'foundational'. Dewey abandoned his attempt to make knowledge foundational and began to describe knowledge as 'justified belief' dropping the 'true' from the well known Platonic definition. Accordingly, Dewey began to emphasis the social aspects of justification rather than the empirco-logical aspects of correspondence between the knowing subject and reality. Wittgenstein began to see language as a tool rather than a mirror. Accordingly, he saw the central problem of the *Tractatus,* that of finding the necessary conditions for the possibility of linguistic representation, as self-delusive. In fact, he described his earlier efforts as: *"buzzing around inside a fly-bottle"*. Finally, Heidegger rejected his existentialist and phenomenological upbringing and began to see the attempt to make the knowing subject the source of necessary truths as one more self-deceptive attempt to substitute a determinate question for *"that openness to strangeness which initially tempted us to begin thinking"*.

At the heart of Dewey, Wittgenstein and Heidegger's later thought was, Rorty argued, the rejection of knowledge as 'accurate representation' or the mind as a 'mirror of nature'. Following the lead of his philosophic heroes, Rorty (1979) made this the centre piece of his book *Philosophy and the Mirror of Nature,* claiming:

[46] The Latter Rorty would probably add Derrida and Davidson to this company.

"The aim of this book is to undermine the reader's confidence in 'the mind' as something about which one should have a philosophic view, in 'knowledge' as something about which there ought to be a 'theory' and which has 'foundations'".

According to Rorty, before Kant there was no conception of philosophy as a distinct discipline from science. However, following Kant, philosophy and science became partitioned by the notion that philosophy's goal was to provide a 'theory of knowledge' that would be distinct from the specific knowledge of the sciences because it was their foundation. Thus, philosophy as a discipline attempted to 'underwrite' the specific claims to knowledge made by other disciplines. According to Rorty, the Kantian conception of the role of philosophy assumes that there is a specific 'problem of knowledge'. This idea finds its source in the assumption that knowledge is an accurate representation of mind-independent reality. One of Rorty's major contributions is the suggestion that **if the traditional way of viewing knowledge is optional, then so is the entire modern epistemological problematic**[47].

Following his rejection of the 'mirror' metaphor, Rorty proposed an alternative characterisation of knowledge to the traditional (modern) one, suggesting that knowledge could be thought of as either:

1. A relation to propositions (and therefore, justification becomes a relation between the propositions in question and other propositions from which the former may be inferred), or

2. A privileged relation to the objects those propositions are about (and therefore justification is the proof of this relation).

If knowledge is construed as (1), then there is no special 'problem of knowledge', because there is no need to end the potentially infinite regress of propositions brought forward in defence of other propositions[48]. However, if knowledge is construed as (2), then there arises a special 'problem of knowledge' (the epistemological problematic). This problem arises because there would be a justifiable desire to go beyond 'argument', to 'compulsion

[47] Rorty argued that the idea of a 'theory of knowledge' grew up from the problems (highlighted by Hume and Kant) of knowing whether our inner representations were accurate. The idea of a discipline devoted to 'the nature, origins and limits of knowledge' – the textbook definition of epistemology – derived from Kant's Critique of Pure Reason. However, if knowledge is no longer defined in representationalist terms, the problems highlighted by Kant cease to be problems at all.

[48] This position is similar to what Putnam (1981) calls the internalist conception of philosophy.

from the object known' such that argument would be impossible because: *"anyone gripped by the object in the required way will be unable to doubt"*. To reach this point is to reach the foundations of knowledge that the epistemologists so greatly desire.

Because of Rorty's insistence that (1) is a more useful way of construing knowledge than (2), his position on knowledge has been described as **'anti-foundational'** and **'anti-representational'**. Indeed, Rorty often quotes Quine's (1969) doctrine of the 'indeterminacy of reference' to argue that there is no self-evidence involved in attributing meaning to utterances and thence to objects. 'Anti-representationalism' accepts there is no way of finding an independent test of accuracy of representation. Or, as Davidson (1984) puts it, *"there is no chance that someone can take up a vantage point for comparing conceptual schemes by temporarily shedding their own"*. According to Rorty, if knowledge is construed as (2) we need to *"climb out of our minds"* in order to justify ourselves. However, if we understand knowledge as (1), it is not judged transcendentally but by the standards of the inquirers of our own day. Accordingly, justification can never be achieved by reference to mind-independent reality but must always be sought by reference to what an individual or community of inquirers already accepts[49].

With the demise of foundational epistemology various attempts have been made to fill the vacuum. One such attempt has been with the 'linguistic turn' and the subsequent study of the philosophy of language. For example, Dummett (1976; 1978; 1981; 1982) sees philosophy of language as foundational in the sense that, following the linguistic turn, epistemological issues are now, at last, being formulated correctly as issues within a theory of meaning. Similarly, some have attempted to ground knowledge in the study of human cognitive practices; the ways people and groups form beliefs. That is, it is transformed into a branch of cognitive psychology. Rorty rejects both of these 'new foundationalisms' claiming that the demise of foundational epistemology did not come about due to the failures of the epistemological program alone, but due to the failures of foundationalist thinking in its own right. Accordingly, Rorty hopes that *"the cultural space left by the demise of epistemology will not be filled"*, claiming that the notion that there is an ahistorical, neutral, framework capable of adjudicating knowledge claims is a fallacy. Such a framework assumes that all possible

[49] Here Rorty suggests that justification is always a matter of coherence with other beliefs. However, these 'other beliefs' are only capable of a similar form of contextual justification.

discourses can be translated into it and hence are commensurable and directly comparable[50].

Much of Rorty's later thought could be characterised as providing an insight into what scientific, philosophic, political and cultural life could possibly look like with the rejection of the mirror metaphor of knowledge (Rorty, 1999). He describes this new outlook as one *"in which the demand for foundations is no longer felt"*, arguing that:

> *"The dominating notion of epistemology is that to be rational, we need to find agreement with other human beings. To construct an epistemology is to find the maximum amount of common ground with others. The assumption that an epistemology can be constructed is the assumption that such common ground exists. Sometimes this common ground has been imagined to lie outside us – for example in the world of Being as opposed to that of Becoming, in the Forms which both guide inquiry and are its goal. Sometimes it has been imagined to lie within us, as in the seventeenth century's notion that by understanding our own minds we should be able to understand the right method for finding truth. Within analytic philosophy, it has often been imagined to lie in language, which was supposed to supply the universal scheme for all possible content. To suggest that there is no such common ground seems to endanger rationality".*

Amongst traditionally minded philosophers, Rorty's anti-foundationalism has been met with considerable ire, as it seems to endanger the idea of the philosopher as the guardian of rationality. Rorty understands this and suggests that the role of the philosopher since Plato be re-cast. According to Rorty, the dominant conception of the philosopher is the Platonic 'philosopher-King', whose job it is to oversee all other areas of inquiry and who knows what everyone else is 'really doing' (whether they know it or not) because s/he knows about the ultimate area of inquiry, the foundations of knowledge, in which everyone else's work is grounded. Finding an appropriate characterisation of the foundations of knowledge is the role of this 'philosopher-King', thus providing the spark for the traditional (modern) epistemological problematic. However, Rorty offers an alternative image to that of the philosopher-King. With the demise of foundationalism, the philosopher could become what Rorty has variously described as the 'informed dilettante', the 'multi-lingual', the 'poly-pragmatic' or the 'Socratic intermediary between various discourses'. Under the Socratic

[50] By 'commensurable', Rorty means "to be brought under a set of rules which will tell us how rational agreement can be reached or what would settle the issue on every point where statements seem to conflict".

intermediary's 'therapy', Rorty sees monistic thinkers *"charmed out of their self-enclosed practices and disagreements between disciplines and discourses compromised or transcended in the course of the conversation"*.

This alternative conception, Rorty labels the **'hermeneutic'** approach, as opposed to the **'epistemological'** approach, to philosophy[51]. According to Rorty, the hermeneutic approach sees the relations between various discourses as strands in a conversation that pre-supposes no disciplinary matrix. Whereas agreement is the goal of the epistemological approach, understanding is the goal of the hermeneutic approach. According to the 'epistemologist', agreement is a token of the existence of some antecedently existing common ground that unites human kind in a universal rationality. Thus, agreement is something to be sought after, as it is the best guide we have of closeness to truth. Similarly, to be rational is to: *"find the proper set of terms into which all contributions should be translated"*. The 'hermeneutic', on the other hand, sees agreement as simply 'agreement', neither intrinsically superior nor intrinsically inferior to disagreement and definitely not a token of some transcendental truth. Thus, according to the hermeneutic, agreement has no inherent superiority over *"exciting and fruitful disagreement"*. Furthermore, to be rational is to: ***"be willing to refrain from thinking there exists a special set of terms in which all contributions to a discourse must be put"***.

In his later writings, Rorty swaps the specific labels of 'epistemology' and 'hermeneutics' for the more general **'systematic'** philosophy, which is foundational (i.e. epistemology) and **'edifying'** philosophy, which is anti-foundational (i.e. hermeneutics). In these writings Rorty argues that whereas systematic philosophy builds towering edifices of ideas aimed at solving the problems of the current generation, edifying philosophy tries to dissolve the problems of the current generation by deconstructing the towering edifice of ideas that created them in the first place. Indeed, Rorty, more than any other pragmatist, has argued that the role of pragmatism is to dissolve difficult problems (such as the nature of truth), thus enabling us to break free from unsuccessful discourses by focussing on issues of more interest. Whereas the vast majority of Western philosophy is systematic, in the sense that it searches for water-tight theories, the work of the American pragmatists Pierce, James and Dewey, and the post-Nietzschean European philosophers Wittgenstein, Heidegger and Derrida is edifying. These edifying thinkers are 'pragmatic' in the sense that they by-pass unsuccessful discourses and 'sceptical' in the sense that they distrust the systematic philosopher's projects of universal commensuration.

[51] Because Rorty's struggle against the foundationalist idea of universal commensurability is akin to the struggle contemporary hermeneutics has undertaken against reconstruction (Gadamer, 1960).

According to Rorty, the point of edifying philosophy is to *"keep the conversation going"*. Indeed, he argues that philosophy fades into self-deception when it attempts to do more than send the conversation off in new and interesting directions. Such new directions may seed new sciences and/or new systematic philosophies, but they are beside the point. The point is always the same: to perform the social function which Dewey called *"breaking the crust of convention"*.

Rorty's pragmatism has significant implications for the scientific enterprise. Much of his writings can be seen as deconstructing traditional accounts of truth and knowledge, and by way of implication, science's claim over them. Accordingly, Rorty argues that **we should give up the idea that the goal of science is to produce models that correspond with reality**. Instead, we should understand that so-called 'scientific' models are but one vocabulary among many. Our question then ought to be: *"which vocabulary works best for our purposes"*?

10.1.3 The Pragmatism of Churchman

As we have seen, in the hands of James and others the consequences of Peirce's subtle reforms have been unfolded over the past century, culminating in Rorty's (1982) *The Consequences of Pragmatism*. According to Rorty, the American pragmatic tradition is **anti-foundational** (in the sense that justification can never be proven because in every chain of argumentation there exists a background context of unprovable *a priori* assumptions), **anti-representational** (in the sense that the goal of inquiry should never be to develop an accurate representation (model) of reality) and committed to a **critical** (or 'edifying') mode of inquiry that seeks to dissolve the problems brought about by foundationalist and representationalist assumptions.

Undoubtedly, **the early systems theorists were uncritically committed to both foundationalism and representationalism**. They aimed to produce models that corresponded with reality (representationalism) and, moreover, assumed that it was feasible to justify the outcomes of their studies by claiming to always model the 'whole system' (foundationalism). As such, by the mid 1960's there was much that the systems tradition could learn from the pragmatic tradition and none was better qualified in both than Charles West Churchman.

Having completed his doctorate under the tutelage of the pragmatist philosopher Edgar Singer, Churchman took up a teaching position in the Philosophy Department of the University of Pennsylvania. In his first year of teaching, Churchman met a young architecture student, Russell Ackoff, who after four years of prescribed architectural studies was taking his first elective. According to Ackoff's (1988) account of this first meeting, Churchman's course in Modern Philosophy was *"intellectually the most exciting course [Ackoff] had taken at the university"* and despite the *"five year difference in [Ackoff's and Churchman's] ages, they hit it off and became friends"*. Much has been written about the intellectual partnership that this fateful encounter spawned. Over the next four decades, the two friends were to have a profound effect on the emerging fields of Operations Research and Systems. However, all of this had to wait, as following the bombing of Pearl Harbour, both of them were immediately seconded to help in the war effort.

During the war, Churchman headed the Mathematical Statistics section of the Frankford Arsenal. It was here that he was introduced to the use of mathematical techniques to help solve real world logistics and inventory problems. Impressed with the practical applications of mathematics and statistics for solving real world problems, upon completion of the war, Churchman took it upon himself to explore some of the methodological issues arising from the use of these analytical techniques. Accordingly, once back at the University of Pennsylvania, and re-united with Ackoff, the two colleagues attempted to establish an Institute in Applied Philosophy. Around the same time, they published their first book together, *Methods of Inquiry: An Introduction to Philosophy and Scientific Method*, which set forth both their philosophical position and their hopes for the application of philosophy to structuring and thence solving complex real-world problems (Churchman & Ackoff, 1950). In *Methods of Inquiry*, Churchman and Ackoff (1950) argued that pragmatism (like Kantian criticism) should be seen as a synthesis of modern rationalism and modern empiricism. **Pragmatism accepted the interplay of observation (empirical datum) and concepts (*a priori* thought), however, it gave neither an autonomous, fixed, or foundational character**[52]. As a consequence, pragmatism rejected the notion of the 'immediacy' of empirical sense-datum as well as the 'provability' of scientific theories. It was, as Rorty (1991a) would have, anti-foundational.

Even at this early stage in his career, Churchman could see that there were important issues to be sorted out in the use of 'scientific' or 'analytical' techniques for solving management and planning problems. Of particular importance was that these new techniques should not be used uncritically in a foundationalist or representationalist manner. Employing them as such

[52] Whereas the modern rationalists were arguing that some knowledge cannot be proven but must be accepted on faith (e.g. metaphysical positions in domains of ontology, epistemology, ethics and aesthetics). On the other hand, the positivists were claiming that 'knowledge' of this sort was meaningless and should be expunged from scientific discourse altogether.

would, according to the pragmatic tradition, lead to self-deception. However, exactly how they should be employed was a difficult question and was set to become the main research agenda of the proposed Institute. Unfortunately, however, the Institute did not receive the support of the University of Pennsylvania and, as a consequence, both Churchman and Ackoff moved to Wayne State University and, following a failed attempt to establish it there, the Case Institute of Technology which finally agreed to support the idea.

At Case, Churchman and Ackoff established a multidisciplinary faculty group (complete with postgraduate programs for Masters and Doctoral students) for specifically studying the philosophy, methodology and practice of tackling complex organisational and societal problems. By borrowing the appellation, Operations Research, from the wartime British and American analysis units set up to help plan operational tactics, Churchman, in effect, established the first post-war academic OR group in the US. Over the next seven years Churchman found numerous sponsors willing to finance research into the sort of complex industrial, governmental and societal problems he had been trying to work on ever since his stint at the Arsenal and, perhaps even more importantly, promoted the new approaches he was developing far and wide by publishing the very first OR text book (in conjunction with Ackoff and Arnoff (Churchman *et* al (1957)) and educating a new generation of Masters and Doctoral students in the new inter-discipline. During his time at Case, Churchman became one of the most prominent early OR practitioners whilst at the same time continuing to excel in his philosophical pursuits[53]. In particular, he established himself as one of the most outspoken advocates for a typically 'pragmatic' epistemic self-understanding within OR.

After the publication of *Introduction to Operations Research* in 1957, Churchman accepted a Professorship at the Graduate School in Business Administration at the University of California, Berkeley, where he remained, Ackoff-less, until his retirement in 1981. Around the same time, OR began to develop rapidly throughout the US. However, this was a bitter-sweet result for Churchman. His hopes of OR being a pragmatic, interdisciplinary approach for tackling big-picture societal issues waned as he saw the discipline gradually adopt a mono-disciplinary culture of mathematical modelling. OR was becoming a 'cookbook' discipline of standard techniques for solving standard problem types. As such, it was thoroughly embracing a representationalist view of methodology and, at its worst, a naively foundationalist view of justification. Accordingly, from the late 1960's Churchman began associating himself less with OR and more with what was becoming known as **'the systems approach'**.

[53] Churchman was, for a time, editor of both Philosophy of Science (1948-1958) and Management Science (1955-1960).

As he had done previously with OR, Churchman became amongst the most outspoken advocates (at times) and critics (at other times) of the new systems approaches. The hope of a philosophically sound, interdisciplinary approach to thinking about complex societal problems, once held for OR, was transferred *en masse* to the fledgling systems community. Both he and Ackoff argued that the systems approach should set out to do what OR failed to. That is, *"advance the big picture societal issues such as poverty, war, health, education etc"* (Ackoff, 1974). To this end, Churchman began work on the research project that he is most famous for: the introduction of critical ideas from the American pragmatic tradition into the systems community. In particular, Churchman sought a **critical understanding** of the related ideas of **holistic inquiry** and **systemic improvement**[54].

Embedded within the American pragmatic tradition as he was, Churchman was not well predisposed to accept uncritically the claims to comprehensiveness of the early systems theories. According to pragmatism, there are limits to what science can know and even the knowledge that it has is seen to be 'floating', as it were, on unprovable metaphysical assumptions. Knowledge, as Churchman was aware, is always infused with the *a priori*. As such, it is always **partial, contextual** and **contingent**. However, systems thinking sought to distinguish itself from alternative ways of thinking by focussing on **'the whole system'** – a claim that did not sit well with the pragmatists understanding of knowledge. To Churchman (1968b), this represented a fundamental epistemic problem. Expressed as:

> *"How can we design improvement in large-scale systems without understanding the whole system, and if the answer is we cannot, how is it possible to understand the whole system"?*

According to Churchman the answer to the second part of this question is that **it is impossible to apprehend (non-contextually) the whole system**.

Within any specific problem one finds the connectedness to all other problems (Churchman, 1982). Such interconnectedness forces systems theorists to bound their systems of interest in ways that are intrinsically distortive to the wider context[55]. Because of the unavoidability of these

[54] An overview of Churchman's contribution to this project can be found in The Systems Approach (Churchman, 1968a), Challenge to Reason (Churchman, 1968b), Design of Inquiring Systems (Churchman, 1971), The Systems Approach and its Enemies (Churchman, 1979) and Thought and Wisdom (Churchman, 1982).

[55] Uncritical adoption of boundaries leads to what Muller-Merbach (1988) describes as the 'given-the-problem' school of thought of OR. This school of thought sees mathematical theorems, proofs and algorithms as the raison d'etre of OR and begins with the assumption that 'the problem' (usually stated in mathematical terms) is 'given', by whomever. It also aptly highlights the difference between measures of performance (MoP's), which attempt to measure the efficiency with which a system **performs** its function and measures of

boundary judgements, Churchman argued that we have a responsibility for the whole-of-system implications of our attempted systemic improvements. Thus, for Churchman, **there is a moral imperative at stake in the systems approach**.

Churchman's writings implicitly reject the notion that the boundaries of a system are given by the structure of reality in favour of a pragmatic understanding that what is 'given' and what is 'constructed' are irreducibly intertwined. Accordingly, traditional distinctions between subject and object (and for that matter ontology and epistemology) are undone and boundary definition becomes an issue not of systems modelling but of practical philosophy. That is, it becomes an ethical issue. Something that appears to be an improvement from a narrow point of view may not be seen as such if the boundaries are extended or arranged in a different way[56]. According to Churchman, systems approaches too often have us analyse 'the problem' as if it represented the total system. However, since these 'as if's' might be sources of deception he claimed that:

> *"The ultimate meaning of the systems approach, therefore, lies in the creation of a theory of deception and in a fuller understanding of the ways in which the human being can be deceived about his world".*

It is the last part of this sentence that Churchman claimed was often forgotten by the early systems theorists. Accordingly, he argued that the systems person has a moral responsibility to:

---

effectiveness (MoE's), which attempt to measure the **effectiveness** of these efficiencies within the broader context.

56 Churchman (1971; 1979) uses the example of the simple inventory problem. Traditional characterisations of this problem bound the system of interest by making strong assumptions about factors such as the development of demand, capital costs, availability of natural resources, manpower etc. For example, Churchman argues that most texts suggest that the appropriate database for estimating future demands is the record of past demands. What these texts fail to understand is that in so doing they are making strong systemic judgements, namely, that the component which generates the demand for inventory is properly designed. However, if there are awkward seasonal variations in demand, then **before** designing an inventory policy, you should **first** see if pricing or advertising can smooth out the demand. This automatically expands the boundaries of the system of interest to include the demand-generating component. According to Churchman, such systemic judgements account for far more serious errors than those typically associated with statistical analysis which formal education spends a disproportionately amount of time on. Furthermore, the real questions to be answered in determining a company's inventory holding costs are: "what is its optimal use of funds?" and "what application of resources can best achieve the purpose of the organisation". These are questions of **value** not **fact** (Mason, 1988).

1. Sweep in as much information about the broader system(s) in which the system of interest is situated[57].

2. Embrace critical thinking (in the Kantian sense)[58].

Churchman's way of fulfilling the dual nature of this moral responsibility was through dialectics. That is, he suggested pitting alternative options (based on alternative *a priori* metaphysical assumptions) against each other[59]. By listening to the arguments of our 'enemies' we become aware of the assumptions in our own thinking and both are better for it. Ulrich (1994) summarises this line of thought by arguing:

> *"What the systems designer needs beyond ever new analytical techniques is a dialectic framework that would enable him to enter into a discourse with these other rationalities and to learn to understand them as what they are: mirrors of his own failure to live up to the systems idea".*

With this thought in mind, Churchman developed an educational program at Berkeley in Social Systems Design (SSD).

One of the aims of SSD was to steer the systems thinker away from the sole reliance on a single set of *a priori* concepts, which may act as an unnecessary restriction to critical thought. Accordingly, SSD aimed at 'surfacing' the implicit worldviews (*a priori* assumptions) of the systems designer and/or decision maker. Once these assumptions are brought to the surface an alternative set of assumptions are developed. From this alternative set, different proposals (courses of action, decisions, systems designs etc) are derived that, because of their different foundational assumptions, challenged the former ones. The aim is to develop a more critical understanding of the complex problem (or system) by seeing aspects of the problem that would have remained hidden by the uncritical implementation of policy founded on a single world-view. The whole process is an exercise in dialectics with three simple steps:

[57] The 'sweeping in' process of Churchman was derived from his philosophical teacher Edgar Singer. According to Singer, Peirce's pragmatic maxim required a systematic attempt to 'sweep in' ever more aspects of a problem situation. In Churchman's language, Singer's 'sweep in' process becomes a process of re-examining the system of interest in the light of the larger system in which the system of interest is situated. However, as both he (1968a,b) and Ulrich (2001) have noted, there is no definitive way in which to bound this larger system. Accordingly, the systems thinker's quest for comprehensiveness is potentially endless.

[58] That is, attempt to understand the influence (and implications) of our implicit, a priori, metaphysical assumptions on our knowledge and actions.

[59] The alternative a priori metaphysical assumptions are more often than not ontological assumptions about what exactly is the object (system) of interest (i.e. where the appropriate boundaries lie). However, metaphysics is not just limited to ontology and could incorporate alternative epistemic, ethical or aesthetic assumptions as well.

Thesis: understanding the decision-makers proposals, *P*, and the world-view, *W*, in which these proposals are sensible[60].

Antithesis: put forward an alternative world-view, *W'*, and develop an alternative set of proposals, *P'*, based on this world-view.

Synthesis: reapproach the problem context in the light of both *W* and *W'* and develop a richer appreciation of the situation.

Churchman's philosophical position and subsequent development of SSD have influenced a generation of systems thinkers. Once the claim to be modelling some real world 'out there' is abandoned (i.e. once the traditional aims of systems modelling are abandoned) the systems approach begins to focus on developing heuristics, or rules of thumb, for helping structure 'messy' problems by encompassing multiple perspectives. Almost no part of the systems movement has remained unchallenged by the Churchman corpus. Following Churchman, a whole host of planning methodologies have emerged that, like SSD, take as a given that objectification (boundary definition) is not a value-free, 'objective' affair, but influenced by all manner of *a priori* concepts.

10.2 CHURCHMAN'S INFLUENCE ON THE SYSTEMS COMMUNITY

10.2.1 Churchman's Influence on Russell Ackoff

As has already noted, Russell Ackoff was not only Churchman's first doctoral student, but his intellectual partner over some four decades. Following their unsuccessful attempts to establish an Institute at both the University of Pennsylvania and Wayne State University, both Churchman and Ackoff moved to the Case Institute of Technology to establish the very first OR group in the US. Here Churchman and Ackoff worked together for seven years, helping to establish the new inter-discipline and educate a new generation of inter-disciplinary thinkers. However, from the mid-sixties onward, it became clear to Churchman that OR was evolving into something entirely different from what he and Ackoff had hoped it would be. Accordingly, his publications increasingly focused on the problems with OR (Churchman, 1966; 1967; 1968b; 1970) and the opportunities afforded by the emerging

[60] Here, world-view is used as a 'catch all' term for a priori metaphysical concepts, including ontology (boundary of the system/problem – what is said to be the object of inquiry), epistemology (what we can know and the nature of our knowledge of the system/problem/object), ethics (the purpose, moral responsibilities involved) and aesthetics.

systems approaches (Churchman, 1968a; 1971; 1974; 1979; 1980; 1981). As such, Churchman contributed to a growing realisation that OR had failed to capitalise on its wartime promise of informing national-strategic planning decisions.

By the mid 1970's Russ Ackoff (now on the opposite side of the continent to Churchman) began to follow his erstwhile mentor into the critique of OR (Ackoff, 1973; 1974; 1977). Indeed, it could be argued that Ackoff took Churchman's critique a step further than Churchman himself was willing to go by publicly discontinuing his association with OR, which he believed was trapped within a false paradigm of the possibility of objectivity and optimisation (Ackoff, 1977). In a paper entitled *"Optimisation + Objectivity = Opt Out"*, Ackoff (1977) argued that those who work under the false assumption of objectivity and optimisation inevitably opt out of tackling the important social and environmental issues of the age. Following this polemical proclamation, Ackoff ceased publishing in *Operations Research* (the US journal) and began publishing in the context of a debate currently underway in the British journal (*The Journal of the Operational Research Society*), criticising the development of the discipline on both sides of the Atlantic.

Ackoff had already articulated his vision for the development of OR in a 1961 publication stating *"OR should be applied effectively to broad problems, as for example, to an industry as opposed to a bench, to a transport system as opposed to a railway track. Incidentally, I can see no reason why OR should not be applied in the future to studies concerned with national planning and national policy"* (Ackoff, 1961). However, by the mid 1970's he was deeply disillusioned with the course and conduct of the discipline. In a double address given to the UK Operational Research Society, entitled *"The Future of OR is Past"* and *"Resurrecting the Future of OR"*, Ackoff (1979a,b) noted that the current problem-solving paradigm that was dominant among OR theoreticians, with its emphasis on optimal solutions, was in most cases irrelevant to senior management. Instead, senior managers found themselves in *"dynamic situations that consist of complex systems of changing problems that interact with each other, thereby rendering optimal solutions redundant"* (Ackoff, 1979a). Ackoff called for a new paradigm whereby OR would replace its problem-solving orientation with one that focussed on planning and system design (Ackoff, 1979b). In his view OR could only save itself by *"adopting a systems orientation and serving executives rather than playing with mathematics"* (Ackoff, 2001).

In one of his most controversial criticisms, Ackoff referred to industrial managers' description of OR as *"mathematical masturbation"*, and argued that OR had been gradually taken over by second-generation academics with no practical experience of real-world industrial problems at the strategic level (Kirby, 2000). Indeed, the mounting obsession with mathematical models and algorithms had been instrumental in eroding one of the original strengths of early OR: its interdisciplinarity (Bonder, 1979; Mueller-Murbach, 1984). This view was later adopted by Bonder (1979), who in complete accord with Ackoff, argued that:

> *"The relevance of current mathematical developments in OR is continually being questioned and, perhaps more significant, the techniques and methods are being developed by individuals who have more of a disciplinary allegiance to mathematics than to operational research".*

The following year Samuel Eilon (1980) published a paper on the changing role of OR within industry, which confirmed the concerns of Ackoff, Bonder, Mueller-Murbach and others. Eilon observed that:

> *"The OR group reports at a level well below the board, its work concentrates on tactical problems, often with limited terms of reference, its contribution to the evaluation of strategy is minimal, its approach is technique-orientated and the competition from other functional areas is real and sometimes threatening: the uniqueness of the OR approach is not seen as indispensable, its methodology is challenged, it is regarded as a narrow specialist discipline, a suitable sanctuary for mathematicians, its involvement in implementation is tenuous and its general impact somewhat limited" (Eilon, 1980).*

All of this came to a head when a number of prominent OR groups were unable to survive various organisational changes in the 80's and 90's and were subsequently closed down (Fildes & Ranyard, 1998). Ackoff, of course, had already predicted the demise of OR arguing that:

> *"Managers are not confronted with problems that are independent of each other, but with dynamic situations that consist of complex systems of changing problems that interact with each other. I call such situations messes. Problems are abstractions from messes by analysis ... Problems may be solved; messes need to be managed. If we insist on the solution mode, analysis will be relegated to those relatively minor problems which are nearly independent, while messes go inadequately managed" (Ackoff, 1979a).*

Accordingly, he discontinued his association with OR and followed Churchman into the fledgling systems community:

> *"In sum, I have discontinued my association with Operations Research because it has increasingly devoted itself to doing the wrong things righter. Systems thinking and practice are what OR could and should*

have become ... However, it is not too late for OR to leave this path and go in a different direction. To do so requires that it give up its devotion to tools and techniques of analysis and the search for problems that these tools fit or for problems that can be distorted to fit the tools. It must become manager- and executive-oriented: identify the system of interacting problems, messes, that preoccupy upper levels of management (usually strategic and seldom operational) and bring to bear all the intelligence, science, art and humanities it can muster. In doing so, it will have more impact and deliver more fun than most Operations Researchers have dreamed of" (Ackoff, 2001).

Once there, he began to work on **Interactive Planning** (IP), a systems approach, designed to deal with the 'messes' that OR had left unmanaged. Accordingly, IP adopts what Ackoff calls the **principle of continuity**. A principle which states that because 'messes' are characterised by dynamism, in the sense that their principle issues and problems are constantly changing, plans need to be constantly revised. To this principle, Ackoff invokes the systems idea in the **holistic principle**, which states that because 'messes' are also characterised by interactiveness, in the sense that their principle issues and problems are all related, planning should be conducted simultaneously and inter-dependently for as many parts of the organisation as possible. Finally, Ackoff adds the **participative principle**. Following Churchman, and indeed the broader American pragmatic movement, Ackoff understood that all understanding of (and solutions to) management problems is infused with the *a priori*. Different *a priori* assumptions lead to different problem characterisations and thence different solutions. Moreover, given the pragmatic dissolution of the rationalist / empiricist debate in the denial of any foundations for knowledge, no set of *a priori* assumptions could be easily legitimated above all others. Accordingly, planning and design must be 'participative' in the sense that it sweeps in the views of a wide variety of stakeholders by allowing those affected by the implementation of plans to plan for themselves.

10.2.2 Churchman's Influence on Richard Mason and Ian Mitroff

Richard Mason and Ian Mitroff are perhaps best known within the systems community for their development of Strategic Assumptions Surfacing and Testing (SAST). What is not as well known is that SAST adopts the Churchman corpus in its entirety. Like Churchman, Mason and Mitroff (1981) argue that policy decisions are always underpinned by a world-view in which the decisions make sense. Like Churchman, Mason and Mitroff suggest that no world-view is ever representative of 'absolute' truth and hence, objectification (the creation of ontological 'objects' – in this case systems) is a relative affair (relative to the *a priori* world-view). Like Churchman, Mason and Mitroff, claim that policy developed on the basis of a single world-view is restrictive, possibly authoritarian and most likely doomed to failure. Accordingly, Like Churchman, Mason and Mitroff suggest that **the assumptions 'underpinning' policy decisions should be 'surfaced' and thence set against counter assumptions.** Thus, for Mason and Mitroff, presuppositional tension is a necessary condition for dealing with complex problems (Jackson, 2000). Accordingly, SAST attempts to surface underlying presuppositions and only arrive at judgements after due consideration of a number of possible opposing perspectives. To do this SAST involves four steps:

1. **Group Formation:** The aim of this stage is to structure groups such that consensus is likely to occur within groups but not across groups. As wide a cross section of individuals and organisations as possible should be involved. These should be organised to increase the likelihood of each group developing different courses of action based on different foundational assumptions.
2. **Assumptions Surfacing:** During this stage each group is asked to 'surface' the most significant assumptions underpinning their preferred policies and strategies. Various 'surfacing techniques' have been suggested to help with this, a review of which is beyond the scope of this work. The aim is to surface the key assumptions upon which the success of each group's strategy rests.
3. **Dialectical Debate:** Here the groups are encouraged to pit their cases against each other in an open and honest manner (i.e. by way of debating the assumptions behind the alternative policies proposed). It is hoped that this will precipitate a closer inspection of the various world-views of the groups and the viability of the various assumptions that these world-views generate. At the end of this stage, each group should have developed a richer appreciation of the world-views and assumptions of other stakeholders and, by way of implication, the incompleteness of their own positions.
4. **Synthesis:** The aim of this stage is to achieve a compromise on assumptions. Often a list of 'agreed' assumptions is drawn up and the inclusions of marginal assumptions further debated. Once the groups feel that a rich enough understanding of the problem context has been generated, a new strategy / policy is suggested based on the synthesised set of assumptions. If no synthesis can be achieved, points of disagreement are noted and the question of what research

might be required to resolve the differences discussed (Jackson, 2000).

Accounts of SAST 'in action' are given in Mason and Mitroff's (1981) *Challenging Strategic Planning Assumptions*. The authors suggest that the main weakness of SAST is its dependence on the *"willingness of participants to lay bare their assumptions"*. Kilmann (1983) agrees with Mason and Mitroff's estimation and argues that SAST *"assumes that the participants want their assumptions exposed"*. It seems that within certain 'contexts of practice' the process of SAST does not provide the free interplay of assumptions that its philosophical foundations demand. Of these contexts, Jackson (2000) rightly suggests that coercive contexts, where the powerful are unlikely to want their assumptions revealed, are perhaps the most abundant. Thus, whilst SAST no longer succumbs to the well-known epistemic problems associated with O.R. (e.g. sub-optimisation and/or naïve representationalism) it nevertheless suffers from a number of practical problems associated with the use (and abuse) of power. Accordingly, of the practical difficulties that arise when applying SAST, none is more discussed in the literature than the difficulties of achieving 'emancipation' from human power relations in the planning process. Happily, another of Churchman's doctoral students, Werner Ulrich, has made this problem the core of his research.

10.2.3 Churchman's Influence on Werner Ulrich

The influence of Churchman on Werner Ulrich is unmistakeable. After completing his doctorate in Economics and Social Sciences at the University of Fribourg in 1975, Ulrich studied at the University of California (Berkeley) under Churchman for five years. In Ulrich's (1999) own words, he moved to Berkeley to: *"become West's disciple"*. Churchman's work was particularly exciting to Ulrich because of its 'critical' and 'moral' implications. As we have seen, Churchman looked at the critical implications of the systems idea on management and planning. According to Churchman, whenever we apply the systems idea to some part of the 'real' world we necessarily make strong assumptions about what belongs to the system of interest and what is part of its environment. Moreover, these boundary judgements inform the systems models subsequently developed and, therefore, our understanding of the relevant portion of the 'real' world. The critical implication of this, is that **the systems concept, by virtue of the a *priori* nature of its boundary judgements, constantly reminds us of its own relative character.** That is, it is clear that what belongs to the 'whole system' is entirely dependent on (or relative to) the inquirer's choice of conceptual boundary. As such, whenever we speak of a system, it should be obvious that we are not speaking of transcendent reality, but of immanent metaphysics (Ulrich, 1983).

Ulrich agreed with this insight and suggested that **whole systems judgements, therefore, represent justification break-off points with regard to the context of application**. That is, when it comes to justifying the 'normative' implications of plans, systems planners can only ever refer back to the context of application – the original whole system judgement[61]. No purely rational justification beyond this point is possible. As Ulrich (1983) states:

> *"It is not the reality 'out there' that determines the boundary between the system and the environment, but rather the inquirers standpoint, the purpose of his mapping effort, his personal preconceptions of the reality to be mapped and the values he associates with it".*

According to Ulrich, justification break-off points occur at the point where the systems planner's *a priori* normative presuppositions cannot help but influence essential planning decisions. These decisions can be as broad ranging as: the boundary between system and environment; the set of stakeholders to be included in the decision making process; the measures of improvement; the appropriate consideration of past or future generations[62]; and the exact role of planners, experts and authorised decision-makers in the process. Such questions are clearly normative and are likely to have an enormous impact on the recommendations that emerge from plans[63].

With these thoughts in mind, Ulrich reviewed the systems approaches to planning currently on offer in the literature and found them wanting. Too often, the problems associated with justification break-off points are entirely ignored. Moreover, where boundary judgements are explicitly discussed,

[61] By 'normative', Ulrich (1983; 1987) refers not only to the normative premises (e.g. design models, planning standards, measures of effectiveness etc) but the normative implications (i.e. the consequences of these premises on others when used as the basis for system designs and plans)

[62] Ulrich (1983) provides the following example questions as possible considerations for past and future generations:
Is intergenerational transfer of costs and risks acceptable?
Should the goals, dreams and traditions inherited from past generations be relevant?

[63] In a more recent offering, Ulrich (2002) relates justification break-off points to systemic boundary judgements, normative evaluations and empirical observations, stating: "Whenever we propose a problem definition or solution, we cannot help but assert the relevance of some facts and norms as distinguished from others. Which facts and norms we should consider depends on how we bound the reference system, and vice-versa; as soon as we modify our boundary judgements relevant facts and norms are likely to change … Systems thinking, despite its desire for comprehensiveness, cannot help but be selective with respect to all three".

they are often only done so in terms of the formal criteria of models (i.e. boundaries judgements are discussed in terms of the availability of data and/or the imperatives of theoretical modelling techniques) and not in terms of the justification break-off points associated with their normative content (Ulrich, 1983; 1987).

As we have seen, Churchman's approach to the inevitability of boundary judgements was twofold:

1. Following his pragmatic schooling under Edgar Singer, Churchman argued that systemic inquiry needed to continually expand the boundaries so as to 'sweep in' as much information about the broader system(s) in which the system of interest was situated.

2. Following neo-Kantian thought in general, Churchman claimed that systemic inquiry needed to embrace critical thinking (in the Kantian sense) by recognising the *a priori* content of all knowledge. In this respect, he proposed a dialectical process of seeking out alternative planning proposals and entering into debate over the presuppositions and worldviews in which each proposal made sense.

Ulrich, who had already been influenced by Churchman before moving to Berkeley, argued that Churchman's 'sweep in' process was potentially unending. At some point in time, it is important to take action in the full knowledge that this action will be infused with *a priori* assumptions and normative (value-laden) baggage. As Kant had put it some 200 years earlier:

> *"We cannot, by complaining about the narrow limits of our reason, escape the responsibility of at least a critical solution to questions of reason*[64]*"*.

Or, as Ulrich put it in the context of the systems approach:

> *"The systems approach, because it strives for comprehensiveness, must learn to live with its own unavoidable incomprehensiveness and must draw the necessary conclusions from this insight"*.

Ulrich's research question, therefore, became **'how is it possible to act rationally, in light of this unavoidable incomprehensiveness'?**

Characteristic of post-Churchman systems thinking in general, Ulrich proposed that the way forward was to 'surface' or 'render explicit' the boundary judgements that caused justification break-off points. According to

[64] According to Ulrich: "a critical solution does not yield any objectivist justification of normative validity claims. It prevents us, rather, from submitting to an objectivist illusion regarding such claims".

Churchman, what was needed was 'critical thinking': thinking that attempts to expose the influence (and implications) of our implicit *a priori* assumptions on our knowledge and actions. Within the systems community, argued Ulrich, this implied a need for **critical (systems) thinking**: systems thinking that sought to render all *a priori* judgements explicit and allow for their critical inspection. In regard to this need, Ulrich initiated his **'critically heuristic' turn**, which sought to answer the question, inherited from Churchman, 'what kind of *a priori* boundary judgements need to be critically reflected upon in order to uncover the justification break-off points of systems designs and plans?'

In formulating his answer to this question, Ulrich reconstructs Kant's Transcendental Doctrine of Elements. To begin with, Ulrich undertakes a reconstructive effort in regard to the Kantian *a priori* forms of pure intuition (The Transcendental Aesthetic). Specifically, he argues that social systems cannot be experienced (phenomenologically) without being related to purpose. Following this, Ulrich moves on to a reconstruction of Kant's *a priori* concepts of understanding (The Transcendental Analytic) by recalling Churchman's (1971; 1979) conditions that must be fulfilled for a system to be capable of purposeful activity. In the hands of Ulrich, these twelve conditions are reformulated into twelve critically heuristic categories. Finally, Ulrich suggests a reconstruction of Kant's *a priori* concepts of pure reason (The Transcendental Dialectic) whereby Ulrich, once again, leverages from Churchman in presenting the systems idea, the moral idea and the guarantor idea as critical standards for reflection and argumentation about the justification break-off points in systems designs and plans.

The critically heuristic turn, as the reader will have picked up by now, is deeply embedded within the structure of Kantian philosophy and, as such, it would be inappropriate to reproduce here in its entirety. Notwithstanding this, the centre-piece of the critically heuristic turn (the twelve critically heuristic categories which Ulrich employs in the form of questions for rendering explicit the justification break-off points associated with boundary judgements) should be briefly discussed.

As we have noted, Ulrich's twelve critically heuristic categories, constitutive of the justification break-off points associated with boundary judgements, are to a large degree informed by Churchman's twelve conditions that must be fulfilled for a system to be capable of purposeful activity. Churchman (1979) arranged these conditions into four groups of three categories such that the first category stood for a social role, the second for role specific concerns, the third for problems surrounding the determination of boundary judgements associated with the role and the fourth group for the critical (self-reflective) aspects of social system design. These twelve categories are shown in Table 1 below:

Social Role Role Specific Concern Boundary Judgement	1. Client 2. Purpose 3. Measure of Performance
Social Role Role Specific Concern Boundary Judgement	4. Decision Maker 5. Components 6. Environment
Social Role Role Specific Concern Boundary Judgement	7. Planner 8. Implementation 9. Guarantor
Social Role Role Specific Concern Boundary Judgement	10. Systems Philosophers 11. Enemies of the Systems Approach 12. Significance

Table 1: Churchman's Categories (a priori concepts) Constitutive of the Possibility of Knowledge of Social Systems (In the sense that knowledge of social systems requires synthetic a priori judgements based on them)

Using each of these *a priori* concepts in an effort to trace the normative content of system designs in the domains of economic, health and environmental planning, Ulrich became dissatisfied with the 'heuristic value' of categories 8,10,11 and 12. According to Ulrich, the significance of these four categories: *"consists in helping us to reflect upon our basic philosophy of planning rather than to map some section of social reality"*. That is, they served a critical, rather than a constitutive role. Accordingly, Ulrich reformulated Churchman's categories in an attempt to trace a link between each category and specific synthetic *a priori* boundary judgements constitutive of knowledge of systems.

In terms of social roles, Ulrich distinguished between the **involved** (in the planning process) and the **affected** (by the plans)[65]. In terms of the involved, Ulrich once again looked to Churchman and distinguished between three sources of influence:

1. **Sources of Motivation** (Clients): Who contributes (ought to contribute) the necessary sense of direction? Whose purposes are to be served?
2. **Sources of Control** (Decision Makers): Who contributes (ought to contribute) the necessary means, resources and decision authority? Who has the power to decide?

[65] These roles are not necessarily mutually exclusive. It is possible to be both affected and involved. Accordingly, Ulrich redefines the affected (so as to make each group distinct) as: those affected **but not involved.**

3. **Sources of Expertise** (Planners): Who contributes (ought to contribute) the necessary skills and knowledge? Who has the know-how to do it?

Obviously, these three groups are not intended to be mutually exclusive. However, according to Ulrich, they are intended to be exhaustive of the sources of influence of the involved.

In terms of the affected, Ulrich acknowledges that the sheer size of this group and the breadth of their concerns makes it impossible to grasp their influence by means of a detailed list of heuristic categories. Accordingly, he limits the claim of heuristic necessity to one category, which he labels **'witnesses'**. The essential aim of this category is to force the planner to adequately trace the normative content of alternative boundary judgements by specifically including the judgements of those who are affected but not involved. Given that they are not involved, by definition, their contribution does not consist in influencing the construction of plans and proposals. Rather, their role can only be understood in terms of a critical consideration of the sources of legitimacy of plans. Thus, a fourth source of influence arises:

4. **Sources of Legitimation** (Witnesses): Who contributes (ought to contribute) to the necessary sense of legitimation? Who has the right to counter?

Having distinguished between four social roles (the 'who' of planning), Ulrich turned his attention to determining their corresponding role-specific concerns (the 'what' of planning) and the effect of these concerns on boundary judgements (the 'so what' of planning).

In regard to the involved (clients, decision makers, planners), Ulrich once again turns to Churchman. The clients, of course, are concerned with their **purposes** for undertaking the social action. Accordingly, their purposes will have an effect on boundary judgements by way of their monological understanding of what constitutes an **improvement**. The decision makers, on the other hand, are concerned with the decisions required to achieve these purposes. As such, they have an interest in the **control** of those aspects of the system that they can change (what Churchman calls components). Accordingly, their area of control (components) will have an effect on boundary judgements by way of their monological understanding of what constitutes their **decision environment**. Finally, the planners are concerned with the skills and knowledge necessary for achieving the client's purposes. As such, they have an interest in the **expertise** required. Moreover, the planner's determination of what expertise is required will have an effect on boundary judgements by way of their monological understanding of how the various expertise required can **guarantee** improvement.

In regard to the affected (witnesses), Ulrich turns to Habermas' (1968) understanding of the emancipatory interest (a.k.a. his theory of knowledge constitutive interests) and Churchman's understanding of the effects of *a priori* worldviews on systems design. The witnesses, as we have seen, are concerned fundamentally with the sources of legitimation of plans. As Ulrich (1983) puts it:

> *"In a democratic society in which every citizen is sovereign, only the voluntary consent of those affected but not involved can possibly legitimise the costs and risks imposed upon them ...the essential point is that the affected must be given a chance of emancipating themselves from being treated merely as means for the purposes of others".*

As such, the witnesses have an interest in their own **emancipation** from the imposition of plans. Accordingly, their determination of what is in their interests will have an effect on boundary judgements by way of monologically imposing their **worldviews** in critiques of proposals.

From the above, Ulrich settles on twelve *a priori* concepts of understanding constitutive of social systems design. These are: client, purposes, measures of improvement, decision makers, components, decision environment, planners, expertise, guarantor, witnesses, emancipation, worldviews. Table 2, below, summarises these twelve categories.

Social System (To Be Bounded)	Involved	Sources of Motivation	Clients
			Purpose
			Measure of Improvement
		Sources of Control	Decision Makers
			Components
			Decision Environment
		Sources of Expertise	Planners
			Expertise
			Guarantee of Improvement
	Affected	Sources of Legitimation	Witnesses
			Emancipation
			Worldview (*Weltanschauung*)

Table 2: Ulrich's Twelve Critically Heuristic Categories (*A Priori* Concepts of Understanding) For Surfacing the Normative Content of Social System Design

Moreover, Ulrich (1987) suggests that these categories may be used as a set of **boundary questions** for unfolding the normative content of existing or proposed social system designs. By so doing, Ulrich provides the means (heuristic support) for planners and concerned citizens to reflect upon the presuppositions that enter into social system designs and thereby aids the development of reflective practice and research within the systems community. Furthermore, in a move reminiscent of Kant's distinction between pure (what is) and practical (what ought to be) reasoning, Ulrich (1987) argues that the critical relevance of these questions is best seen by contrasting them in and **'is' mode** and an **'ought' mode**. Thus, Ulrich helps guide the reflective practice suggested by his twelve concepts towards a consideration of **the ethics of drawing boundary judgements** (a.k.a. Churchman's ethics of the whole system). The twelve boundary questions are given in Table 3 below.

Once the normative assumptions associated with boundary judgements have been surfaced the problem of justifying them remains (especially given Churchman's argument that no 'purely rational', or foundational, justification is possible). This is a very important point! It is often assumed that critical, self-reflection on behalf of the systems expert (i.e. such as that suggested by the self-reflective use of the twelve boundary questions) provides the justification for the normative content of systemic inquiry. However, this is entirely mistaken. The problem of justifying the normative content once it has been self-reflectively (or otherwise) surfaced is quite a different one from the problem of rendering explicit the normative content. In regard to this problem, Ulrich proposed to follow up his critically heuristic turn with a **dialectical turn**.

Whilst not as obvious as in the critically heuristic turn, Churchman's influence is also found in Ulrich's dialectical turn. In a 1988 festschrift paper, Ulrich unfolds Churchman's 'process of unfolding'. According to Ulrich (1988b), what Churchman was concerned with was unfolding the nature of the different boundary judgements of the involved and the life practical implications of them for the affected. That is, he was interested in the dialectical interplay between the systems rationality of the involved and the social rationality of the affected.

	'ought' – mode	**'is' - mode**
Sources of Motivation	Who ought to be the **client** (beneficiary) of the system to be designed or improved?	Who is the actual **client** of the system design (i.e. who belongs to the group of those whose purposes, interests and values are served)?
	What ought to be the **purpose** of the system (i.e. what goal states ought the system be able to achieve so as to serve the client)?	What is the actual **purpose** of the system (as being measured not in terms of the declared intentions of the involved but in terms of the actual consequences)?
	What ought to be the system's **measure of success** (or improvement)?	What, judged by the design's consequences, is its built in **measure of success**?
Sources of Control	Who ought to be the **decision maker** (i.e. have the power to change the system's measure of improvement)?	Who is actually the **decision maker** (i.e. who can actually change the measure of success)?
	What **components** (resources and constraints) of the system ought to be controlled by the decision maker?	What **components** (resources and constraints) of the system are actually controlled by the decision maker?
	What resources and constraints should not be controlled by the decision maker (i.e. what ought to be his/her decision **environment**)?	What resources and constraints are not controlled by the decision maker (i.e. what is his/her decision **environment**)?
Sources of Expertise	Who ought to be involved as the **designer / planner** of the system?	Who is actually involved as the **designer / planner** of the system?
	What kind of **expertise** ought to flow into the design of the system (i.e. who ought to be considered an expert and what should his/her role be)?	What kind of **expertise** is flowing into the design of the system (i.e. who is involved as expert and what role does s/he play)?
	Where ought the involved seek the **guarantee** that their planning (designs) will be successful (judged by the system's measure of success)?[66]	Where do the involved seek the **guarantee** that their planning (designs) will be successful (judged by the system's measure of success)?[66]
Sources of Legitimation	Who ought to belong to the **witnesses** (who represent the concerns of the citizens that will or might be affected by the design of the system)? That is to say, who among the affected ought to be involved?	Who, if anyone, does belong to the **witnesses** (who represent the concerns of the citizens that will or might be affected by the design of the system)? That is to say, who among the affected is involved?
	To what degree and in what way ought the affected be given a chance of **emancipation** from the premises and promises of the involved? Ought the affected be considered a client?	To what degree and in what way are the affected being given a chance of **emancipation** from the premises and promises of the involved? Are the affected being considered as clients?
	What **worldview(s)** ought to underly the system's design? Should the worldview(s) of the affected be incorporated?	What **worldview(s)** are underlying the system's design? Are the worldview(s) of the affected being incorporated?

Table 3: The Twelve Critically Heuristic Boundary Questions for Rendering Explicit the Normative Premises that Inevitably Flow into Social System Design

66 Are the involved seeking the guarantee of improvement in the theoretical competence of experts? In consensus amongst experts? In the validity of empirical data? In the relevance of mathematical models or computer simulations? In political support on the part of interest groups? In the experience and intuition of the involved? Can these assumed guarantors secure the design's (plan's) success, or are they false guarantors?

The dialectic (or contradiction) between these two groups arises from the different normative assumptions made by them. At the heart of the 'process of unfolding', therefore, is the problem of justifying the normative content of systems designs and plans. According to Ulrich (who, in turn, utilised Habermas), the problem of justifying the normative content of systems designs and plans cannot be 'solved' monologically but can only ever be 'resolved' through some sort of participative debate between the involved and the affected. Ulrich's dialectical turn, therefore, seeks to stimulate a 'process of unfolding' in which the 'systems rationality' of the involved is only ever justified in dialectical dialogue with the 'social rationality' of the affected. According to Ulrich, what is required is for the involved and affected to enter into dialogue, and thereby drive the dialectical 'process of unfolding' through challenging each other to make transparent the normative presuppositions and implications of their respective positions.

The 'process of unfolding', therefore, seeks to arbitrate between the professional competence of planners and the democratic participation of citizens, by following Kant's (1781) suggestion that:

> *"Reason must in all its understandings subject itself to criticism; should it limit freedom of criticism by any prohibitions, it must harm itself, drawing upon itself a damaging suspicion. Nothing is so important for its usefulness, nothing so sacred, that it may be exempted from this searching examination, which knows no respect for persons. Reason depends on this freedom for its very existence. For reason has no dictatorial authority; its verdict is always simply the agreement of free citizens, of whom each one must be permitted to express, without let or hindrance, his objections or even his veto."*

Thus, the dialectical turn is secured by allowing the affected the opportunity to challenge the normative validity of the plans proposed by the involved. Moreover, Ulrich arms concerned citizen's with the ability to employ his boundary questions in a polemical mode in situations where planners are unwilling to take the critically heuristic turn (i.e. render explicit their boundary judgements) or the dialectical turn (i.e. seek justification for these judgements through dialogue). As such, he affords a 'symmetry of helplessness' to both the involved planners and affected citizens with respect to dogmatic pronouncements and, thereby, allows concerned citizens to emancipate themselves from technocratic incursions of the systems rationality of the planners into the social rationality of their lifeworlds.

10.2.4 Churchman's Influence on Gerald Midgley

One researcher who has responded to Churchman's call for systems thinkers to adopt a critically reflective attitude in regard to boundary judgements is Gerald Midge. Following Churchman and Ulrich, Midgley set himself the task of considering what happens in practice when there is a conflict between different groups of people making different boundary judgements. To begin with, Midgley considers the scenario that unfolds when one group set their boundaries narrower than another. In such a situation, the gap between the two boundaries will furnish elements considered important to the problem (or decision) at hand by one group and irrelevant by the other. According to Midgley (2000), conflict over the inclusion or exclusion of these **'marginal elements'** is often unresolvable, as the respective positions are driven by different value systems. What tends to happen is that the conflict creates a situation where marginal elements are either radically valued (by those arguing for the broader scope) or radically de-valued (by those arguing for a narrower scope). Thus, by the very existence of boundary conflict, marginal elements tend to take on an importance above and beyond that which they would have taken if no conflict existed. Borrowing his terminology from anthropology, Midgley (2000) labels the radically valued marginal elements **'sacred'** and the radically devalued marginal elements **'profane'**. "Moreover, he argues that **when the broader scope prevails, marginalised elements become sacred** by assuming a disproportional importance, whilst **when the narrower scope prevails, marginalised elements become profane** and are actively disparaged.

An example offered by Midgley (2000) to clarify the above, is the process of marginalisation that takes place due to conflict over the normative bases of **democracy** (informed by social liberalism) and **capitalism** (informed by economic liberalism)[67]. According to Midgley (2000), whilst the capitalist and democratic value systems are, in the main, mutually supportive, occasionally they come into conflict. One of these conflicts concerns the

[67] This is not the terminology of Midgley (2000). In particular, Midgley refers to what I have called 'democracy' as 'liberalism'. The reason for changing the terminology is two-fold. **First**, it is the opinion of the author that the liberal tradition of John Locke, Thomas Hobbes, Edmund Burke, John Stuart Mill, David Hume and Immanuel Kant (which

status of the unemployed. Whereas democracy values all people equally due to their status as 'citizens', capitalism values people differently according to the 'market value' of their (commodified) labour. Obviously, the value of unemployed citizens remains an area of conflict. According to Midgley (2000), this particular conflict stabilises only when a 'sacred' or 'profane' status is imposed on the unemployed. If the capitalist boundary judgement (i.e. the narrower boundary) prevails, then citizens tend to be valued according to the processes of labour commodification. This justifies regarding the unemployed as 'profane' and promotes programs designed to make the unemployed earn any benefits they may receive (such as the current Australian Government's 'Work for the Dole' program). However, if the democratic boundary judgement (i.e. the broader boundary) prevails, then citizens tend to be valued equally due to their equal status as citizens (according to the one person, one vote rule). This justifies regarding the unemployed as 'sacred' and promotes programs designed to support the unemployed through social inclusion and training.

As the above example indicates, it is often difficult to obtain consensus in conflict over boundary judgements. More often than not, both sides solidify their respective positions through reinforced patterns of behaviour and expression (i.e. social ritual). Ultimately, each group may consider the other group as profane. This can have disastrous consequences when different groups need to cooperate together over a shared problem but bring significantly different boundary judgements to it. Midgley (2000) notes that this kind of situation often obtains in multi-agency working in the public sector where Government's demand a 'whole-of-government' approach to issues and problems they see as pressing.

A prominent example in Australia of conflict over boundary judgements within a multi-agency setting in recent years is the whole-of-Government approach to stopping asylum seekers reaching Australia's shore during the last Federal election campaign. The authoritative account of this series of events is found in Marr and Wilkinson (2003). However, for the interested reader, a brief overview is provided here.

---

emphasised the sovereignty of individual citizens over that of monarchs) can be seen to inform **both** capitalism and democracy. Whereas in the capitalist system of thought, the primacy of the individual is secured by the autonomy of market forces (i.e. individuals free choice to value or otherwise goods and labour), in the democratic system of thought, the primacy of the individual is secured by the equality of citizen's (as expressed in the one person, one vote rule). The **second** reason is that others, such as Habermas (1981a,b), discuss conflict between these two value systems by using the terminology 'capitalism' and 'democracy' (e.g. his theory of the colonisation of the lifeworld by the system).

A few weeks prior to the announcement of the election, the Australian Rescue Coordination Centre (RCC), responded to an SOS from a small Indonesian vessel stranded in International waters close to Christmas Island (Aus) by sending out a call for help to shipping in the region. At this point, however, the Department of Immigration and Multicultural Affairs (DIMA) made an unprecedented attempt to interfere in a search and rescue operation by calling RCC Australia and requesting that *"vessels that respond to Australian search and rescue broadcasts tow the stranded vessel to Indonesia"*. The series of events that followed highlight the potential difficulties (and disastrous consequences) that can unfold in whole of Government approaches to problems that are perceived of remarkably differently by different agencies.

To begin with, the ship that responded to the RCC request, the *MV Tampa*, rescued all 438 passengers of the ailing vessel and made its way to nearby Christmas Island in order to allow for the survivors to disembark. However, upon its approach, the *Tampa* was ordered by Australian authorities not to dock and stand twelve nautical miles off shore, following which a remarkable drama unfolded. Under direction from DIMA and the Department of Prime Minister and Cabinet's (DPMC) 'Unauthorised Arrivals Task Force', The Captain of the *Tampa*, Arne Rinnan, was told that the ship was not allowed to land the survivors in Australian Territory. However, facing a serious health crisis on board, including mass dehydration, diarrhoea, skin disease and several unconscious survivors, Rinnan radioed back that he was not going to move and requested immediate medical aid. As such, the *Tampa* entered into a standoff with DIMA and DPMC.

From DIMA's and DPMC's point of view, the survivors could not, under any circumstances, reach Australia's shore (or be treated by civilian doctors ferried out to the *Tampa*), as this would trigger the machinery of the Migration Act. However, from RCC Australia and the Royal Flying Doctor Service's (RFDS) point of view, the medical situation on board the *Tampa* was the first and foremost problem that needed to be solved, the easiest means of which would be to disembark the survivors at Christmas Island and allow for their treatment at the local hospital. During the subsequent standoff, the survivors were left without medical attention for several days causing their condition to deteriorate remarkably. Eventually, faced with a shortage of essential supplies, Rinnan broadcast a Pan Pan: an urgent message concerning the safety of a vessel and those on board, which requested *"urgent medical assistance"*. The RFDS scrambled, but was informed that Cabinet had arranged for a military solution. Thus, the elite Special Air Service (SAS) were ordered to board the ship. Accompanying them, was a military doctor who treated the sick and arranged for their eventual removal to the troop carrier *Manoora*, which was headed for Nauru and a new Australian built and funded offshore detention camp for would-be asylum seekers. Thus, the 'Pacific Solution' was born.

Shortly before the *Tampa* standoff, the Australian Government had embarked on a significant off-shore deterrent program involving DIMA, the Australian Federal Police (AFP) (in collaboration with their Indonesian counterparts, the INP) and the Australian Secret Intelligence Service (ASIS). Coinciding with DIMA sponsored media releases in Indonesia (advertising the perils of travelling to Australia by boat), the AFP and ASIS recruited local agents to actively disrupt the operations of Indonesian people smugglers. The details of these operations are still clouded in secrecy. However, it has been reported that amongst the more unsavoury practices of the local operatives involved in Australian sponsored disruption operations were the posing as people smugglers and subsequent robbing of asylum seekers of all their possessions (so they could no longer afford to pay a genuine smuggler) and the sabotaging of boats (so they could not embark on the journey, or worse, fail to reach Australia after embarking).

Even if only a small amount of what has been reported to occur actually did occur, then this would amount to a significant conflict over boundary judgements between agencies such as the AFP (which must always operate under local laws) and the Australian Human Rights and Equal Opportunity Commission (which must report all cases of known human rights abuse regardless whether the perpetrators are the Australian Government) on the one hand, and ASIS (which is not limited to operate under local laws) and the local (Indonesian-based) agents recruited for the disruption operations on the other. All of this came to a head when Indonesian officials raided the home of an expat Australian in West Timor working as a local agent. In Marr and Wilkinson's (2003) words:

> *"Suddenly, difficult questions about the Australian operation were being asked at senior levels of the Indonesian Foreign Ministry. These included questions about Enniss [the agent in question], about Australia's aid to select Indonesian police officials and the conduct of the disruption program ... Australia's operations up there were threatening to fall apart".*

To add to the melange of agencies involved in the deterrent program, following the arrival of the *Tampa*, the Royal Australian Navy (RAN) launched Operation Relex. The idea behind Operation Relex was to deploy a significant Naval presence (including three warships, ten patrol boats, a supply ship, a transport vessel, several seahawk helicopters and a P3-C Orion surveillance aircraft) to blockade the Indian Ocean and prevent any further refugee arrivals. Where possible, the RAN were tasked with intercepting, boarding and returning vessels back to International waters. From the beginning, conflicts over boundary judgements between the RAN and the DPMC emerged. Chief of these inter-agency conflicts were the conflicts that arose over the legal requirements set out by the Navigation Act and the RAN's Maritime Commander's Orders which required the RAN to operate at all times under the imperatives set forth by SOLAS (Safety Of Life At Sea) regulations. According to the RAN, the over-riding concern at all times was that no life was to be lost at sea. How people find themselves in trouble was irrelevant. If the RAN were patrolling waters nearby a vessel in trouble they would be compelled to rescue survivors and arrange for immediate medical aid (which would usually involve bringing them back to the nearest Australian port). However, from the perspective of the RAN's political masters (and the Department's of Prime Minister and Cabinet and Immigration and Multicultural Affairs) whilst it was important that the Navy fulfil its obligations under SOLAS, the over-riding imperative was that no refugee land on Australian soil (whether in need of medical aid or not). This inter-agency conflict resulted in the suspension of the usual reporting arrangements. Naval operational information was passed directly to Government and the existing arrangements for the Department of Defence to brief the press were excised.

As the above examples demonstrate, whole-of-Government approaches to complex social problems are not necessarily the panacea that the political classes hope them to be. In fact, they may result in disastrous unforeseen consequences brought about by differences of opinion over boundary judgements. No recent event has demonstrated this more, in an Australian context, than the circumstances that led to the death of 353 refugees on board what became known as SIEV X.

Overcrowded and unprepared for the perilous journey that lay ahead, SIEV X slowly slipped out of the Bay of Lampung on its way to Christmas Island in the early hours of the morning on Thursday October 18 2001. Almost immediately, local AFP operatives alerted Coastwatch that an overcrowded vessel carrying asylum seekers had left Indonesia and that the vessel was considered 'at risk'. In fact, according to sources in Canberra, SIEV X had already been a target of the covert ASIS disruption operation prior to leaving Indonesia (Marr and Wilkinson, 2003). Regardless, Coastwatch was aware of the precarious condition of SIEV X from the moment of its departure. However, due to the nature of the whole-of-Government response to people smugglers, rather than liaising with RCC Australia to ensure the safe passage of the vessel, Coastwatch alerted the Naval commanders of Operation Relex who were, at the time, sending out surveillance flights over the approach to Christmas Island with the intention of intercepting approaching vessels early on.

The chief concern for Relex, as demanded by the Minister of Defence and his cabinet colleagues was that no vessel slipped passed the blockade. Accordingly, surveillance flights focussed heavily on the area closest to Christmas Island. After hearing about the expected arrival of SIEV X from Coastwatch, a P3-C surveillance flight was ordered over the approach, once again focussing heavily on the seas closest to Australian shores. Strangely, the boat was not found to be where it should have been. However, as far as Operation Relex was concerned a late boat was a good sign. They could safely wait until morning in the knowledge that the boat would not be able to reach Christmas Island for many hours yet. At this point it may be useful to point out that if the whole-of-Government approach to people smuggling was not in operation, Coastwatch and RCC Australia would not have passed responsibility for SIEV X over to Operation Relex and would have treated the voyage of the fateful vessel as a search and rescue operation. The fact that the vessel was late, or not where it should have been, would have, under normal circumstances, sparked an intensive search.

The fate of SIEV X will probably go down in Australian political folklore. As is well known, the vessel succumbed to the sea sometime on the morning of Friday October 19, leaving 399 asylum seekers, most without lifevests, at the mercy of the ocean for over 24 hours. Eventually, 46 of the original 399 were rescued by a passing Indonesian fishing boat and taken back to Jakarta. 353 died. Coastwatch and RCC Australia only learned of the disaster when the few that did survive disembarked in Jakarta to tell their stories to the world.

This kind of story brings home the reality of how important boundary judgements can be in a way that the philosophical language of pragmatism could never hope to. **No wonder Ulrich contended that besides 'intellectual honesty', the other defining characteristic of Churchman's life and work was 'moral outrage'.**

10.3 CONCLUSIONS

No thinker has had a greater influence on the systems movement over the course of their intellectual life than Charles West Churchman. Under Churchman's watchful eye, the systems community has adopted many of the epistemic precepts of pragmatism whilst remaining, in the main, unaware about the content of the American pragmatic discourse, or that of the broader epistemic dispute between the rationalists and empiricists, which the pragmatist saw themselves as dissolving. By so doing, Churchman has helped move systems thinking into step with the broader intellectual spirit of our post-modern age. Old precepts such as subject-object dualism, the unity of scientific method and the steady progress of science towards correspondence-truth have been set aside as the systems community has attempted to come to terms with the contextual, contingent and plural nature of systems.

Following Churchman, systems thinking (by virtue of the problematic constitution of system boundaries) has dissolved into (critical) systems thinking. By critiquing its central concept (the concept of holistic inquiry), Churchman challenged systems practitioners everywhere to be critically self-reflective about the *a priori* assumptions they must inevitably make. Anything less than this is dogmatic and naïve. Churchman was perhaps the first thinker to really understand this and by sharing this understanding with the systems community he has influenced an entire generation of systems practitioners. Let's hope that his influence will continue to be felt well into the twenty first century, helping to shape both the **epistemic basis** and **ethical motivation** of generations of systems theorists to come.

REFERENCES

Ackoff, R., 1961, The meaning, scope and methods of OR, *Progress in OR*, 1: 3-34.

Ackoff, R., 1973, Science in the systems age: beyond IE, OR and MS, *Operations Research*, 21:661-671.

Ackoff, R., 1974, The systems revolution, *Long Range Planning*, 7:2-20.

Ackoff, R., 1977, Optimization + objectivity = opt out, *European Journal of Operational Research*, 1:1-7.

Ackoff, R., 1978, *The Art of Problem Solving*, Wiley, New York, USA.

Ackoff, R., 1979a, The future of operational research is past, *Journal of the Operational Research Society*, 30:93-104.

Ackoff, R., 1979b, Resurrecting the future of operational research, *Journal of the Operational Research Society*, 30:189-199.

Ackoff, R., 1988, C. West Churchman. *Systems Practice*, 1:351-355.

Ackoff, R., 2001, OR: after the post mortem. *System Dynamics Review*, 17:341-346.

Bonder, S., 1979, Changing the future of operations research, *Operations Research*, 27:209-224.

Britton, G. and McCallion, H., 1994, An overview of the Singer/Churchman/Ackoff school of thought. *Systems Practice*, 7:487-521.

Churchman, C.W., 1966, Perception and deception, *Science*, 153:1088-1090.

Churchman, C.W., 1967, Wicked problems, *Management Science*, 14:141-142.

Churchman, C.W., 1968a, *The Systems Approach*, Dell, New York, USA.

Churchman, C.W., 1968b, *Challenge to Reason*, McGraw Hill, New York, USA.

Churchman, C.W., 1970, Operations research as a profession, *Management Science*, 17:B37-B53.

Churchman, C.W., 1971, *The Design of Inquiring Systems: Basic Concepts of Systems and Organisation*, Basic Books, New York, USA.

Churchman, C.W., 1974, 'Philosophical speculations on systems design', *Omega (The International Journal of Management Science)*, 2:451-465.

Churchman, C.W., 1979, *The Systems Approach and its Enemies*, Basic Books, New York, USA.

Churchman, C.W., 1980, The status of the systems approach. *Omega (The International Journal of Management Science)*, 8:277-279.

Churchman, C.W., 1981, Trends in systems theory, in: *Measurement for Management Decision*, R. Mason, and E. Swanson, eds., Addison Wesley, Menlo Park, USA.

Churchman, C.W., 1982, *Thought and Wisdom*, Intersystems Publications, Seaside, USA.

Churchman, C.W. and Ackoff, R., 1950, *Methods of Inquiry: An Introduction to Philosophy and Scientific Method*, Educational Publishers, St Louis, USA.

Churchman, C.W., Ackoff, R. and Arnoff, E., 1957, *Introduction to Operations Research*, Wiley, New York, USA.

Davidson, D., 1984, *Inquiries into Truth and Interpretation*, Clarendon Press, Oxford, UK.

Dewey, J., 1909, *The Bearing of Pragmatism Upon Education. In The Middle Works: 1899 – 1924*, Vol. 4, 1976, J. Boydston, ed., Southern Illinois University Press, Carbondale, USA.

Dummett, M., 1976, What is a theory of meaning, in: *Truth and Meaning*, G. Evans and J. McDowell, eds., Oxford University Press, Oxford, UK.

Dummett, M., 1978, *Truth and Other Enigmas*, Duckworth Publishers, London, UK.

Dummett, M., 1981, *The Interpretation of Frege's Philosophy*, Harvard University Press, Cambridge, USA.

Dummett, M., 1982, Realism, *Synthese*, 52:55-112.

Eilon, S., 1980, The role of management science, *Journal of the Operational Research Society*, 31:17-28.

Fildes, R. and Ranyard, J., 1998, The foundation, development and current practice of OR: an editorial introduction and overview, *Journal of the Operational Research Society*, 49:304-306.

Flood, R. and Jackson, M., 1991, *Creative Problem Solving: Total Systems Intervention*, Wiley, Chichester, UK.

Gadamer, H-G., 1960, *Truth and Method*, Translated by G. Barden, and J. Cumming, Cross Road, New York, USA. This edition first published, 1984.

Habermas, J., 1968, *Knowledge and Human Interest*, Translated by J. Shapiro, Beacon Press, New York, USA. This edition, 1972.

Habermas, J., 1981a, The theory of communicative action, Volume 1, *Reason and the Rationalisation of Society*, Translated by T. McCarthy, Beacon Press, Boston, USA. This edition, 1984.

Habermas, J., 1981b, The theory of communicative action, Volume 2, *Lifeworld and System: A Critique of Functionalist Reason*, Translated by T. McCarthy, Beacon Press, Boston, USA. This edition, 1984.

Jackson, M.C., 2000, *Systems Approaches to Management*, Kluwer Academic/Plenum Publishers, New York, USA.

James, W., 1897, *The Will to Believe: And Other Essays in Popular Philosophy*, Harvard University Press, Cambridge, USA. This edition first published in 1956.

James, W., 1902, *The Varieties of Religious Experience*, Harvard University Press, Cambridge, USA. This edition first published in 1975.

James, W., 1907, *Pragmatism: A New Name for Some Old Ways of Thinking*, Harvard University Press, Cambridge, USA. This edition first published in 1975.

James, W., 1909a, *The Meaning of Truth*, Harvard University Press, Cambridge, USA. This edition first published in 1975.

James, W., 1909b, *A Pluralistic Universe*, University of Nebraska Press, Lincoln, USA. This edition first published in 1996.

Kant, I., 1781, *Critique of Pure Reason*, Translated by N. Kemp Smith, Macmillan, London, UK. This edition first published in 1929.

Kilman, R., 1983, The costs of organisation structure: dispelling the myths of independent divisions and organisation-wide decision making, *Accounting, Organizations and Society*, 8:341-357.

Kirby, M., 2000, Operations research trajectories: the Anglo-American experience from the 1940s to the 1990s. *Operations Research*, 48:661-670.

Kirkham, R., 1997, *Theories of Truth: A Critical Introduction*, MIT Press, Cambridge, USA.

Marr, D. and Wilkinson, M., 2003, *Dark Victory*, Allen & Unwin, Crows Nest, Australia.

Mason, R., 1988, Exploration of opportunity cost and consideration for future generations: two lasting contributions to systems thinking, *Systems Practice*, 1:367-375.

Mason, R. and Mitroff, I., 1981, *Challenging Strategic Planning Assumptions*, John Wiley & Sons, Chichester, UK.

Matthews, D. 2004, Re-thinking Systems Thinking: towards a postmodern understanding of the nature of systemic inquiry, PhD Thesis, University of Adelaide, Australia.

Menand, L., 2001, *The Metaphysical Club*, Flamingo Press, London, UK.

Midgley, G., 2000, Systemic intervention: philosophy, methodology and practice, *Contemporary Systems Thinking Series*, Kluwer Academic/Plenum Publishers, New York, USA.

Muller-Merbach, H., 1984, Inter-disciplinarity in operational research: in the past and in the future, *Journal of the Operational Research Society*, 35:83-89.

Muller-Murbach, H., 1988, Mutual understanding, revisited after nearly 25 years, *Systems Practice*, 1:385-397.

Peirce, C., 1877, The fixation of belief, *Popular Science Monthly*, 12:1-15.

Peirce, C., 1905, What pragmatism is, *The Monist*, 15:161-181.

Peirce, C., 1932, *Collected Papers*, C. Hartshorn, P. Weiss, and A. Burks, eds., Harvard University Press, Cambridge, USA.

Putnam, H., 1981, *Reason, Truth and History*, Cambridge University Press, Cambridge, UK.

Quine, W.V.O., 1969, *Ontological Relativity and Other Essays*, Columbia University Press, New York, USA.

Rorty, R., 1967, *The Linguistic Turn: Essays in Philosophical Method*, University of Chicago Press, Chicago, USA.

Rorty, R. 1979. *Philosophy and the Mirror of Nature.* Princeton University Press, Princeton, USA.

Rorty, R., 1982, *Consequences of Pragmatism*, University of Minnesota Press, Minneapolis, USA.

Rorty, R., 1991a, *Objectivity, Relativism and Truth: Philosophical Papers, Volume 1*, Cambridge University Press, Cambridge, UK.

Rorty, R., 1991b, *Essays on Heidegger and Others: Philosophical Papers, Volume 2*, Cambridge University Press, Cambridge, UK.

Rorty, R., 1999, *Philosophy and Social Hope*, Penguin Books, New York, USA.

Schiller, F., 1903, *Humanism: Philosophical Essays*, Macmillan, New York, USA. This edition, 1912.

Schiller, F., 1907, *Studies in Humanism*, Macmillan, New York, USA.

Strawson, P., 1964, A problem about truth: a reply to Mr Warnock, in: *Truth,* G. Pitcher, ed., Prentice-Hall, Englewood Cliffs, UK.

Ulrich, W., 1983, *Critical Heuristics of Social Planning: A New Approach to Practical Philosophy*, Haupt Academic Publishers, Bern, Switzerland.

Ulrich, W., 1987a, Critical heuristics of social systems design, *European Journal of Operational Research*, 31:276-283.

Ulrich, W., 1988a, C. West Churchman – 75 years, *Systems Practice*, 1:341-350.

Ulrich, W., 1988b, Churchman's process of unfolding: its significance for policy analysis and evaluation, *Systems Practice*, 1:415-428.

Ulrich, W., 1994, Can we secure future-responsive management through systems thinking and design?, *Interfaces*, 24:26-37.

Ulrich, W., 1999, *An Appreciation of C. West Churchman.* http://www.isss.org/lumCWC.htm

Ulrich, W., 2001, The quest for competence in systemic research and practice, *Systems Research and Behavioural Science*, 18:3-28.

Ulrich, W., 2002, Boundary critique, in: *The Informed Student Guide to Management Science*, H. Daellenbach and R. Flood, eds., Thomson, London, UK.

Wittgenstein, L., 1922, *Tractatus Logico-Philosophicus,* translated by C.K. Ogden, Routledge, London, UK.

Wittgenstein, L., 1953, *Philosophical Investigations,* translated by G. Anscombe, Blackwell Publishers, Oxford, UK.

Chapter 11

CHURCHMAN AND MEASUREMENT

TIMOTHY L.J. FERRIS

Abstract: Churchman's main contribution to the theory of measurement flows from bringing together the work of a number of scholars. This action brought a multidisciplinarity to the theory of measurement that was new, and led to many later scholars viewing measurement more holistically than was previously the case. "In both the UK and USA, scientists from a wide variety of disciplinary backgrounds came together during the second World War to support the planning of military operations. They mainly applied mathematical modelling techniques to find optimal solutions to complex problems. Churchman (1987) provides a typical example: he describes how he applied mathematics to the problem of determining an optimal quality control procedure in the manufacture of ammunitions. After the war some OR [operations researchers] continued to work in defence, but most went into civilian positions in the public and private sectors...First OR came to be seen as largely synonymous with 'management science'. (Midgley, 2004:4). But this is no longer the case as critical questioning has become central to both better risk management and greater accountability.

Key words: wholistic, multidisciplinary, measurement.

11.1 BACKGROUND

C. West Churchman completed a PhD in philosophy prior to World War II and during the war was engaged in statistical work related to the production of small arms ammunition. At the same time the field of Operations Research was opened up by others to provide a means to address issues in the optimisation of logistics and production processes (Churchman,

Ackoff & Arnoff, 1957). Churchman's work placed him near to the development of Operations Research. This position gave him significant knowledge of the work, practices and effectiveness of Operations Research practitioners but also placed him outside their professional community of practice. As an outsider to the founding community of practice he had a freedom to express divergent views, which he often did. Churchman's background contributed to his ability to position himself as one of the early authors in Operations Research (Churchman, Ackoff & Arnoff, 1957), a leader in the Operations Research, but also led him to have an uneasy relationship with many practitioners of the field (Churchman, 1970).

Churchman, drawing on his own experience, made much of the observation that during World War II the professional contributors to solutions of production and logistics problems often made significant progress when they asked 'stupid' questions (Churchman, 1968). These 'stupid' questions are questions that appear to have answers rooted in the structure of the issue but which often had been left simply unaddressed by the principal stakeholders as the assumptions underlying the approach taken to the issue. This observation and the decision to communicate it by Churchman is evidence of his belief that one must not take any methodologically uncritical approach to addressing issues (Matthews, 2004).

The issue of 'stupid' questions that prise open the intellectual framework surrounding a problem space is important in the development of a truly systemic approach to issues. The traditional analysis of any issue comes from a framework that has appeared useful in the past and has become embedded as the means of viewing the particular issue space. It therefore addresses the nature of a model of the space and over time grows to become the paradigm, using the analysis of Barbour (1974).

Initially the perception of the issue is a means to try to explain observations and thus has the descriptive function of a model. However, over time, and particularly with demonstrated apparent effectiveness of the model in assisting reasoning about the matter, the model is transformed from a construct that is perceived as a useful attempt at description of the matter to an expression of the truth about the matter. Further along, that which is perceived as the truth about the matter is transformed in thought to become the presuppositions about the matter that it is unimaginable to challenge. When this has happened one has a paradigm. The paradigm governs thought about the matter concerning which it makes assertions. It is not possible for the mind to escape the bondage imposed by the paradigm without some kind of externally sourced challenge. The paradigm becomes sufficiently deeply embedded in the person's perception of the world that new information obtained is filtered and rearranged so as to be made compliant to the original paradigm. It is unusual for new data to sufficiently challenge the existing paradigm to have the intellectual leverage to prise open the seal imposed by the paradigm.

An example of such a struggle with a paradigm can be found in the late nineteenth and early twentieth century problem of 'ether' in physics. The discovery that electromagnetic radiation propagates as a wave led to the assumption that, consistent with the experience of other wave phenomena and the paradigm that had developed around the understanding of those phenomena, there was necessarily some medium the electromagnetic waves required for their support and propagation. This medium was named 'ether'. Various experimental observations and thought experiments were made that raised challenges to the assumption that there was such a medium, but for a long time the paradigm held firm. The contrary observations and reasonings led to contortions to enable the maintenance of the original concept of the ether as a necessary medium. The contortions made electromagnetic theory complex and complicated. The breakthrough came when Einstein asked a fundamental question concerning the nature of time, which eventually led to the realisation that electromagnetic waves do not require a medium for their support but rather are self-supporting.

The role of the 'stupid' questions to which Churchman referred was one of challenging the paradigm by being the external idea that provides a key to identify and examine assumptions underlying the current approach to issues. This is a role that can usually only be played by the outsider who can see and understand enough of the issue to be able to ask questions that are sufficiently insightful as to challenge assumptions. However, the process of using questions rather than assertions to challenge assumptions is important because of the social manner in which the discussion is conducted. The questioner is the outsider, who is regarded as naïve concerning the issue, and therefore as not qualified to make challenging assertions. However, the naivety of the questioner may act as a factor that enables the questions to be effective, since the answerer, who is expert in the field wishes to display knowledge and may then discover limitations of the paradigm in the process of trying to give explanation of the model.

Churchman's interest in pointing out that the 'stupid' questions asked broke through the barriers created by the rigidity of thought developed about an issue area in a specialist's mind. Questions are valuable in concerning the nature of thinking systemically about issues. In turn, this perspective is linked to his view of Operations Research as being primarily a systems oriented field, rather than as primarily an application of mathematics. As a systems field Operations Research should involve consideration of issues from a full range of perspectives reflecting the concerns of all the possible stakeholders. The emphasis of Churchman on this point was because of his speaking within a polemical situation in which he was aware of Operations Research having been shaped as a quantitative field, contrary to his understanding of its nature and relevance as a systems science. This point is conveyed in relation to Operation Research program structures by Ackoff, (2001). The difference between Operations Research as a systems science and as a quantitative science was the point of Ackoff's (1987) contrast of market oriented and input-oriented professions. Ackoff's lament was that Operations Research had lost its way and shifted its focus from producing solutions to real and important problems to methodology development and application.

Churchman contrasted the "rational" story of precision and certainty with the "irrational" story of ambiguity and inconsistency (Churchman 1970). The contrast is important in identifying the motivations of participants in Operations Research and Management Science. The 'rational' were those interested in providing precise and accurate analyses of situations to guide actions. This group, naturally, were interested in the quantitative mathematical approaches to Operations Research because the mathematical approaches provide the possibility of giving complete and accurate solutions to problems. The 'irrational' were prepared to tolerate ambiguity and inconsistency because their goal was to develop insight into the issues and to obtain guidance for action even in considerable uncertainty.

11.2 THEORY OF MEASUREMENT

Churchman's interests in Operations Research, even though diverging from the interests of many other practitioners, meant that he addressed a field that is inherently numerical, and requires that one represent the variables affecting conclusions numerically in order to perform analysis to guide decisions. This presents the core problems of the theory of measurement: how states of nature may be represented on scales of measurement; the nature of the scales of measurement; and the consequent range of valid operations that may be performed on those measurements. The theory of measurement is the field of study in which the fundamental nature of measurement is explored. For this reason Churchman was drawn towards the study of the theory of measurement. His training in philosophy enabled him to make some distinctive contributions to the field, which had been dominated by physical scientists and applied mathematicians prior to World WarII, and in the early post-war period was in a state of transition as people in other fields sought to measure new kinds of quantities.

Throughout the history of the theory of measurement there have been contributions by practitioners of several disciplines. The major disciplinary roots of contributors to the theory of measurement are physical sciences and engineering: Helmholtz, Campbell, Bridgeman, Finkelstein, and others; social sciences, particularly psychology: Stevens, Suppes, Tversky, Krantz, and Luce; mathematics: Roberts, Pfanzagl; philosophy: Ellis, Kyburg and others; and economics and management. The reason for theory of measurement attracting people from diverse fields to contribute has been that measurement is a fundamental *a posteriori* epistemic process dealing with the means by which knowledge of reality is obtained. Knowledge of reality is necessary in the various sciences because the subject matter of interest is contingent upon states of nature. Consequently, the analysis of the means of obtaining knowledge of states of nature is critical to the success of the sciences because their success depends on the amount of confidence that can be placed in the conclusions that they derive.

Mathematicians are interested in measurement because the naïve approach to measurement asserts that the numbers obtained in measurement processes can be operated upon as numbers following the usual rules of arithmetic with the result that *a priori* valid constructs upon the *a posteriori* data obtained through measurement may be performed. The naïve assume that this process facilitates valid assertions about the state of nature observed, or the state of nature that would be established if action were taken corresponding to the *a priori* steps applied to the data. Mathematicians are interested in the extent to which the usual rules of arithmetic are valid when applied to measurement data to either make assertions about states of nature that have not actually been observed, or to make plans for future action and manipulation of states of nature to bring effect to situations that support certain desired states. Engineers are concerned with the theory of measurement because engineering is the purpose driven analysis of possible changes to the state of nature that will effect outcomes that are the goal of the engineering development process.

Philosophers are interested in the theory of measurement because measurement is a process by which people obtain knowledge of the world that is measurement is one of the epistemic processes. Since epistemology is one of the on-going interests of philosophers the particular subset of epistemology that pertains to assisted or structured observation of states of nature and the nature and quality of the knowledge that it provides is a natural interest of philosophers.

The earliest commonly cited contribution to the theory of measurement is Helmholtz's lecture series (Helmholtz, 1881). Helmholtz was a physical scientist. He approached the theory of measurement by analogy to Euclidean geometry, observing that there are dimensions in observed entities, and that each dimension can be represented by a number, but that each number can only represent one dimension, with the results that entities cannot be reduced to a single number, and that entities can only be compared with respect to particular dimensions, not as wholes. This work was a valuable first formal discussion of measurement.

The first formal definition of measurement was offered by Campbell (1920):

> "Measurement is the process of assigning numbers to represent qualities: the object of measurement is to enable the powerful weapon of mathematical analysis to be applied to the subject matter of science."

This definition reflects the interest in representing states of nature with numbers in order to permit the application of the full power of mathematical methods to the analysis of the states of nature so represented. The principle advance presented by Campbell's work was to distinguish between nominal, which classifies states of nature, and quantitative measurement, which permits at least some numerical relations to map the physical relations of the entities represented.

Helmholtz' concept of dimensionality was advanced by Bridgman (1922) who originated the method of dimensional analysis. Dimensional analysis is a method used in sciences to ensure that equations purporting to represent relationships between measures of particular aspects, dimensions, of an entity balance in terms of the units, and provide representations of physically meaningful operations.

During the 1930's Campbell chaired a British Association for the Advancement of Science committee investigating whether measurement could be applied to sensory events, such as perceived loudness of sounds, and whether such a scale could be a scale for measurement (Stevens, 1946). The committee provided a report that reflected disagreement, with one contributor rejecting the idea of a scale of sensory perception "until a meaning can be given to the concept of addition as applied to sensation" (Stevens, 1946). The reflection of Stevens on this issue led to the seminal paper already cited in this paragraph, in which Stevens proposed four kinds of measurement scales: Nominal, Ordinal, Interval and Ratio, each of which corresponds to a particular relation between the observed manifestation and the numbers used to represent it, and the structure of relations within the manifestation. This work was important in opening the way for measurement of qualities that had previously been considered as unsuitable for measurement for reasons such as cited in the committee report, and for providing a foundation for integrating the concept of measurement across the physical sciences and the social sciences. Stevens' work led to the development of the Representational Theory of Measurement, which provides a mathematical description of the relations that exist within the observed and in the numbers representing manifestations in a broad range of measurement situations. This work has been extended by researchers of various discipline backgrounds, including notable contributions from Krantz *et al* (1971), Pfanzagl *et al* (1971), and various papers by Ludwik Finkelstein. These contributors span many of the discipline areas with special interest in the theory of measurement. Another major contribution that extended Stevens work was made in philosophy by Ellis (1966). This work is important because it provide a sound elaboration of the nature of the various kinds of scale identified by Stevens.

Stevens' defence of the idea of scales of sensory perception and development of the foundations of the representational theory of measurement led, by the mid 1950's, to people constructing scales of diverse kinds of manifestation. An example is the interest of market researchers to develop a scale of consumer preferences (Coombs, 1954), and the realisation that such a scale would be better if it relied on a direct measure of choice rather than on an indirect measure relying on a model of the market (Papandreou, 1954; Tobin, 1954). This was the broad scheme of the theory of measurement when Churchman entered. This account of the development of the field has prepared the way to discuss the specific contributions of Churchman, and then to look to the future directions that arose from his work.

11.3 CHURCHMAN'S CONTRIBUTION

In 1959 Churchman co-edited, with Ratoosh, a conference proceedings containing a set of contributions to the theory of measurement, *Measurement definitions and theories*, and contributed one chapter to this work (Churchman 1959). This book brought together the work of people with diverse approaches to the matter of measurement, including philosophy, physical science, management science and psychology and included contributions from two seminal thinkers in the field, Stevens (1959) and Suppes (1959), the former already recognised as a major contributor and the latter rising prior to his major achievements.

In Churchman's chapter he commenced by observing that measurement is a fashionable process because it is used by many to demonstrate that their activity is a worthwhile scientific endeavour. This is also linked to the observation that measurement results in a quantitative approach to the subject matter, in contrast to the qualitative approach that is forced when one makes non-quantitative observations. Churchman also states that there was, in 1959, no theory of measurement that could be used as a basis for decision making. Such a theory was still lacking in 1992 (Ferris). The primary reason such a theory continued to be absent was that the majority of work in the theory of measurement was devoted to the development of a clear understanding of the traditional concern of theory of measurement, the relation of states of nature to measures of those states of nature and the kind of operations on the measures that are consistent with the operations that occur within the observed. This work, whilst very important, left a gap in the development of the theory that meant that the theory of measurement had not yet been transformed from a scientific theory to an engineering methodology. The essence of Churchman's concern was to transform the theory of measurement from a field with primarily theoretical and scientific concerns, such as presenting a clear exposition of the nature of measurement because that knowledge is inherently valuable to the engineering concern of using knowledge to support and enable purposeful activity. In this particular case Churchman's concern was the design of measurement systems that would support the making of sound decisions concerning the manner in which one should interact with the world as a result of certain observations. Churchman's concern for the purpose driven issue of guiding decisions reflects his background in the Pragmatist tradition that encouraged the view that all knowledge should be developed and used for a purpose. The only field in which there is a well-developed theory of the interaction of decisions and measurements is feedback control systems. This field involves the use of measurement processes that are, normally, directly connected to the inputs of some control system with the result that the monitored output directly affects the input to the system for the purpose of adjusting the output to be closer to the desired output.

Churchman continued his discussion of measurement systems pointing out that the specification of measurement systems involves determining what is to be measured and what method is to be used to perform that measurement. These problems sound elementary, but are not. In the context of measurement providing quantitative description of states of nature and a culture that values quantitative description above qualitative description it is easy for people to assume that a better understanding of the phenomenon will be obtained through measuring more variables, and this can lead to obtaining large quantities of data but not developing a strategy for the analysis of the data that makes the data of significant value to the holder of the data. At this point, Churchman again revealed his engineering concern, stating that the design of a measurement system must also consider the cost of obtaining particular measures and balance that cost against the benefit obtained through the insight concerning the observed obtained. Another aspect of the theory of measurement to which Churchman referred was the theory of error. The theory of error has developed in parallel with other aspects of the theory of measurement since the origin of the field because it has been recognised that no measurement is perfectly accurate. The purpose of the theory of error is to provide a foundation for determining the confidence with which one can hold that the measured value is within a certain range around the value that was measured, and therefore the reliance that one may put on the measure when making decision about action based on the measure.

Among the other contributors to this book Caws presents the physicists view of measurement as the essential foundation that provides the link and the separator between physics and mathematics (Caws, 1959). It is by measurement that, following Campbell's definition of measurement, mathematics can be used to describe the physical world, but it is also measurement which maintains the separation of kind between mathematics, a set of processes that can be investigated as an *a priori* conceptual structure, and physics which concerns the physical world which can only be known *a posteriori*.

Stevens' contribution (Stevens, 1959) was largely an expansion of his earlier paper (Stevens, 1946). Stevens' notes that definitions of measurements had been progressively broadened in response to the effort to measure a growing variety of kinds of manifestation, and particularly named his own contribution in the area of measurement of subjective magnitudes and the work of others in the measurement of 'utility', the subjective scale of the value of things. Kircher (1959) approached measurement from the viewpoint of management, a starting position close to that of Churchman. He stated that the first step in establishing a measurement process is to define the objective of the management process to which the measurements will contribute. This point is close to Churchman's concluding point that ultimately measurement is about obtaining information that will be used as the foundation of control of some outcome situation (Churchman, 1959).

Suppes' (1959) contribution emphasized the need for measurement to involve a comparison between the observed and either a unit of magnitude or some reference magnitude to be meaningful. A measurement must be meaningful in order to contain information that is useful to the holder of that information. Since measurement concerns quantitative description of the observed it is necessary for that quantitative description to contain reference to some magnitude of reference. It is not particularly important whether the reference magnitude is a publicly accepted reference magnitude, a unit, or a private reference, because, at least in principle, the private reference could be calibrated to a public reference unit.

Pap considered whether physical magnitudes are operationally definable (Pap, 1959). In the theory of measurement 'operationalism' refers to the concept that the measurement process is established first and when applied produces outputs, the measures of states of nature, which are defined to be whatever comes out of the specified measurement process. Such an approach permits arbitrariness in the definition of physical manifestations because they are defined in a manner dependant on the measures of them that are obtained. The alternative view is that manifestations are inherently something, and are defined first. The process of measurement then requires the design of means to objectively and meaningfully obtain measures of the manifestation as defined. The difficulty with this issue, ultimately, is circularity. If we claim to use a non-operational approach to measurement, then the basis on which we define quantities that we then seek to determine means to measure is our knowledge of other quantities. However, our knowledge of those other quantities is developed through processes of observation and measurement accepted as consistent with the scientific method.

The conference proceedings edited by Churchman and Ratoosh brought together the work of people from diverse backgrounds. This was a significant achievement because previously the contributions to the theory of measurement coming from the various disciplines were published in discipline specific venues and rarely read across disciplines. The achievement of bringing together practitioners of the disciplines with an interest in the theory of measurement enabled the beginning of the recognition of the theory of measurement as a multidisciplinary field of endeavour. The contributions of each of the authors made modest advances in the state of the disciplinary knowledge of measurement, as discussed above. The major value of the work was in having these contributions brought together in a single volume, thereby providing the reader with the sense that the contributions belonged together as parts of a field of discourse and providing a state of the art account of the knowledge within each of the contributing areas. This important contribution of Churchman and Ratoosh has not previously been highlighted, because as is common in disciplinary discourses, the later contributions to the theory of measurement have tended to look to the seminal works of Helmholtz (1881), Campbell (1920), Bridgman (1922) and Stevens (1946). These authors have a rightful place as seminal contributors to the theory of measurement and fully deserve the recognition that they receive. But Churchman's contribution as the person, who motivated through his systemic perspective on analysis of issues, worked to bring together the diverse threads of the theory of measurement and so contributed to the development of the theory of measurement as a unified and multidisciplinary field of endeavour, deserves greater recognition than just citations of the individual works in the proceedings volume. Since Churchman the theory of measurement has maintained the kind of disciplinary cross-fertilization that this book stimulated.

Churchman provided a philosophical discussion of the issues arising from Operations Research (Churchman, 1961). A fundamental issue presented by Operations Research is the problem of measuring human values because it claims to provide a means to produce the best possible, optimal, outcome. The difficulty with providing what is claimed to be the best possible outcome is that situations are multidimensional, and so cannot be simply or trivially analysed to provide a uni-dimensional measure of desirability of outcome that would enable a direct comparison of all possible outcomes to determine the best. The main additional point presented in this book compared with Churchman's earlier work (Churchman, 1959) is that measurement is a theoretical construct that enables relation of the observed to a theory of the observed. This point was made but not developed much by Churchman, and by no-one else until1997 (Ferris, 1997a). This history of the idea of measurement as a theoretical construct is evidence of the strength of Churchman's thought in being able to see a characteristic of measurement that remained unexplored for so long. A contributory aspect to the long delay is the deep embeddedness of the idea of measurement in the scientific worldview. The reconsideration only became possible with the general willingness to question the assumptions underlying strongly held positions in relatively recent times. The delay was associated with the relatively small number of people actively engaged in the theory of measurement at any time.

Churchman (1961) also observed the fundamental circularity of the process of measurement, which arises regardless of whether measurement is defined operationally, in which the circularity is clear, or otherwise, in which the circularity is hidden but still real. Churchman contended that the important issue is for one to determine a way to use the circularity implicit in measurement creatively rather than to have it undermine the value of one's investigations.

11.4 CONCLUSION

Churchman's direct contributions to the theory of measurement was to point out that measurement requires understanding what one is one is trying to measure and the many intervening variables that impact on the observation and representation process. He also stressed that the purpose of measurement is to provide a basis for decisions, and that measurement is a theoretical construct. Since his time there has been very little work performed to solve the problems raised by the issues that he identified. In itself this absence of follow-up work by others reflects that the discipline as a whole has not looked sufficiently to Churchman as a seminal thinker.

Discussions[68] with major scholars of the theory of measurement demonstrate that practitioners of the field did not pursue the questions arising out of Churchman's work, and did not perceive Churchman to be a major contributor. This work has more recently led to a definition of measurement that is consistent with the nature of measurement as a theoretical construct:

> "Measurement is an empirical process, using an instrument, effecting a rigorous and objective mapping of an observable into a category in a model of the observable that meaningfully distinguishes the manifestation from other possible and distinguishable manifestations. "(Ferris 2004)

In summary, Churchman's main contribution to the theory of measurement is flows from his collaboration with Philburn Ratoosh in bringing together the work of a number of scholars of the theory of measurement of various disciplines and collecting their work into a single volume publication. This action brought a multidisciplinarity to the theory of measurement that was new, and led to many later scholars viewing measurement more holistically than was previously the case.

REFERENCES

Ackoff, R.L., 1987, President's symposium: OR, a postmortem, *Operations Research.* 35(3):471-4, viewed May-June.

Ackoff, R.L., 2001, OR: after the post mortem, *Systems Dynamics Review.* 17(4):341-6, viewed Winter.

Barbour, I.G., 1974, *Myths, Models and Paradigms: The Nature of Scientific and Religious Llanguage*, S.C.M. Press, London.

Bridgman, P.W., 1922, *Dimensional Analysis*, Yale University Press, New Haven.

[68] As a result, at the beginning of my own doctoral studies my supervisor, Peter Sydenham, could say that there was no substantial work available in the field of the epistemology of measurement, which he suggested would be a fruitful area on which to spend some time. I pursued the question and wrote my thesis (Ferris 1997b). He said this on the basis of his own knowledge of the field, and did not specifically introduce me to the work of Churchman, who as was stated above, identified the gap but did not develop a sustained investigation of the matter. Towards the end of my candidature, my associate supervisor, Ludwik Finkelstein was able to find a freshness in the issues that I was addressing, leading to the writing of (Ferris 1997a). The fact of the freshness being found in my work addressing the epistemological issues of measurement by the then editor of "Measurement"[68] was evidence of the absence of other work stimulated by Churchman's in the intervening interval.

Campbell, N.R., 1920, *Physics the Elements*, Cambridge University Press, Cambridge, UK.

Caws, P., 1959, Definition and measurement in physics, in: *Measurement Definitions and Theories*, C.W. Churchman and P. Ratoosh, eds., John Wiley, New York, pp. 3-17.

Churchman, C.W., 1959, Why measure? in: *Measurement Definitions and Theories*, C.W. Churchman and P. Ratoosh, eds., John Wiley, New York, pp. 83-94.

Churchman, C.W., 1961, *Prediction and Optimal Decision Philosophical Issues of a Science of Values*, Prentice-Hall Inc., Englewood Cliffs, New Jersey.

Churchman, C.W., 1968, *The Systems Approach*, Delacorte Press, New York.

Churchman, C.W., 1970, Operations research as a profession, *Management Science*, 17(2):B37-B53, viewed October.

Churchman, C.W., Ackoff, R.L. and Arnoff, E.L., 1957, *Introduction to Operations Research*, John Wiley and Sons, Inc., New York.

Coombs, C.H., 1954, Measurement of preferences. in: *Consumer Behaviour the Dynamics of Consumer Reaction*, L.H. Clark, ed., New York University Press, New York, p. 95.

Ellis, B., 1966, *Basic Concepts of Measurement*, Cambridge University Press, London.

Ferris, T.L.J., 1992, Purpose of clinical measurement - distinguishing disorders in patients, *Australasian Instrumentation and Measurement Conference*, The Centre for Continuing Education, University of Auckland, Auckland, New Zealand, pp. 225-30.

Ferris, T.L.J., 1997a, The concept of leap in measurement interpretation, *Measurement*. 21(4):137-46.

Ferris, T.L.J., 1997b, *Foundation for medical diagnosis and measurement*, PhD thesis, University of South Australia, Adelaide.

Ferris, T.L.J., 2004, A new definition of measurement, *Measurement*, 36(1):101-9.

Helmholtz, H., 1881, *Popular Lectures on Scientific Subjects*, vol. 2, Longmans, Green, and Co., London.

Kircher, P., 1959, Measurements and managerial decisions, in: *Measurement Definitions and Theories*, C.W. Churchman and P. Ratoosh, eds., John Wiley, New York, pp. 64-82.

Krantz, D.H., Luce, R.D., Suppes, P. and Tversky, A., 1971, *The Foundations of Measurement*, vol. 1, Academic Press, New York.

Midgley, G. and Ochoa-Arias, A.E., 2004, *Community Operational Research. OR and Systems thinking for Community Development*, Kluwer, New York.

Pap, A., 1959, Are physical magnitudes operationally definable, in: *Measurement Definitions and Theories*, C.W. Churchman and P. Ratoosh, eds., John Wiley and Sons Inc., New York, pp. 177-91.

Papandreou, A.J., 1954, An experimental test of a proposition in the theory of choice, in: *Consumer Behaviour The Dynamics of Consumer Reaction*, L.H. Clark, ed., New York University Press, New York, pp. 90-1.

Pfanzagl, J., Baumann, V. and Huber, H., 1971, *Theory of Measurement*, 2nd ed., Physica-Verlag, Wurzburg-Wien.

Stevens, S.S., 1946, On the theory of scales of measurement, *Science*, 103(2684):677-80.

Stevens, S.S., 1959, Measurement, psychophysics, and utility, in: *Measurement Definitions and Theories*, C.W. Churchman and P. Ratoosh, eds., John Wiley, New York, pp. 18-63.

Suppes, P., 1959, Measurement, empirical meaningfulness, and three-valued logic, in: *Measurement Definitions and Theories*, C.W. Churchman and P. Ratoosh, eds., John Wiley and Sons Inc., New York, pp. 129-43.

Tobin, J., 1954, Introduction, in: *Consumer Behaviour the Dynamics of Consumer Reaction*, L.H. Clark, ed., New York University Press, New York, pp. 88-90.

Chapter 12

JUSTIFYING KNOWLEDGE CLAIMS

MIKE METCALFE
University of South Ausralia (City West)

Abstract: This paper will present a means of justifying[69] knowledge claims (research findings, conclusions) about technology-assisted social systems. It uses the ideas developed by the systems thinker C. West Churchman (1914-2004), so individual experience and experimentation are seen as important, but insufficient for dealing with the design of social systems. Richer criteria for justification are required, including consideration of stakeholders' perspectives and alternative explanations. This paper will explore the conjecture that Churchman's five 'guarantors' of knowledge can provide systems designers and researchers with criteria for justifying their findings, their claims to know something. The evidence to support this argument will be drawn from Churchman's books, 'Systems Thinking and Its Enemies' and 'The Design Of Inquiring Systems' as well as the argumentative inquiry and perspectival thinking literature. A list of 'justification' criteria for researchers is provided as a summary.

Key words: knowledge claims, justification

12.1 INTRODUCTION

"There is only perspective seeing, only perspective 'knowing'; the more effects we allow to speak about one thing, the more eyes, different eyes, we observe one thing, the more complete will our 'concept' of this thing, our "objectivity," be. (from Nietzsche's Genealogy of Morals (Kaufmann, trans).

[69] The terms 'guarantor', 'justification', 'warranted', and 'validated' are seen as fairly synonymous. The term 'guarantor' was used by Churchman. The term 'justification' aligns more with the rational argument perspective relevant to this paper.

Those of us involved in research into the design of equitable 'technology-assisted' complex social systems require more than the methodologies of the physical sciences to evaluate our findings, our knowledge claims. Claims of the form 'that system X is preferable to system Y' require subtle interpretations of complex social issues. For the non post modern systems designer, such interpretations need to be justified to a knowledgeable audience; argued in preference to disputed alternatives. As Churchman (1971) put it, there is a requirement for some 'guarantors of truth'. While both the words 'guarantee' and 'truth' seem excessively 'objective' nowadays, they are useful to underline Churchman's approach to justifying knowledge claims in the context of conflicting social tensions. The word 'truth' is being interpreted very liberally to include the objective 'one truth' as well as the interpretive approach of community-accepted knowledge claims. These are justified explanations of the world. This aligns with Churchman's broad definition of truth, so the two phrases 'guarantors of truth' and 'justified knowledge claims' would be seen as synonymous. Personal knowledge is being excluded.

This paper, therefore, agrees with Churchman that we need somehow to construct a means of justifying knowledge claims about complex social choices, in order to be able to distinguish 'ignorance' or 'mistaken' opinion from '*Knowledge*'. A discipline is defined by its episteme. If system designers determine that the scientific methods are sufficient to justify knowledge claims then it will be returning to engineering and dismissing the opportunity to include interpretive knowledge claims with its domain. If systems designers determine that knowledge can be personal, and not be required to be justified to a wider audience then they will be aligning with some post-modernist, and perhaps putting lives at risk from new technologies. A middle road approach is to encourage critical argument using rational evidence between stakeholder groups so as to enable them to justify their interpretation of a situation.

Designers of technology assisted social systems need to be able to distinguish 'justified' knowledge from error and ignorance, knowledge that goes beyond a definition which only allows for the scientific criteria of measurement and repeatability, knowledge that includes well argued alternative interpretations of phenomena. It will be suggested that this be done by using agreed upon means of justification, for identifying the difference between unsupported hunches, and what is to be treated as justified interpretations (i.e. knowledge claims) validated by reasoning and empirics, to the satisfaction of a knowledgeable community. There have to be some minimum criteria about the justification process. We need to be able to distinguish the justified from unjustified if our discipline is avoid very real dangers of institutionalised ignorance. Such ignorance allowed the horrors associated with the Nazi science of race, their persecution of "Jewish physics" and Stalin's enforcement of "Marxist biology." In both cases 'ignorance' could not be distinguished from justified knowledge and as a consequence millions of people died. Writers are now calling for "Feminist algebra," and "New Age Physics" (Stove, 1998). Some justified middle ground between only recognising scientific, objective, or physical, knowledge and the 'anything goes' of post modernism may be appropriate for a discipline responsible for designing life-support and community wealth-creating technologies. This concern for a set of acceptable criteria for justifying knowledge claims about social systems needs to be adopted by consultants, text-book writers, lecturers, and researchers in systems design. While somehow incorporating their years of experience, it needs to challenge them as to how they know that what they are advising is 'justified knowledge'.

The aim, therefore, is to explore the conjecture that *Churchman's hierarchy of five 'guarantors' can be used as pragmatic (works in practice) means of justifying interpretive knowledge claims (research conclusions, findings, results*). Apart from needing to interpret Churchman's guarantors into a knowledge justification method, these need to be interpreted into day to day systems design and research methods. They need to provide a richer yet pragmatic alternative to objective (evidence based) knowledge and to one not requiring any justification of what they claim to 'know' to others.

First, the paper will provide a brief look at Churchman's background is required to explain his assumptions about science, argument, systems and perspective seeking. This is followed by a more specific review of Churchman's book, *The Design Of Inquiring Systems,* which will be presented in order to discuss how his ideas might be seen as a way to justify knowledge claims. Last, the implications of these justifications will be discussed.

12.2 RATIONAL JUSTIFICATION, SYSTEMS AND PERSPECTIVE

Churchman's background as a philosopher of science, as a scientific engineer designing gun sights and as a writer on the mathematical techniques of operations research and mathematical programming, suggests he was anchored in the rationality of science. It is fairly safe to say he would have agreed with the famous quote from Hegel, 'that rationality is real' and that knowledge claim need to be argued before a sceptical audience as the basis required from which to define knowledge. Rational argument is the mechanism to produce justified knowledge. Irrational ranting or un-argued belief is an unlikely source of justified knowledge, although they may be the primary source of innovative conjectures that will later need to be rationally justified. The process of trying to persuade others of a new knowledge claim involves rational argument (debate, not quarrelling), with the presentation of conflicting evidence in order to learn (for more on argumentative inquiry see for example Crosswhite (1996); Eemeren, et al (1987); Meyers and Seibol, (1989); Niederman and DeSanctis (1995); Perelman and Olbrechts-Tyteca (1969); Popper (1972); Tracey and Glidden-Tracey (1999); Walton (1998). The argument needs a boundary.

Churchman developed into a system thinker. His enrichment of the criteria for justifying knowledge related to technology-assisted social systems seems to have started from the issue of the centrality of measurement as knowledge, followed by the issue of universal boundaries. As Ackoff (2000), Churchman's PhD student, constantly points out, the essential difference between systems thinking and scientific thinking is that of boundaries. In science there is one boundary for everything, our universe; within that universal truth. What is true in one place, generally speaking hopes to be true everywhere, over all time. System thinking uses the unit of analysis of a mentally imagined, interconnected, bounded, *system* as defined by a system designer. It may be a justice system, a payroll system, a defence system, a town, an organisation, a hen or a nation. It is bounded; changing that boundary changes the system within a context. For example, the production system becomes the organisational system by including the other organisational functions. Knowledge claims about a system do not have to be universal, only to be relative to the system under consideration. What is true for one system, and its occupants, may not be for another.

Claims have to be agreed upon by all the elements of the system; stakeholders in a community. If the doctor element of a health system claims to have done a wonderful job of containing a contagious disease outbreak, this is only 'justified' if the finance, patient and drug supply elements of the hospital system agree. If there were insufficient notice from doctors for drug production rescheduling to fill the orders then the doctors have not done a wonderful job. Said differently, the systems perspective also addresses the generalisation requirement for justified knowledge claims. To claim that 'John should not marry' is not as general as to claim, 'all men should not marry'. The greater the generalisability of a knowledge claim the greater the knowledge claim. Using systems thinking concepts, the generalisability needs to extend through the whole system under study but not necessarily all systems. The more 'universal' the systems the more generalisable the knowledge claim.

As Churchman (1971) keeps pointing out, the fact that systems design is dealing with 'design' also makes it very different to the sciences which can only be interested in explaining (theory) about 'why' physical events occur. Scientists are not able to be interested in re-designing nature. To design is to decide what is preferred; in other words, which of numerous alternatives is the one chosen? The systems design literature (text books especially) is full of knowledge claims that design X is better than design Y. Is this design knowledge claim 'justified'? This is not an issue for scientists as their methods of validating knowledge are insufficient. The design issue connects systems management to that part of critical social theory (the Frankfurt School etc.) which aims to redesign social systems to emancipate the disadvantaged. Critical social theory has long claimed that social research needs to result in improved social action leading to emancipation. As will be discussed later, for Churchman this means that justifying knowledge claims needs to include hearing from all those involved with the re-designed system, their interpretations somehow need to be included in any claim about how 'best' to re-design a system.

Hornsby's poem (1997) explains that knowledge claims may be divided into three; objective, interpretive and critical, and can be viewed systemically. Rather than set these three against each other they need to be inter-related. The physical world is measured and tested using scientific, objective methods. So for example, when trying to first appreciate, and then design a response to, the complex social system that delivered the Bali bomb attack, there will be some physical facts. When, where, and what? Rigour rules. Objects can be measured using the classic scientific research methods which are convincing because they can be repeated by the fiercest critic. When all agree on these measurements, perhaps of the number of deaths, explosion source and fire temperature, then a justified knowledge claim can be by arguing about these physical aspects of the bombing. The measurements provide convincing evidence in support of the rational argument. The next fact to be assessed involves the interpretations. Why did they do it? Should they have done it? How could it have been done better? The scientific measurement methods based around the convincingness of repeatability are no longer of any use. Insights, ideas, and different perspectives are sought. How convincing will these be? Are they not just opinions? Justification still needs to use rational argument but now argument for alternative interpretation need to be encouraged. Multiple insights need to be justified to a sceptical audience. The requirement of argument helps distinguish the 'insane' from the insightful. Moreover, these insights need to be justified within the context of the entire system being studied.

12.3 CHURCHMAN'S FIVE GUARANTORS

Looking now more closely at Churchman's book (1971), *The Design of Inquiring Systems*, the title is problematic[70]. He suggests an alternative, "The Design Of An Inquiring Society", which hints at his grander intentions. But the introduction makes it clear he is exploring the extent to which research can be computerised, something very 1970s and something his Nobel Prize competitor Herbert Simon also focused upon. Churchman seems to be trying to identify how much we can computerise human-like inquiry so distinguish what can and cannot be mechanised. He seems to trying to bring out the issues, and thus improve our appreciation of human inquiry. His is a multiple perspective approach as opposed to Simon's earlier mono-decision tree approach. Churchman identifies three ways of seeing the world that need to be capable of being mastered by human inquiry skills. These are:

1. the world as composed of atoms (from Democritus),
2. the world as composed of purposeful entities (Aristotle) and
3. where events in the world are only known with a certain degree of probability (Carneades).

It is the second one that seems most relevant to systems thinkers given the identification of the emergent property of purposeful action in human activity systems (see Ackoff (2000), Ulrich (1983) and Checkland (2000)). As Ulrich, (1983) Metcalfe and Powell (1995) and many others have pointed out, purpose defines what is knowledge (relevant knowledge). Purpose causes problems to exist (a difference between what is sought and what exists) and inquiry is an attempt to solve that perceived problem. A tree does not 'know' anything, nor does it want to or need to know anything. Only a system with purpose can know things. A community with the purpose of defending itself from nature needs to 'know' many things.

Churchman can be interpreted as providing five criteria for justifying whichever of the three sorts of knowledge is being assumed. That is, he designs a hierarchical function of human thinking into a fairly arbitrary five layers of justification required to deal with these three ways of seeing the world. These he gets from his own background – the history of the theory of knowledge. The base level he labels Leibnizian, one of the founders of calculus. Logic, and algebra provide a means of justifying knowledge. Its basis is repeatable thinking steps, logic; an in-the-head repeatability; a critical argument with yourself. This knowledge could be generated and validated by someone in a sensory-deprivation tank. It involves processing memory facts rather than sensory inputs. Given the mind's need to consider thoughts in a linear fashion, it is a completely mental process of associating

[70] With thanks to Chris Stewart for alerting me to this.

a string of propositions using 'if-then' functions as in: If Socrates is a man, and all men are mortal, then Socrates is mortal. After some processing in this way, a whole series of facts can be thought of as connected by a sort of network of justified and related facts. Socrates is connected to mortality, Ancient Greece, public debate and so on. Ancient Greece is connected to Plato, the Iliad, and democracy and so on….

Practical application of this Leibnizian level of justifying knowledge claims may be that, "leadership is a communication role". The justification may include logical/algebraic reasoning along the lines "if a group has 3 members, then there are 3 communication channels; if there are 5 people then there are 10 communication channels [(n/2)(n-1)]; there is an exponential growth. When a group has more than 10 people, more than 45 communications channels, then a different communication system is required. If all the communications are directed through one person, then there will only be 10 communication channels. This communication role is sometimes called leadership. This is a logical justification of the knowledge claim, but not sufficient.

This then provides a way of justifying knowledge claims even about non physical phenomena such as the concept of leadership and communication. Another example would be for a preference for one technology-assisted system, over another. However, the justification can only be as good as the quality of the reasoning that formed the connection links. So for example, in the example the need for less than 45 communication channels may be questioned. It was Russell who in the mid twentieth century rather ended people's trust in the sole use of logic to justify claims when he showed that many of the so called logical steps in most mathematical proofs were more one of faith than pure logic. That said, logic is still considered a good starting point for justifying knowledge claims but perhaps as a Popperian conjecture that needs to be tested further through a process of it being argued to a sceptical audience. Anyway, this form of inquiry is not expected to be a vibrant source of innovative and creative knowledge claims that providing insightful interpretations useful for solving complex social problems.

Leaving aside the logic level for justifying knowledge claims, Churchman then discusses the other way to justify knowledge claims which he names after the empiricist Locke. Justification may be through the senses, personal experience, including all sensory input (sight, touch, smell, and hearing). This is 'out there' inquiry, outside the head. In the simple Lockean form, we wander about the planet absorbing experiences in the form of sense data which we store as personal knowledge In this way, we continually expand what we think we know but as anyone who watches films knows, our senses are easily deceived. Both what we see and how we interpret what we see can deceive us. Experience becomes unjustified knowing, the more something appears to happen the more certain we become we have command of some knowledge when perhaps we are not testing it sufficiently. Justification comes from wanting to convince others of what we think we know, and being able to do. This justification takes the form of saying, go ahead repeat my sensory experience for yourself, then you will experience what I did and be convinced I am correct. For example, if you wanted to convince someone that a speech contained mention of some issue then you could say to a sceptical audience, "listen for yourself, this is the original tape recording of her speech, what she actually said was..." The justification for the knowledge claim is through repeatable sense experiences. This tends to exclude spiritual knowledge claims.

Therefore, with the example of a knowledge claim that "leadership is a communication role" an empirical justification is required. This may take the form of a sociometric analysis where members of a group are asked who they talk to about work related issues. If everyone mentions one person, then the sociometric network will show this person as a central communications node; the hub. If this person had a 'leadership' title like, manager, then this would be empirical justification of the knowledge claim. This could be *repeated* for many different groups, and repeated by many different researchers. It could be falsified, and I suspect it would be due to our social habits of breaking up into 'family size' sub groups. Social networks tend to look like a series of loosely connected clusters. However, note that repeatable evidence, both in terms of logic and empirics, makes for very convincing justification.

The use of empirics to justify what it can, i.e. physical events, encourages an ever increasing demand for more accuracy of the sensory input; the precision of the observation. This leads to measurement and experiments which utilize careful observation under careful conditions. Convincing others of the accuracy of our empirical experiences becomes based upon the accuracy of recording the circumstances of our observations. Cohen (1994) argues that the unique quality of science is this rigor, the 'precision of empirics'. Given the ease with which people's senses can be deceived, sensory experiences collected under controlled experimental conditions provides preferable justification to everyday experiences. Thus the experimental scientist's knowledge claims are preferred over the experienced person. However, in many cases repeatable laboratory experiments are not possible, such as with the effectiveness of a military system in certain combat conditions. In these situations justification is sought from the reasoning (logic) of the 'experienced' person but again this must be open to challenges. There is a large step from recounting what happened and explaining why it happened that way. In court 'eye witnesses' are allowed to repeat what they thought they saw. Only very credible 'expert' witnesses are allowed to interpret *why* something might have happened.

This begins to overlap with Churchman's next level of thinking about inquiry, which is based on Kant's ideas on the need to combine explanation with empirics. Observations have to be explained. Theory, an explanation why, is sought. For example, why does a stick appear bent in water? One well justified explanation (theory) is that light travels more slowly in denser mediums. But these explanations need to be logical, satisfactory to a knowledgeable sceptical audience. A justification needs to include an explanation of why the phenomenon is how it is. It is not sufficient to say, "Hey it works, see for yourself, I do not know why it works but it always does". Justification that it works needs to include 'why'.

Therefore, the example claims that, 'leadership is a communication role' needs to be justified by explaining 'why' leadership is a communication role? The logic/algebraic justification, while open to pragmatic criticisms (does it work in practice), is inherently explanatory. It explains that leadership needs to be a communication role because of the exponential growth of channels. The empirical justification is not so explanatory. If shown a sociometric network with one central node, then the question, 'why' do people organise themselves like that, still needs to be asked and answered. Churchman's Kantian justification requires that an explanation be provided for these epirics. The intent is to encourage critical argument between explanations.

Figure 1: Churchman's Guarantees

Name	Guarantor	Comments
Singer	'Sweeping in' all interpretations	An ethical justification is required to give all stakeholders a 'voice', an interpretation.
Hegel	Competition	Rational argument for alternative interpretations.
Kant	Rational explanations	Justifications need to include explanations (why) using rational reasoning to a universal audience. This explanation therefore needs to draw on the prior experiences of the audience.
Lockean	Empirics, sensory experiences	Interpretive knowledge claims can be justified with empirics (experience through the senses).
Leibniz	Logic, Reasoning	Interpretive knowledge claims can be justified with logic (rational reasoning).
Assumptions	Rationality and Community	Knowledge is agreed by a knowledgeable audience, not personal. Rationality is real.

The next level of justification (see figure 1) Churchman suggests for a knowledge claim about a socio-technical system he calls Hegelian. This justification starts to become separated from the simplistic view of science, and move towards a more Kuhnian view of science, that is, one of competing explanations. Explanations need to be made in the context of alternative explanations why, alternative theories. The argument is not 'that X is (or is not) proven' but that explanation X is preferable to explanation Y. The idea of seeking alternative explanations of the same phenomenon starts from Hegel's separation of the phenomenon under study from the perspective being taken on that phenomenon. An example of the phenomenon may be a technology-assisted social system such as a national health system. The different explanations for its failure may be seen as organisational, technical or personal.

For the leadership is communication knowledge claim, alternative explanations may be that humans are a hierarchical and need to be led, that we break into family size sub groups to automatically counter the problems of number of channels, that the sociometric network merely reflects power through the control of resources, that in a complex world having one dominate worldview keeps things understandable, or that modern electronic communications mean the number communication channels can grow enormously. The original knowledge claim will be better appreciated if it explicitly competes with these sorts of alternative explanations.

At the fifth level, the justification for a knowledge claim need to include what Churchman calls Singerian justifications. This requires the justification of a claim to seek stakeholders' alternative interpretations of the phenomenon (e.g. socio-technical system). All possible rival perspectives need to be given an opportunity to be justified. This is an ethical imperative. Stakeholders' interpretations need to be 'swept in' and allowed for in the justification of a knowledge claim, if it is to be an ethical claim. At the start of this paper the example was given of doctors in a health system trying to claim they had done a wonderful job. If other parts of the system (e.g. drug suppliers) have a different interpretation that is not allowed for, then the doctor's claim cannot be considered ethically justified. In the leadership example, this ethical justification may be achieved by ensuring a voice has been given to those with experience in leadership and follower roles, both good and bad. Group communication facilitators and those involved in group communications technology also need to be given a voice.

12.4 RESEARCH CRITIQUE

Having very briefly outlined Churchman's five levels of justification for knowledge claims about socio-technical systems, these can be used to critique a piece of research. The critique questions below are only intended to be a guide, but they do attempt to include all of Churchman's five levels.

1. What is the knowledge claim?
   Has the research produced a knowledge claim? Has it produced something new, novel, or surprising to the majority of whoever is your knowledgeable audience? Is it falsifiable? Is it at least possible to find empirical evidence that counters the knowledge claim?
2. Logical Justification
   Has the knowledge claim been justified to a knowledgeable audience using algebraic, logic, or mathematical reasoning, possibly in the form: 'if…then'? In the classic layout of a research article or thesis this may well be achievable as a result of using previous research and discussion (literature review) to reason for a knowledge claim. Alternatively it may suggest the use of logic in the form of a mind experiment: if X and Y then Z.
3. Empirical Justification
   Can the knowledge claim be justified empirically? Is there any supporting evidence that can be seen, heard, smelt, touched or tasted? Can this experience be repeated by sceptical others? In situations where the experience cannot be repeated due to its context, e.g. the installation of the first computer), can those recounting the experience (expert witnesses) provide reliable repeatable, precise 'eye witness' reports?
4. Explanation as Justification
   Can you provide rational explanations to a knowledgeable audience as to ***why*** some phenomena are preferably explained using the knowledge claim rather than some other explanation? The justification for the knowledge claim needs to include more than "because it works in practice". Why does it work?
5. Justification Through Countering of Alternative Explanations
   Does the evidence in support of the knowledge claim acknowledge and counter alternative explanations?
6. Ethical Justification
   The justification of the knowledge claim needs to include consideration of all elements in the system; all stakeholders need to be given a voice; for their perspective to be justified.
7. Actioning the Knowledge Claim
   As a result of justifying the knowledge claim what is recommended; so what? It is considered inadequate for social research to merely explain or describe, it needs to suggest a design for an improved world. Moreover, these actions should be actionable in every day decision making, i.e. they should not be too abstract.

12.5 IN SUMMATION

Churchman provides us with a means for uniting interpretive research with science and with systems thinking. The systems thinking stream being drawn upon here is that of shifting boundaries; that is, seeing things from different but still rational interpretations or perspectives. Justification is also about bounding those perspectives to one system. Research should result in a knowledge claim justifiable to and by all parts of that system. This justification needs to:

- be logical and empirical,
- be able to provide explanation,
- to consider alternatives,
- to include all stakeholders and
- to result in improvements to humanity.

REFERENCES

Ackoff, R., 2000, *Making a Difference, Systems Thinking/Systems Change*, cited 8/9/03, http://www.judgelink.org/Presentations/GirlsLink/

Checkland, P., 2000, Soft systems methodology: a thirty year retrospective, *Systems Research and Behavioural Science,* 17(1):S11-S58.

Churchman, C.W., 1971, *The Design of Inquiring Systems*, Wiley, New York.

Cohen, H.F., 1994, *The Scientific Revolution*, University of Chicago Press.

Crosswhite, J., 1996, *Rhetoric of Reason*, University of Wisconsin Press.

Eemeren, F.H., Grootendorst, R. and Kruiger, T. 1987, *Handbook of Argumentation Theory*, Foris Publications, Dordrecht.

Hornsby, R., 1997, Critical thinking: a poem, *System Research and Behavioural Science*, 14(4):277-278.

Linstone, H.A., 1999, *Decision Making for Technology Executives: Using Multiple Perspectives to Improve Performance*, Artech House, Norwood, MA.

Mason, R.O. and Mitroff, I., 1981, *Challenging Strategic Planning Assumptions*, John Wiley and Sons, New York.

Metcalfe, M. and Powell, P., 1995, Information: a perceiver-concerns perspective, *European Journal of Information Systems*, 4:121-129.

Meyers, R.A. and Seibold, D.R., 1989, Perspectives on group argument, *Communications Yearbook*, 14, 268-302.

Niederman, F. and DeSanctis, G., 1995, The impact of the structured argument approach on group problem formulation, *Decision Sciences*, 26(4):451-475.

Perelman, C. and Olbrechts-Tyteca, L., 1969, *The New Rhetoric: A Treatise on Argumentation,* University of Notre Dame.

Popper, K.R., 1972, *Conjectures and Refutations*, Routledge and Paul, London.

Stove, D., 1998, *Anything Goes,* Maclay Press, New South Wales.

Tracey, T.J.G. and Glidden-Tracey, C.E., 1999, Integration theory, research design, measurement and analysis, *Counselling Psychologist*, 27(3):299-324.

Ulrich, W., 1983, *Critical Heuristics of Social Planning*, Wiley, New York.

Walton, D., 1998, *The New Dialectic*, Toronto University Press, Toronto.

Chapter 13

BUSH TUCKER, CONVERSATION AND RICH PICTURES

Capacity Recognition and Entreprenuership Through Systemic Praxis with Neporendi Forum Inc.

JANET McINTRYE-MILLS AND DOUGLAS MORGAN
Flinders Institute of Public Policy and Management , University of South Australia

Abstract: Aboriginality is a matter of family connections and a history of marginalisation and disadvantage. Aboriginal health and social inclusion are intractable problems for government. This project contributes to redressing family violence and social inclusion issues through participatory action research (PAR) with Neporendi Forum Inc (an Aboriginal organization) to enhance social justice. The research enables an integration of activities across networks that span researchers based at two universities, local youth organizations, schools and local government (see Fals-Borda & Rathman, 1991; Edgar 2001; Duhl 2001). PAR builds the capacity of participants to work holistically across the organizations and community settings. It is sustainable, addresses socio-economic concerns and uses an intergenerational and interactive design.

Key words: Participatory design, Indigenous, strategic partnerships, well-being

13.1 INTRODUCTION

Poor outcomes in Indigenous health and social exclusion are the result of a history of dispossession (McIntyre-Mills, 2003). It has been identified, as a result of similar work in Central Australia (McIntyre-Mills 2003c) and part of an ongoing collaboration between Neporendi[71], Flinders University and University of South Australia that there is a need to address health outcomes through PAR in a community settings. The 'can-do' PAR approach (learning by doing, creating new ways, keeping the best of the old ways, and networking with others) can be helpful, albeit by no means an answer to all problems. It is merely a strand in a skein of much needed governance interventions to balance both social rights and responsibilities across the state (and wider regions) the market and civil society (of Indigenous first Australians and non Indigenous Australians).

The participatory action research comprises researchers, community leaders and participants at Flinders University, the University of South Australia and Neporendi Aboriginal Forum Inc.[72] The aim of the project is to address complex social inclusion issues in partnership. This approach is more than a cynical Third Way approach to link business, government and community, orchestrated to bring state or market together merely to save costs. Instead it was designed and owned within the community context from the start and then we applied for the funds together.

The project is culturally appropriate as it is delivered by Aboriginal community leaders. For instance, a senior female (whose family deferred to her) and a young male in his early thirties with considerable technical and business skills (who was seen to be a vital resource for the Neporendi network). PAR offers a practical approach to address the communication and policy issues relating to Indigenous family violence and to social inclusion. Aboriginal health and social inclusion are intractable problems for the Commonwealth Government. This research makes a contribution to redressing social health problems by designing and developing a process with Neporendi, an Aboriginal organization.

The elegance of the project is that is designed and run by and for the project participants. The participatory action research facilitator's role is to

[71] Neporendi Aboriginal Forum Inc is a not for profit organization with the primary objective of providing an informed referral service, which, although specifically established for Indigenous people, is used by both Indigenous and non-Indigenous people seeking assistance with food, accommodation, employment, training, health, well being, legal issues, the aged and youth requiring assistance. The service coverage is from Glenelg to Victor Harbor, with most service assistance being delivered to residents in the Reynella, Port Noarlunga, Christies Beach area. Approximately 600 families, consisting of 1,560 Indigenous Australians are in the service area. The social context of these Indigenous Australians reflects the circumstances of Indigenous Australians in general, suffering from poor health arising from an historical context, including dislocation from communities and separation from families. The impact of which has contributed to mental health problems, high rates of substance misuse, family violence, self-injury and suicides.

[72] Indigenous people from seeking health care and other needed services from mainstream providers. The southern metropolitan area is a region where the population is growing and the demand for services is increasing for both Indigenous and non-Indigenous users.

work with the Neporendi participants to assist them in networking, accessing resources (White 2001, 2002) and recognition of their own capacity (Foley, 2002, 2003).

Reworking governance can make a contribution to redressing social exclusion if practical co-ordinated projects and service delivery (McIntyre-Mills[73]) are used as a basis for extending existing networks to include local and state government representatives, non government organisations and businesses.

13.2 BACKGROUND, AIMS AND OBJECTIVES

This is a two-part project comprising governance, community building and research on social inclusion with the Department of Human Services. The multimethod participatory action research project supports better life chances for young people and their families. The longer-term research agenda for this project grew directly out of the needs identified in our current research. This project reworks the notion of knowledge management and develops approaches to address the conceptual, practical and policy issues relating to social inclusion and addressing complex problems in a holistic manner.

The project provides a case study of how so-called 'Third Way' (Giddens (1998) approach could be re-worked if they are owned and directed by the local people and for themselves (see McDonald and Marston, 2003 for a critique of neo-conservatism and the retreat from universal welfare rights). Governments use the rhetoric of 'steering not rowing' (See Osborne and Gaebler, 1992:20 cited by Rhodes in Pierre, ed 2000) in order to make a distinction between policy directions and the delivery of services. Unfortunately, it can also lead to what Jamrozik (2002) has called the 'postwelfare state' (also see Rhodes in Pierre, ed., 2000:56; Rhodes, 1997; Peters and Savoie, 2000).

[73] "Participatory computer design and strategic partnerships to address complex health, housing and social inclusion issues with indigenous people". This is work in progress funded by an Australian Research Council Linkage Grant for 2005-2007. This research will contribute to redressing social inclusion issues by designing, developing and testing a dynamic management tool together with Neporendi Forum Inc, an Aboriginal organization, Aboriginal researchers and the South Australian Department of Human Services. The research enables an integration of services to maximize provider effectiveness and user outcomes and could achieve better solutions to complex social health problems, build the capacity of an Aboriginal PhD student and the experience of the participants in the West Churchman Research Network who apply critical and systemic design processes.

Entrepreneurship and capacity recognition is not used as a means to justify limiting the role of the state, or offloading the responsibility for some citizens (whilst privileging others). It is a means to span the social divides between the Aboriginal and non-Aboriginal community and to use dialogue and practical actions to empower social agents to remake their life chances (through working and re-working the boundaries[74]).

The project was an outcome of friendship networks across two universities, the community members, local government and community organizations and supported by a state government, Human Services and a Community Benefit SA Grant. This first Community Benefit South Australia funded project provided the basis for the Australian Research Council Funded Linkage Grant with the South Australian Department of Human Services that will enable the past president of Neporendi Forum Inc to be mentored through a postgraduate degree at Flinders University with a team of supervisors spanning informatics and engineering, public health, public policy, management and Aboriginal history, education and politics.

The two phases of the project aim to:

- Develop the capacity of participants with complex problems (such as homelessness, drug misuse, domestic violence) through participatory action research.
- Make a contribution to theory and practice.
- Build capacity across service users and providers to improve health, housing and welfare (to achieve greater social inclusion) requires a paradigm shift in the way social problems are understood and treated.
- Address problems systemically across sectors and disciplines.

13.3 RATIONALE AND POLICY CONTEXT

Neporendi is representative of the issues faced by Indigenous communities in other areas of Australia, such as: high levels of health-related issues, unemployment, homelessness, family violence and reduced education opportunities. The effects of family violence are wide ranging, for instance: the criminalisation of the offenders has financial, physical and psychological consequences for the women, children and men involved. It also has intergenerational consequences. It disrupts families, destroys assets and creates a demand for emergency housing and assistance and disrupts the education of children. Social inclusion, homelessness, unemployment,

[74] Governance is a construct that plays out in the way people think about the world and the places they live in (see McIntyre-Mills "Systemic Governance: Working and re-working the conceptual and geographical boundaries of governance and international relations", forthcoming).

gambling, family violence and drug misuse are facets of a complex, interrelated problem[75] that requires a coordinated governance response across departments in the public, private and non-government sectors. However, current compartmentalised thinking in respect of some aspects of human services has led to disciplinary specializations. There is a need to address ways of developing the capacity to work across disciplines to understand better the nature of "joined up" social problems (as they relate to social well-being and governance) (Davies and Kelly, 1993; Ackoff and Pourdehnad, 2001; Duhl, 2002; Gaventa, 2001). This is not merely a change to policy and practice, but a move away from the perception and definition of issues in separate compartments. A review was undertaken in 2003 by Berry (a Masters student, Primary Health Care – Flinders University) and this project builds upon this work to assist Neporendi's move towards implementing its vision of achieving better health outcomes. So the project is a shared

[75] This is clearly demonstrated by the twenty year gap in life expectancy (Steering Committee for the Review of Government Services, 2003). Indigenous Australians experience unemployment levels 2.8 time higher than non-Indigenous Australians (if CDEP 'work for the dole' scheme participants were include the rate would be much higher) and have average incomes significantly lower than non-Indigenous Australians across all regions (ABS, 2001; Steering Committee for the Review of Government Services, 2003). In South Australia 39.4 per cent of Indigenous Australians are employed however, 22.8 per cent are employed via CDEP (ABS, 2001). Home ownership levels are also affected with just 31.9 per cent of Indigenous Australians purchasing or owning their own homes compared to 69.5 per cent of non-Indigenous Australians (Steering Committee for the Review of Government Services, 2003). Around 2 per cent of Indigenous Australians live in caravans, cabins, tents or improvised housing (ABS 2001). In South Australia 67 per cent of housing is rental accommodation (though this is affected by use of community housing and the relative youth of the Indigenous Australian population – home ownership being a product of increased age) (ABS 2001). Indigenous Australians are more likely to be subject to violence (murder, assault, sexual assault and family violence) with hospitalization rates of 13.3 per 1000 compared to 1.0 per 1000 for non-Indigenous Australians (Steering Committee for the Review of Government Services, 2003). Homicide rates are also higher with Indigenous Australians representing 15 per cent of the victims despite accounting for only 2.4 per cent of Australia's population (Steering Committee for the Review of Government Services, 2003). Indigenous Australians are 15 time more likely to be in prison than non-Indigenous Australians with juveniles 19 times more likely to be detained (Steering Committee for the Review of Government Services, 2003). In South Australia 17 per cent of the prison population were Indigenous Australian (ABS 2001). The Steering Committee for the Review of Government Services (2003) identified three priority areas for action to address these appalling statistics. They are creating 'Safe, healthy and supportive family environments with strong communities and cultural identity'; 'Positive child development and prevention of violence, crime and self-harm' and Improved wealth creation and economic sustainability for individual, families and communities' (Steering Committee for the Review of Government Services, 2003:1).

initiative with Neporendi Forum Inc, a voluntary Aboriginal Community Organization that services the Southern Metropolitan region of Adelaide. The extended Neporendi family members and the community participate as volunteers. The Board of Neporendi facilitates human services for Aboriginal peoples in the Noarlunga area. Its governing committee has representatives from universities, local government and many areas of the Department of Human Services. These partnerships with a range of government and non-government service organizations provide integrated services across Local Government, human services, Healthy Cities Noarlunga, the Southern Youth Exchange, the Living Guarna Centre and businesses prepared to market their products.

The participatory action research project was undertaken at a time when Aboriginal politics was at a low ebb.

> "As Aboriginal people we are saddened, alarmed and disgusted by the radical welfare plans proposed by the Howard government in the article 'Sit –down cash ends for blacks' (11/11). Racism caused the disadvantage across every social indicator (health, employment, education, housing) that Aboriginal people experience today. This plan will only result in further deterioration and dispossession. It is racist...The National Aboriginal Community Controlled Health Organization represents 131 Aboriginal health services across Australia. What we are hoping for is proper consultation with Aboriginal people, and with the key organizations that represent Aboriginal interests and adequate resources committed to implementing these plans...This plan was devise behind closed doors, apparently without consultation with the very people it will affect most. It fails to address the underlying issues, and it too will fail." (Tony McCartney, Nov 13th, 2004)[76]

This statement stresses that Noel Pearson's (1999) responsibility approach to 'make a go of things' has been taken too far, as the approach is potentially punitive and racist if it is applied only to Aboriginal welfare recipients. It is yet another example of the erosion of the welfare state (Jamrozik 2001). The Aboriginal leadership of Noel Pearson, Patrick Dodson and Michael Long stresses that rights and responsibility based on reciprocity has resonance in the Aboriginal community[77], but the devil will be in the detail and the way that it is interpreted for men, women and children.

At the theoretical level, it fills a void in the literature by investigating the influence of PAR and action learning on empowering Aboriginal Australians

[76] Chairman, National Aboriginal Community Controlled Health Organisation, The Weekend Australian 13-14th, 2004, page 18.

[77] Kelley, P. 'Black leaders offer new accord'. and Rintoul, S. 'Long Walk over but longer journey ahead' The Weekend Australian December 4-5.

in practical ways. The outcome of this research will form the basis for developing an intergenerational model of capacity building. At the practical level, insights into PAR are provided that will lead to more effective processes for trans cultural community consultation, that is not about appropriation, the denial of the value of diversity, but instead it attempts to co-create meaning for the purpose of social and environmental justice. It effectively turns the tables on the design process and empowers those who are marginalised (see Ackoff in Midgley 2004, Ackoff and Pourdehanad 2001).

13.4 PARTICIPATORY DESIGN

Poverty and a sense of trying to catch up from a position of disadvantage remains the challenge. This is a capacity building program using a participatory action research approach to assist communities to identify opportunities and funding and to help the participants to apply entrepreneurship and a 'can do' attitude to addressing social disadvantage. This project uses social/cultural discourse management to address complex problems and capacity building to systemically address problems in community settings.

This project is a contribution towards our program of 'breaking the deficit cycle' (Doug Morgan, Chairperson of the Neporendi Forum Inc, pers comm. 2004) and enabling the Neporendi participants to enhance their own life outcomes through identifying opportunities and trying out or piloting possibilities that could be developed into self-employment.

A participatory design approach (Banathy, 1996) to address complex social problems (Rittel and Webber, 1984; McIntyre, 2004) can provide more effective outcomes, both in treatment and community settings (Roche and MacDonald, 2001; McIntyre, 2002, 2003a, b, and c). At a family day for conversation (Churchman, 1982; Warren, 2004) and planning is was clear that finding the resources to enable families to get together for a good time in pleasant surroundings is important. Providing food, green, shaded areas for children to play and enabling adults to network with family and friends to share their stories is helpful. Through recreational activities families enjoy themselves. The catering is shared and there is time to talk. Violence seems a world away. Their daily challenge of not being able to pay utility bills, the rent or unemployment can be discussed within pleasant surroundings.

PAR that supports participatory governance (McIntyre 2003a,b, c, 2004) through listening and learning from others experience enhances the capacity of community to design services and processes to meet the perceived and expressed needs of the Aboriginal service users and thus to improve user outcomes.

13.5 PHASE 1: PRACTICAL ENJOYABLE ACTIVITIES ENGAGE FAMILY MEMBERS OF ALL AGES

The approach focuses on positive ways to address challenges. A constructivist approach was used, to model 'positive *can do* approaches' to turning lives around. A talented senior Aboriginal woman and her unemployed son- in- law, a skilled computer technician, continue (at the time of writing) to give voluntary time to the forum and work for social security funds via a community development employment project (CDEP). They lead the project work to develop a community art project with Wirreanda School children and parents. They designed the artwork used to package home made biscuits and Christmas /greeting cards. Family days were held at which children and family worked on a design. Members of the Neporendi team who specialize in health matters were present at the event. Children and families were asked to draw pictures of a) What makes them happy. b) What helps to draw family members together? The design was used as a basis for the biscuit box packaging and the greeting cards. Community days were held at which biscuit making was the main activity, supervised by L the senior Aboriginal grandmother, daughter and daughter – in –law.

Aboriginal 'local knowledge' (of their social, cultural -including age[78] and gender[79]- political and economic) circumstances is central to the success of the project (See McIntyre-Mills 2003c, Edgar, 2001; Gaventa, 2001; Geertz, 1973) and the approach is directed at two practical concerns: building community supports through art with children, parents and grandchildren, cooking using bush herbs and then packaging the material in boxes designed by the children and their parents. The designs for packaging focus on things that make them happy as families. The biscuits are marketed, the proceeds paid into a revolving fund. One of the goals is to develop a small Aboriginal business producing Aboriginal products.

[78] McIntyre, J., 2003a, 'Yeperenye dreaming in conceptual, geographical and cyberspace: a participatory action research approach to address local governance within an Australian Indigenous Housing Association', Systemic Practice and Action Research, 16(5):309-338.

[79] Fonow, M. and Cook, J., 1991, Beyond Methodology, Feminist Scholarship as lived Research, Indiana University Press, Bloomington.

13.5.1. An Intergenerational and Systemic Approach to Social Inclusion

Art and cooking activities provided a means to address greater understanding of definitions, causes, consequences and ways to address family violence with different age and gender groups who form part of Neporendi. The art activity aimed to a) communicate the ideas of different age and gender groups and to b) promote family well being through recreational activity that enhances communication across interest groups. Participants were asked to plan ways to build strong families and communities. Men, women and young people worked in different and shared groups. Creativity, recreation, capacity building were the focus of the community activity that aimed to enhance communication across age and gender groups.

The first phase of the intergenerational project is based on creating products for sale. Aussie damper and myrtle biscuit recipes using bush tucker ingredients were the focus of the participatory process around which conversations were held. 'Learning by doing' is the focus of the project (Freire, 1982).

> "Where do you find the myrtle for the biscuits?" I asked whilst being instructed on the number of spoonfuls of dry ingredients to add to the baking bowl.
>
> "We buy it", said my biscuit making partner, dryly, adding with a laugh, "we are not hunter gatherer's any longer!"

This comment (encapsulating self determination and a postcolonial mindset) together with the conversation on cable television and entertainment options and the need for more cash, underlined the worldview and the aspirations of Neporendi members. It also made me reflect on the nature of our diverse overlapping and changing identities as participants (See McIntyre-Mills, volume 2: *Molar and molecular identity and politics*).

Families assisted with baking and decorating the packaging and stationary with drawings by children and parents. The younger children were occupied with Aboriginal dance practice and drawing whilst the rest of the family contributed to the project.

The pictures were of what makes families happy together, in other words, a version of: "these are my favourite things" and "from small things, big things grow". Painting pictures was not the only focus of the activity. The artwork and cooking were the means to bring the families together, rather than focusing on the negative or normalizing problems (McIntyre-Mills, 2003). Instead the focus was positive: "We use our creativity and culture in a sustainable project that leads to earning some money, celebrating our talents and strengthening the community." The products generated to support a family well-being project were sold at local fetes, schools, through personal networks and at the time of writing were in the process of being marketed via local government contacts to ensure that a) the project was sustainable, b) skills could be learned by the participants and c) some entrepreneurial spirit would add zest to the community participation through having fun together as families!

13.5.2. Reflection on Facilitation Processes

Enthusiasm stems from the root conceptualization being undertaken by the participants. Processes for cross-cultural work need to be based on facilitating the facilitators, rather than playing a central role (McIntyre, 2004, 2003). This was learned through experience with a previous project with Aboriginal organization in Central Australia. This important for self-determination and two-way learning about facilitation processes based on the pooling of skills. Respectful relationships are based on acknowledging one's different ways of knowing. Identity and meaning cannot be factored out or 'bracketed out' of any relationship.

Doug Morgan, as chair of Neporendi provided the governance in terms of the Neporendi constitution. Along with Doug, my role was providing some project planning and implementation skills and building ties with local and state government. D. C., a community leader provided the technical skills in art design and small business computing skills, whilst his mother-in law, a senior Aboriginal woman and L, a woman in her early twenties provided the cooking skills and dance skills. Members of the Neporendi board and participants played a role in drawing the members of the community together to support the project.

13.5.2.1 Critical and systemic underpinnings

Systemic approaches to the management of complex problems (McIntyre–Mills, 2003 draws on the insights of many critical and systemic thinkers in the areas of sociology, cultural studies, management, development studies and education). The basic assumption is that bureaucratic and compartmentalised responses are inadequate to address complex multifaceted problems. The paradigm shift is from the machine metaphor of natural sciences (and the attendant command and control and lineal thinking) to the complexity metaphor (of interrelated systems) and the attendant open communication and matrix responses shape PAR. Methodologically the project is based on participatory action research.

Theoretically it draws on community of practice theory and 'third way' governance approaches to build links within and across community and local government to support disadvantaged families.

The aim and challenge was to address family violence, enhance involvement of the stakeholders and achieve practical sustainable outcomes. The objective of this research is to investigate the extent to which entrepreneurial capacity recognition could empower Indigenous Australians.

Conversation that strives to generate shared understanding across generations is central to the project. The I- thou relationship (to cite Buber, 1947, in Buber, 1965) across generations was developed within the context of enjoyable activities at which the community excelled. Celebrating the positive and using this as a basis for developing better understanding and resonance (West Churchman, 1982; Banathy, 1996, 2000; Freire, 1982) that energizes the participants to work together. Open communication is mindful of multiple viewpoints and meanings associated with different cultural maps. This approach to communication draws on the work of Habermas (1984) and takes into account the so-called objective, subjective and intersubjective worlds of participants. The research process is mindful of the need to take cognisance of power (Foucault and Gordon 1980) when listening to narratives, recording rich pictures and developing areas of convergence and divergence in soft systems maps. Problem appreciation is an iterative process of taking into consideration many variables and exploring the meanings of many interest groups through critical systemic thinking techniques (Churchman, 1982; Jackson, 2000; Flood, 2004; Midgley, 2004). Rich pictures are conceptual maps drawn by the participants. Using conceptual maps is culturally familiar and based on Aboriginal art. Nevertheless the strategy was informed very loosely by the work of Institute of Development Studies in Sussex (for example Chambers, 1997) and the work of Checkland and Scholes (1990) who developed soft systems and rich pictures for problem definition and problem solving in organizational/ community management contexts. Perceptions are mapped using soft systems modelling. This means showing the detailed connections in the thinking and perceptions. It is a well-known technique used in management, policy and human services/organisational research. The more details of what works and why, the more likely the partners are to be able to address the problems and to identify entrepreneurial opportunities.

The conversations about what makes the participants happy were the starting point of PAR. New technology such as digital cameras, tape recorders and cable television featured highly in discussions, along with dance, music and food. Drawings were used to decorate greetings cards and packaging for home made biscuits. An Aboriginal community leader with computing expertise gave his time. A sense of purpose was generated and a pride in community generated products that could be marketed. Social capital, according to White (2002:268) is a concept that needs to be considered critically. Conceptually it can mean different things to different interest groups. Thus merely studying social networks as if they were objective indicators of something uniform and meaningful for all the participants is mistaken from this critical and systemic point of view. Building networks of trust is indeed a worthwhile goal for enhancing civil governance, but it is by no means unproblematic from the point of view of the participants with unequal power (See McIntyre Mills 2003c). Age, gender and level of education lead to differing perceptions and awareness of perceptions is central aspect of the project.

Problems of organizational and social complexity occur in combination and are interconnected (see Rittel and Weber, 1973) such as polydrug use (use of more than one drug in a range of combinations and for varying lengths of time and varying amounts), homelessness (sleeping rough, sleeping with friends, sleeping in shelters, not belonging to any place, feeling that one is less than welcome), domestic violence (including emotional and physical dimension as they impact on both genders and all age groups), unemployment, low self esteem associated with addictive or risk taking behaviour.

Participatory Governance (McIntyre, 2003 a, b, c, 2004) requires working with diverse people and ideas. This requires working with different ideas and arguments (or discourses). Discourse analysis strives to understand patterns in arguments and the rationale underpinning arguments. It strives to understand areas of agreement and disagreement. Knowledge management (KM) can be a process of working with different domains, discourses or areas of knowledge, rather than in specialized areas, in order to understand the ways in which problems are defined and perceived and to innovate ways to solve problems with the participants. It is more than information management using computers; it is based on understanding the value of diverse ways of knowing (through narratives) for problem solving with the stakeholders. This version of KM strives to improve better communication and better understanding of what works, why, how and to what effect in a community.

a) Problems were mapped as tasks and processes to "what" , "how" and "so what will be the benefit and to whom ?" questions[80].

[80] Building a sense of both rights and responsibilities was central and it was agreed at our meetings that D.C. and L. H. would take responsibility, with advice from the chairperson. The Neporendi Board agreed that systemic approaches fitted well with their understanding of the world. The following tasks were undertaken by the organization:

Drawing up a budget and a time line for the use of 5 000 dollars. The budget comprised costs for materials for art project- paper and paint

Costs for materials for biscuit making.

Costs for marketing. Projected income and price for biscuits and that the profit would allow the biscuit making and packaging to continue in a sustainable manner.

b) Soft systems thinking focuses on tasks, processes and the rationale for the way we think about problems. Soft systems refer to the "why" questions or rationale for thinking and practice. The use of conversation, art and cooking was thought to be useful as a means to create opportunities to work with participants by building on activities that families identified as enjoyable.
c) Critical systems thinking is systemic and works with knowledge, rather than within knowledge parameters. It is open to the ideas of many participants. Participants disagreed with one another and moved into or out of the project, for personal reasons, mostly to do with power and status differences associated with being employed within Neporendi or being senior in terms of age (and therefore having more control over the resources). This had to be managed by some informal conflict resolution, such as "mini meetings" at family or recreation days.

13.6 PROGRESS TO DATE

Aboriginal people of all age groups, including people with disabilities participated. At least 50 people will be assisted over a six-month period and as a result of the support of Australian Research Council, Human Services and Flinders University the project has been extended to three years. Neporendi currently has 30 participants involved in a Women's Program and 20 participants in its CDEP family violence project who have been affected by family violence. Practical projects engage participants and give hope. People feel that they want to devote time to a project that will lead to multiple outcomes, namely: some income, some fun and some sense of collaboration and community. The initiative to work on mutual capacity building between Human Services and Neporendi Board was considered particularly helpful by all the stakeholders. It also provides an opportunity for practical learning, such as learning skills in the kitchen, following recipes and identifying ways to apply learning in a friendly environment. These friendships will provide the basis for the more formal research on social inclusion that will follow in the second phase of the project. Humour, used gently, provides a way to address problems of literacy and numeracy. The challenge was to address the desire to keep the profits for individuals rather than for the community.

Organize the packaging and marketing.

Organize the design project with the children and families

Allocated leaders to organize the community gatherings and biscuit making.

Agreement that everyone would work according to the budget once it is agreed. The cards and biscuits will be ready for marketing according to an agreed schedule.

Debate ensued over who owned the biscuit recipe and how to divide up the profits[81]. A participant who saw the initiative only in terms of personal advantage chose to opt out. This was partly, because of the conflict between a senior and junior Neporendi members and it is an area that needs to be carefully considered for social inclusion. People chose to participate as a result of their extended family links with the Neporendi board. A critique of power, social capital and networks is relevant. One of the reasons for remaining affiliated with Neporendi was because it provided community development employment training and the powerful personal and political links with community decision makers associated with the state and commonwealth government who disburse grants.

The participants across the different age and gender groups understand the nature, causes and consequences of family violence. Circumstances were described as being the causes of conflict. Being unemployed and at the behest of paid employees lead to clashes. Two of the volunteer participants were in competition with paid staff and board members. Power lay with the board and the staff, even though the rhetoric of participation continued and the volunteers saw it in their interests to go along with the board and staff in the hope that employment opportunities would eventuate. Competition over recipes and resources was evident.

Nevertheless the PAR process built capacity using community development networks. It addressed i) improved family well being outcomes, iii) better understanding across age and gender groups and iv) improved quality of life through capacity building and greater respect across age and gender groups.

The project was possible because of the volunteer work by members of the Neporendi Board and members of the Community Development Employment Program. Members of an existing Family Violence Support Network provided the core of the program. At least 10 paid staff and volunteers were involved.

It can be concluded that the members of Neporendi Forum Inc were encouraged to work across organizations to enhance governance and community development.

13.7 PHASE TWO OF THE PROJECT

Phase two addresses social inclusion issues pertaining to health, education and employment. The complex systemic problem is a priority

[81] At the time of writing the biscuits and cards were to be marketed to a list of potential customers. The profits of the exercise are to accrue to a) Neporendi, in order to sustain their funds and to build community and b) to a small business bank account managed by the Neporendi.

policy area. It builds on the art activities to make pictures that are meaningful and directed to a particular purpose. Both the service providers and the service users will draw the rich pictures. The questions will be extended from 'what makes us happy as families?', to encompass questions about 'what services and interventions work, why and how?'.

Health, development and wellbeing are associated with life chances such as access to education and employment opportunities within a healthy environment. Self-expression and being heard is the starting point of good governance (see Bevir et al 2004) and wellbeing.

Data will be collected from Neporendi and from Service providers on what works, why and how. Rich pictures are to be used as a means to understand perceptions. They are conceptual maps drawn by the participants. The process is adapted from the work of Bourdieu (1977) on structures, habitus and power. It also draws on Walsh and Mitchell (2002) whose work in Central Australia was conducted with some of the same participants in my own work (see McIntyre –Mills 2003c) and the Institute of Development Studies in Sussex (see Chambers 1997) and the work of Checkland and Scholes (1990) who developed soft systems for problem definition and problem solving in organizational/ community management contexts. Perceptions are mapped using soft systems modelling. This means showing the detailed connections in the thinking and perceptions. It is a well-known technique used in management, policy and human services/ organisational research.

We anticipate that the computer model will enable users and providers to work across organisational areas. It will enable them to translate the current understanding that social health problems do not limit themselves to the boundaries of any one discipline into the practical management reality of working in partnership for better housing /social inclusion outcomes. The project will enable an Aboriginal student to develop workforce capacity to think more systemically about interrelated or 'complex' problems. Collaboration will help to build skills and could advance governance across organizations and across cultural maps of Indigenous and non-Indigenous participants as they pertain to the complex problem of housing and homelessness that have not responded to departmentalized approaches to problem solving. The social justice benefits of knowledge management to the organization; staff and service users are:

- Responsiveness based on matching the needs of the user in terms of age, gender, cultural requirements for Indigenous and non Indigenous users and health needs and the service providers in a recursive feedback system, based on soft systems modeling.

- Timeliness, the quick identification, contact and access to most or all appropriate service providers, (not to be confused with the capacity of the each service involved).
- Minimum backtracking, the individual service pathway can be mapped out to ensure continuous progress towards desired outcomes.
- Minimum staging, the client isn't moved from one service to another in sequential stages with needs evaluation processes (often duplicating previous similar processes) occurring at each stage.
- Organizational data that can be analyzed to determined network gaps or inefficiencies, patterns of use, social and operational costs.

We[82] will research the following hypotheses:

- The greater the use of a) participatory design processes, b) cross cultural knowledge management systems and c) Indigenous – non Indigenous capacity building partnerships to address complex problems (such as homelessness, family violence, drug use, unemployment and social inclusion issues) the better the problem solving outcomes for both human service users and providers.
- The greater the use of participatory design to build capacity to articulate partnerships, the more effective the management of complex problems. This project recognizes that the chances for a healthy start in life are

[82] McIntyre with Professor Roche and Professor Roddick will supervise the PhD student. Dr McIntyre, a senior lecturer at the Flinders Institute of Public Policy and Management and Dr Douglas Morgan, will contribute to the research process by undertaking research alongside the student. They will be supported by Roche (Director of NCETA) who will provide guidance on health related matters and knowledge management. McIntyre will act as the overall project leader. McIntyre is a sociologist / social anthropologist with over 20 years experience as an academic, researcher, community development specialist and consultant. Roddick will give detailed advice on the computer programming and will be supported by Morgan (who Chairs the Neporendi committee) and Metcalfe with computing expertise and soft systems modelling. Since the late 1980's Professor Roddick has contributed to the area of intelligent databases including the development of techniques for data summarization, spatio-temporal databases, query languages, evolution and change in data and metadata management, information semantics and, data mining and knowledge discovery. Morgan, previously a senior lecturer and the Co-ordinator of Yunggorendi Student Services at Flinders University and currently a senior research fellow at University of South Australia will act as a mentor for the researcher and assist with the research. His areas of expertise are Indigenous transition education, transferal of academic skills in higher education and cross-cultural and diversity training. He has a track record of publications in this area and in Indigenous knowledge management and has a wealth of experience with Indigenous organizations, Indigenous transition education and cross-cultural and diversity training. He has undertaken research and consultancies and managed projects for ATSIC examining CDEPs in South Australia and evaluating the effectiveness of cultural and diversity training for the South Australian Justice Portfolio.

embedded within a complex cycle of a) socio-economic status, b) employment, c) housing, and d) education opportunities and that unless these issues are addressed the well-being objectives cannot be met.

13.7.1 Potential outcomes and outputs

- Provide a computer model of knowledge management that places Indigenous knowledge at the forefront and that could be applied in a range of human service settings.
- Enhance workforce capacity building and problem solving to span organizations and based on working with, rather than within knowledge areas (see Van Gigch 2002).
- Demonstrate systemic collaboration.
- Create management links across areas of knowledge relevant to achieving better outcomes (Habermas 1984).
- Identify the extent to which collaborations can meet current needs.
- Map expressed need, normative need and perceived need, the referral traffic and the refusals in a dynamic evaluation and management model.
- Share the findings and its relevance to policy through workshops with a range of stakeholders.

The resulting flow-on effects of adopting an holistic and systemic approach (Roche & McDonald, 2001) for promoting intergenerational health and well-being will: a) achieve better role modelling across men and women of all age groups and b) lead to greater control over employment opportunities, socio-economic status, and improved health (Warren, 2004).

13.8 CONCLUSION TO WORK IN PROGRESS

The participatory design approach addresses Indigenous human service and social inclusion issues through capacity building to develop and strengthen partnerships and entrepreneurship.

Capacity building to improve workforce management skills across health, housing and welfare are needed to achieve greater social inclusion requires a paradigm shift (Roche and McDonald, 2001) in the way social problems are approached and managed.

Problems need to be addressed systemically by developing ways to manage knowledge, in order to address complex challenges across sectors and disciplines.

REFERENCES

Australian Bureau of Statistics (ABS) 2001, Census of Population and Housing.

Ackoff, R.L. and Pourdehnad, J., 2001, On misdirected systems, *Systems Research and Behavioural Science*, 18(3): 199-205.

Ackoff, R., 2004, A black ghetto's research on a university, in: *Community Operational Research*, G. Midgley and Ochoa-Arias, Eds. Kluwer, London.

Barton, J, Emery, M., Flood, R.L., Selensky, J. and Wolstenholme, 2004. A maturing of systems thinking? Evidence from three perspectives, *Systemic Practice and Action Research*, 17(1): 3-36.

Banathy, B., 1996, Designing Social Systems in a Changing World, Plenum, New York.

Bevir, M, Rhodes, R.A.W. and Weller, P., 2003 Traditions of governance: interpreting the changing role of the public sector. *Public Administration.* Vol 81 No 1., 1-19.

Buber, M., 1965, *Man and Man*, Macmillan, New York.

Bourdieu, P., 1986, The forms of capital. Reprinted from Richardson, J. in: *Education, Culture, Economy and Society*, A.H. Halsey, ed., 1997, Oxford University Press, Oxford.

Bourdieu, P., 1977, *Outline of a theory of practice. Cambridge studies in social and cultural anthropology*. Cambridge University Press, Cambridge.

Chambers, R., 1997, Whose Reality Counts: Putting the First Last, ITDG, Bath.

Churchman, C. West, 1982, *Thought and Wisdom*, Intersystems Publications, California.

Checkland, P. and Scholes, J., 1990, *Soft Systems Methodology in Action* Wiley, Chichester.

Duhl, L., 2001, *Systems and Service at School of Public Health*, International Systems Sciences, Berkley.

Edgar, D., 2001, *The Patchwork Nation: Rethinking Government-Rebuilding Community*, Harper, Sydney.

Fals-Borda, O. and Rathman, M.A., 1991, *Action and Knowledge: Breaking the Monopoly with Participatory Action Research*, Intermediate Technology, London.

Maturing of Systems Thinking? Evidence from Three Perspectives

Foley, D., 2003, An examination of indigenous Australian entrepreneurs. *Journal of Developmental Entrepreneurship*, 8(2).

Foley, D., 2002, Indigenous standpoint theory, *Sharing the Space Conference*, Flinders University.

Foucault, M. and Gordon, C. eds., 1980, *Power/Knowledge: Selected Interviews and Other Writings 1972- 1977*, Harvester, Brighton.

Freire, P., 1982, Creating alternative research methods: learning to do it, in: Creating Knowledge: a monopoly? Society for participatory research in Asia, B. Hall, A. Gillette and R. Tandon, New Delhi: 29-37 in: *The Action Research Reader*, Deakin University publication, Victoria.

Gaventa, J., 2001, Towards participatory local governance: six propositions for discussion, Paper presented to the *Ford Foundation, LOGO Program with the Institute of Development Studies*, June.

Geertz, C., 1973, Thick description: towards an interpretive theory of culture, in: *The Interpretation of Cultures. Selected Essays*, C. Geertz, Basic Books, New York, pp.3-32.

Giddens, A. 1998, The third way, Polity, Cambridge.

Habermas, J, 1984. *The theory of communicative action: reason and the rationalization of society*, Beacon, Boston.

Jackson, M., 2000, *Systems approaches to Management*, Plenum, New York.

Jamrozik, A., 2001, *Social Policy in the Post Welfare State: Australians on the Threshold of the 21st Century*, Longman, NSW.

McCline, R.L., Bhat, S. and Baj, P., 2000, Opportunity recognition: an exploratory investigation of a component of the entrepreneurial process in the context of the health care industry, *Entrepreneurship Theory & Practice*, Winter.

McDonald, C. and Marston, G., 2003, Fixing the niche? Rhetorics of the community sector in the neo liberal welfare regime, *Just Policy: A Journal of Australian Public Policy*. 27 August, VCOSS, pp. 3-10.

McIntyre, J., 2002a, A community of practice approach to knowledge management and evaluation re-conceptualised and owned by Indigenous stakeholders, *Evaluation Journal of Australasia*, 2(2): 57-60.

McIntyre, J., 2002b, Critical systemic practice for social and environmental justice: a case study of management, governance and policy, *Systemic Practice and Action Research*, 15(1): 3-35.

McIntyre, J., 2003a, Yeperenye dreaming in conceptual, geographical and cyberspace: a participatory action research approach to address local governance within an Australian Indigenous Housing Association, *Systemic Practice and Action Research*, 16(5): 309-338.

McIntyre, J., 2003b, Participatory democracy: drawing on C. West Churchman's thinking when making public policy, *Systems Research and Behavioural Science*, 20:489-498.

McIntyre Mills, J., 2003c, *Critical Systemic Praxis for Social and Environmental Justice: Participatory Policy Design and Governance for a Global Age*, Kluwer, London.

McIntyre, J., 2004, Facilitating critical systemic praxis by means of experiential learning and conceptual tools to enhance our capacity to design systemic policy and praxis that promotes understanding through better governance, *Systems Research and Behavioural Science*, 21:37-61.

Peters, G. and Savoie, D., 2000, *Governance in the twenty-first century: Revitalising the Public Service*, Canadian Centre for Management Development.

Pierre, J. ed., 2000, Debating *Governance: Authority, Steering and Democracy*, Oxford University Press. New York.

Rhodes, R.A.W., 1997, *Understanding Governance: Policy Networks, Governance, Reflexivity and Accountability*, Cromwell Press.

Pearson, N., 1999, Positive and negative welfare and Australia's Indigenous Communities, *Australian Institute of Family Studies in Family Matters*, 54(Spring/Summer).

Reason, P., 2002, Justice, sustainability and participation inaugural lecture, http://www.bath.ac.uk/%Emnspwr/Papers/InaugralLecture.pdf, published in: *Concepts and Transformations*, 7(1): 7-29.

Roche, A. and McDonald, J., 2001, *Systems, Settings, People: Workforce Development Challenges for Alcohol and Other Drugs Field*, Commonwealth Department of Health and Aged Care.

Van Gigch, J.P. 2002, Comparing the epistemologies of scientific disciplines in two distinct domains: modern physics versus social sciences. Part 1: the Epistemology and knowledge characteristics of the physical sciences. Systems Research and Behavioural Science Vol. 19, No 3: 199-210.

Van Gigch, J. 2003, *Metadecisions: rehabilitating epistemology*. Kluwer.Plenum.

Walsh, F and Paul Mitchell 2002. Planning for country: cross cultural approaches to decision-making on Aboriginal lands. Alice Springs .IAD press.

Warren, L., 2004, A systemic approach to entrepreneurial learning: an exploration using storytelling, *Systems Research Behavioural Science*, 21.

Wenger, E., 1998, *Communities of Practice: Learning, Meaning and Identity*, Cambridge University Press.

White, L., 2001, 'Effective Governance' through complexity thinking and management science, *Systems Research and Behavioural Science*, 18(3).

White, L., 2002, Connection matters: exploring the implications of social Capital and social networks for social policy, *Systems Research and Behavioural Science*, 19(2): 255-270.

Chapter 14

ADDRESSING INDIGENOUS AUSTRALIAN SOCIAL EXCLUSION:

A Strategic Praxis Approach To Research And The Provision Of Social Services

DOUGLAS MORGAN
Chair of Neporendi Forum Inc and Senior Lecturer , University South Australia

Key words: strategic, praxis, services

Indigenous Australians have been the object of much academic interest since Anglo-Europeans first came into contact with them and they have become the subjects of intense research – prompting claims that they are the most researched peoples in the world. Early research portrayed Indigenous Australians as subhuman, bereft of technology, organized social systems, codes of law and any trappings of civilization. The initial focus upon Indigenous Australians was to civilise them and bring them into the 'brotherhood of man' through science and religion, although it was not expected that they would participate equally. They were considered to be unevolved, Aboriginal ('original man'), without the developed mental capacities to take full advantage of the benefits of civilisation and thus for much of the past two centuries they were denied it. During the early twentieth century they became the focus of academic curiosity seeking to juxtapose (and justify) Western civilised social structures with those of 'stone-age' primitive cultures. Since the advent of the 'civil rights' movements of the 1950's and 1960's much of the research has centred upon addressing Indigenous Australian social disadvantage. While Indigenous Australians enjoy many of the benefits that Western civilisation has provided, centuries of research and academic endeavour have done little more than entrench notions of their inability to engage and take advantage of the social benefits on offer in Australia.

Prior to Anglo-European contact and domination, Indigenous Australians had developed a complex social system that was sufficient for their needs and consistent with their lifestyle and beliefs. At the time of the arrival of Anglo-Europeans they enjoyed greater life expectancy than their counterparts from Europe. They had better health outcomes, almost unlimited access to their lands and thus to nutritious foods and through the kinship system experienced social support structures that defined not only their societal roles, obligations and responsibilities, but also confirmed their social and spiritual identity. Over a period of some two hundred years the situation has changed markedly for Indigenous Australians. Their access to their lands and their rights to its bounty have been denied through imposed European laws that failed to recognise Indigenous ownership, their populations have been decimated through introduced disease and maltreatment and they have been largely excluded from the benefits of the social system that replaced their own. Until 1967 Indigenous Australians did not have citizenship rights in their own land and were denied equality of opportunity to participate in and share in the wealth generated from their plundered resources. To facilitate this plunder they were removed to church run missions and state run reserves where their every action was governed by an alien system of rules and conventions that in many instances extinguished the former social support structures that had sustained them for more than 40,000 years. Many were forced to work in menial employment for low wages or where their wages were paid to the state to be kept in trust for them only to see their money stolen by the very apparatus that was supposed to protect it – the state. Despite now having citizenship Indigenous Australians have benefited little. They currently fare worse that Anglo-European Australians on just about every indicator. They have shorter life expectancy, poorer health outcomes, are more likely to be incarcerated for minor offences, experience higher rates of unemployment, have less disposable income, are more likely to live in overcrowded substandard housing, and are less likely to achieve educational standards comparable with Anglo-European Australians. For this Indigenous Australians deserve redress and just compensation. Anglo-European Australians have a moral and ethical duty to right the wrongs of the past and address the current inequality. However, while most Australians are supportive of Indigenous Australians being treated fairly and achieving equality of outcomes they have been largely excluded from the benefits conferred by membership to Anglo-European Australia and its social structures. The 'safety net' of social support systems and structures that have been developed to ameliorate disadvantage for all Australians has failed Indigenous Australians – the most socially disadvantaged group of Australians. This chapter examines the Indigenous Australian social context, explores the issues and suggests ways to address current approaches that have done little more than entrench the disadvantage experienced by Indigenous Australians.

14.1 THE SOCIAL CONTEXT

Until 1967 the lives of Indigenous Australians was regulated by others. They were not able to make decisions about the things that other Australians took for granted. They had no control over who they associated with (on some reserves and missions a husband could not visit his wife and family without permission), were not free to seek employment, change employment or to expend the fruits of their labour without authority (in Queensland, Indigenous Australians' wages and savings were stolen while supposedly held in trust), could not seek medical attention for themselves or their children without permission (often this was not available), were denied more than a basic education (often amounting to as little as 3-5 years of formal education), were forced to live in substandard housing (often without running water and electricity) and were denied political representation (as non-citizens they were not entitled). While many Indigenous Australians earned citizenship through 'being of good character' (renouncing their Indigenous identity, becoming Christianised, or serving in the Australian armed services) the rights that accrued could be rescinded for deviant activity (public drunkenness, associating with community and family who were non-citizens or associating with the wrong political party – the communists). The affects of this, when granted citizenship, were a subsection of Australian society who were ill equipped to take advantage of the social services and structures in place for all Australians.

Indigenous Australians emerged from a period of 'protection', with little or no skills in determining their futures (this had formerly been decided for them by their 'protectors'), and from racist policies (based upon the arrogance of cultural superiority) of 'assimilation' which resulted in many families being devastated by the forced removal of their children into homes and foster care. This has had a devastating impact upon the social and emotional well-being of Indigenous Australians through the loss of their identity, the destruction of Indigenous social support structures, the non-transference of Indigenous parenting skills and the failure to replace this loss with mechanisms and structures to facilitate the transition into mainstream Australian society (irrespective of the arguments for doing so - or not). Subsequent policies have been implemented based upon research that has failed to take into account the impact of past experiences of Indigenous Australians or their commitment to their culture and identity. The ability of Indigenous Australians to be 'self determining' has been limited by the lack of skills for them to do so and by 'tied' funding that counteracts the stated aims of the policies.

14.2 EQUALITY OF OUTCOMES – A MORAL AND ETHICAL RIGHT

Research is not free of moral or ethical considerations – researchers are members of the Australian community who must address the current circumstances of Indigenous Australians if they are to reasonably think of themselves as participating in a fair and just society – a claim that cannot be sustained at present. It is not enough to advocate the removal of barriers that previously excluded Indigenous Australians from equal access to resources. Equal access to the resources of the various social services has not redressed the problems experienced by Indigenous Australians. While one clearly identified subsection of the Australian community remains disadvantaged, Australia's international standing is diminished. Australia, as a nation, cannot question the actions of other nations towards their own citizens while a subsection of the citizens of this land experience health, housing, education and employment outcomes significantly worse than those experienced by Anglo-European Australians. Australians cannot reasonably conceive of themselves as participating in a just society when some of its members are disadvantaged disproportionately to the rest of the members of their community. In a just and humane society, equality of life outcomes and basic rights for all of that society's members is to be expected, and if not then it would be expected that measures be implemented to achieve this end. Australia's record for human rights rates below that of other Western nations with inequalities in health, employment and incarceration suffered by Indigenous Australians being major contributing factors. These issues must be addressed before Australia can have credibility with the international community and be a member of good standing.

14.2.1 The Limits of Individual Responsibility

It is generally believed that moral responsibility is limited to individual action or personal complicity and their specific consequences. It is also believed that as harm is suffered by individuals, only individuals can and should be redressed for injuries and harms that they have individually suffered. Personal responsibility is seen as being limited, as a function of individual choice. If an action is not endorsed then one is not individually responsible for the consequences, nor can one be expected to contribute compensation. In bounding the limits of responsibility to individual actions and responsibilities, most Anglo-European Australians would argue that they are not responsible for the actions of their forebears and thus are not liable or responsible for the current plight of Indigenous Australians, nor responsible for compensation as those who perpetuated or experienced the original harm are no longer around. It is to this criteria and guiding principles that most Australians resort when facing dilemmas such as those posed by the issue of Indigenous Australian disadvantage. It is these criteria and processes of reasoning that prevails in contemporary Australia and underpins the evaluation of which competing actions is appropriate for resolving Indigenous Australian disadvantage. However, these criteria and reasoning are mistaken and ill-informed.

Contemporary Australians are not immune from moral and ethical responsibility for the plight of Indigenous Australians as a society's individuals are not detachable from that society or its history. Each member of a society bears responsibility (to varying degrees) for the actions, both present and past, of their society. Individuals are not limited to their own actions but are morally responsible for the actions of their society. Responsible individuals are a product of their social environment - they do not spring into being *ex nihilo*. As such, they do not choose actions independent of social character-formation and historical conditioning - individuals are products of their society's moral traditions. They are born into particular societies with particular histories and make use of the social assets and resources of that society. When responsible individuals act, they do so from a perspective that is sanctioned by their society and they choose actions for which there are historical precedents. However, while individuals are products of their society and bear collective responsibility they are still relatively autonomous individuals, who come to have, and make, choices. To argue otherwise, would amount to the claim that no one is responsible for his/her actions; i.e., one could not be morally responsible for one's actions because, as an individual, one is a social construction. Such a position would deny any responsibility for the historical inequalities suffered by Indigenous Australians. Collective responsibility rests not with the denial of individual freedom but derives from the relationship between free, though socially informed individuals, and the historically extended social agencies that incorporate them. Society and its social institutions are of value to the individual, and the individual benefits from them - no matter how impoverished the society or its institutions, they are used by the individual to constitute themselves. Societies are constructed for, and by, their members who collectively own their resources - the members of a society are the reason for its composition. There are no other acceptable justifying grounds for the existence and basic activities of collective social agencies than the well-being and demands of its members. Social resources are collectively owned and as such bring collective responsibility. Moral responsibility is not limited to liability for praise or blame, it also incorporates liability for the collective moral debt and thus for social compensation.

14.2.2 A Collective Moral Responsibility

Moral responsibility derives not from personal complicity or wrong-doing but because there is a continuance of harm from an original wrong by a historically extended social entity. Indigenous Australians remain as an impoverished and alienated minority within the broader Australian society – the dominant culture that originated with the Anglo-European invaders (the vertical dimension of morality). Members of a society are morally responsible for the just distribution of that society's resources and are morally accountable for inequalities. The responsibilities of members of oppressed and alienated minorities are attenuated - they are not responsible for the inequalities suffered but will be owed redress from those who benefit (the horizontal dimension of morality).

At the time Australia was 'colonised', Indigenous Australians were deprived of their traditional lands and, in most cases, denied access to their traditional resources in contravention of the moral and social standards of the day. This was done without negotiation. Indigenous Australians have never been compensated for their loss of lands and resources - consent was not given and was never sought. Indigenous Australians resisted dispossession, which was often brutal, and clearly counter to Colonial Office policies from the 1830s. Native title was recognised throughout the British Empire and yet the rights of Indigenous Australians were ignored. The early colonists of Australia had a responsibility to uphold the rules of their society at that time but in failing to follow the directives of the Colonial Office, an institution of their society, they were at odds with their moral responsibilities. Through mostly failing to sanction those who transgressed its rules, the society of the day assumed the moral responsibility for the actions of those members that dispossessed and brutalised the Indigenous owners of Australia.

Contemporary Australians have inherited the assets accrued through the actions of their forefathers and thus their moral debt. These assets are now collectively owned by all Australians and with it the moral responsibility. While no member of the current Australian society can be held individually responsible, or be blamed for what has happened in the past, as collective moral responsibility does not depend on personal blameworthiness or praiseworthiness, they are morally responsible nonetheless. Contemporary Australians are liable for a collective moral debt and thus liable for social compensation. Historical liability is not a result of personal complicity in the original transgression, but a result of an on-going wrong to an identifiable social sub-group of Australian society caused by previous members of that society. Membership of a society is sufficient alone for liability for the acts of past generations. Contemporary Australian citizens are members of a society, a historically extended entity, who are not only constituted by their present society but their past history as well. Thus moral responsibility for the disadvantage and inequality experienced by Indigenous Australians does not derive alone from the actions of those in charge of the present social institutions but also from the past actions of their ancestors. As the members of contemporary Australian society have inherited the assets of their forebears then it is reasonable to expect that they also inherit and assume responsibility for their moral debts.

That individuals can be responsible for actions in which they are not complicit is not without precedent. Precedents exist from which one can argue that responsibility for actions of social agencies extends beyond recent activities. In the 1950s and 1960s the British Government, with the permission of the then Australian Government, contaminated Maralinga in South Australia with plutonium from nuclear weapons testing. Even though neither the present British Government nor a majority of the present British people took an active role in or actively supported the nuclear weapons testing they were held to be responsible for the contamination of the Maralinga site. After ten years of legal action by local Indigenous Australians it was found that both the British and Australian Governments have both the legal and moral responsibility for cleaning up the site and compensating the original owners for their loss - at a cost of 100 million dollars. Not only is one responsible for one's actions and collectively responsible for the actions of contemporary's, but one is also liable for the acts of past generations. It was accepted by the British judiciary that the local Indigenous people have a right to restitution from both the British and Australian people, even if, in strictly moral terms, the current generation is blameless. Personal complicity in wrongdoing is not necessary to establish a collective moral debt, nor is it necessary to establish a direct relationship between one's moral responsibility and that of social agencies. Societies and their agencies are extended historical entities that benefit past, contemporary and future members. At any point on this continuum one is collectively morally responsible for the past and present actions of the social agencies. Moral responsibility, and the liability for compensation, is not contingent on personal complicity in or apportioning personal blame for wrongdoing. The current generation of Australians is morally responsible for the plight of Indigenous Australians, though not personally to blame. It follows then that there is a moral justification for Indigenous Australians to receive compensation for the losses suffered through invasion and for the inequalities that they still experience. There is a special moral imperative to address the unequal social outcomes of Indigenous Australians, which justifies greater than equal distribution of resources to meet their needs, or whatever else is required even if this then disadvantages other Australians.

14.2.3 A Special Moral Imperative to Act

That Indigenous Australians deserve just compensation for past treatment is clear but it is also clear that equality of access to social resources is not sufficient redress. While Indigenous Australians are no longer denied equal access to resources there are many Australians who would baulk at providing greater resources despite the fact that they remain a sub-group of Australian society, which is clearly disadvantaged. They are entitled to a just share of Australia's resources in order that the source of their inequality is eliminated. However, to interpret this as access to a portion of resources in equal measure to that received by every other Australian does nothing but entrench social and economic disadvantage and maintain the *status quo*. Appropriate redistribution of resources in compensation is justified.

The notion of compensation or increased access to resources, to address social inequality is not new. It has been the subject of much debate in modern philosophy. The discourse on the just distribution of social resources usually has two starting assumptions. One is the assumption that all persons are and should be treated equally, unless there are relevant differences between them. The second is the assumption that it is reasonable to assume that all persons are equal. However, it is a mistake to think that it should follow automatically that one requires a justification to treat people differently as the second assumption does not follow from the first. It is not self evident that all persons are alike and can be classed together. Such claims need to be justified, as persons are both alike and unalike, so to claim one or the other requires justification. It is only when the claim, that all persons are alike, is justified that there follows a justification for the claim for equal treatment for all. Arguments against increased Indigenous Australian access to social resources on the grounds that this is in itself iniquitous, are not arguments that can be justified from available evidence. Based upon relative disadvantage, the notion that social resources should be allocated only in equal measure requires as much justification as arguments to the contrary. In a society supposedly founded on principle of fairness, Indigenous Australians' poor outcomes on just about every social indicator provide sufficient justification for differences in their treatment.

Indigenous Australians have a right to expect equality of outcomes in all areas of Australian society. There is no doubt that they were adversely affected by policies, some of which were well intentioned, that were based on mistaken assumptions underpinned by culturally biased research. Many past policies were racist, falsely based and proved ineffective in achieving their stated outcomes. For interaction to be effective in achieving beneficial outcomes for social service delivery it is evident that a basis must be clearly established and agreed by both parties. Past explanations, arguably premised on the dominant cultural viewpoint, have resulted in Indigenous Australians experiencing socially inequitable outcomes disproportionately worse than those Anglo-European Australians, despite structural changes being made to the processes of social and physical interaction. The pragmatic dimension that this presents has been insufficient to motivate action to resolve the resulting paradox – increased resources and recognition of the issues have not resulted in improved Indigenous Australian social outcomes.

Establishing and accepting moral responsibility for the current inequalities experienced by Indigenous Australians does not give a direction, a course of action to follow, in providing redress. Recourse to moral principles requires that one has a choice to make between two or more courses of action. Despite few Anglo-European Australians accepting that they have a moral responsibility for Indigenous Australian social outcomes, there has been an increase in resources in some areas to rectify the current inequalities. As shown above, Indigenous Australians have a right to special treatment that exceeds the duty of social service providers merely to provide equitable services. Indigenous Australians are entitled to special treatment as restitution for their illegal dispossession, and consequent economic loss, and because they continue to experience unequal outcomes on just about every social indicator. Despite increased attention from researchers and academics and additional funding little improvement has occurred.

14.3 THE INDIGENOUS AUSTRALIAN SOCIAL CYCLE

Much of the research pertaining to Indigenous Australians has failed to recognise or properly account for the impact of their social experiences and current circumstances. Despite some increased resources to establish and resolve inequality, and reasonable and logical reasons for doing so, it has not been sufficient and may not be sufficient to resolve Indigenous Australian social disadvantage. Indigenous Australians have consistently rejected Anglo-European social resources where they impacted negatively upon their cultural identity. Thus the paradox remains for service providers - an environment of equality of access to social services has been created but it has not produced outcomes of equality. Responding to issues of inequality to satisfy the dictates of institutions derived from one cultural perspective have failed. New responses are required – responses that include Indigenous Australian perspectives. However, there is sufficient evidence from past interactions to suggest that for areas of conflict to be resolved between Indigenous and Anglo-European Australians requires an approach that has not been sufficiently addressed or has been unrecognised by past Indigenous policies.

14.3.1 The Indigenous Social Experience

A new research and service paradigm is required as Indigenous Australians appear to be trapped in a cycle that entrenches their disadvantage. While there may be reasons for non-participation in Australian society that are 'cultural', many of the barriers to Indigenous Australians' social inclusion are not of their making. Since 1967, Indigenous Australians have been the subject of much research resulting in many initiatives to address their relative disadvantage. There are programs to develop and enhance employment opportunities, to provide affordable housing, to assist and support primary, secondary and tertiary education and a major emphasis upon redressing health inequalities. With a few minor exceptions (infant mortality rates have been significantly reduced, though still far greater than that for all Australians and there have been some improvements in educational outcomes) these have largely failed. For Indigenous Australians the issues of employment, housing, education and health are interconnected and influence the outcomes in other areas, and crucial to the implementation of the services provided is the political climate in which they occur. The resulting social cycle is difficult or impossible for Indigenous Australians to escape as each area influences the other as the diagram below indicates.

14.3.2 Employment

While employment policies have recognised the value of being educated in the skills requisite for this to occur, other crucial factors are often ignored. Indigenous Australians are still being impacted upon by the policies that were in place prior to citizenship. These policies split families, denying many Indigenous Australians the benefit of modelling behaviour from a breadwinner that would assist their transition into the employment market place. A dependence upon welfare was created that has in many cases become entrenched. The control of the fruits of Indigenous Australians labour by the state and their agencies removed the nexus between employment and financial independence and responsibility. Compounding this is the availability and location of housing and the poor health status of Indigenous Australians. Those Indigenous Australians with the closest links to their lands were able to retain those links because the lands were not of economic benefit to Anglo-European Australians. The limited economic activity in these areas reduces employment opportunities for those who wish to maintain their identity upon their own lands and forces those who would participate in employment to break their family and cultural links by moving away to areas where employment is available. Their socio-economic status forces many Indigenous Australians to occupy areas on the urban fringes that are poorly serviced and are often located further from sources of employment. Despite initiatives to address employment issues (many Government and private organisations have programs in place to recruit and retain Indigenous Australians), the ability to respond to opportunities where they present is severely impacted upon by the availability of disposable income, the requisite skills and the health status of those concerned. With effective unemployment rates of around sixty percent, there are many in the Indigenous Australian community who put the case that if every Indigenous Australian had employment that was meaningful and well remunerated then the issues of their disadvantage would be resolved within a few generations.

14.3.3 Housing

The types and availability of housing for Indigenous Australians result from a number of interrelated factors – historical, socio-economic and cultural. These factors affect the ability of Indigenous Australians to avail themselves of the benefits offered in other areas of the social cycle. Historically the reserves and missions were located away from major Anglo-European settlements. Often these were located within the lands of the original inhabitants. Those lands that had economic benefit to Anglo-European Australians were quickly subsumed and thus denied to Indigenous Australians. Given the strong attachment of Indigenous Australians to their country many have chosen to stay on the missions and reserves despite the lack of suitable housing and the lack of employment opportunities. In some of these communities the housing available was built during the nineteenth century for populations that were half what they are today leading to overcrowding with the resultant impacts upon health outcomes. While there have been some major initiatives to address this often the responses have been inappropriate or inadequate. In many cases the housing provided does not account for Indigenous Australian family structure but is designed for the typical Australian 'nuclear' family or it is inappropriate for the location and climate and often lacks infrastructural support such as maintenance services for sewerage, plumbing, electricity and rubbish removal. For those who, by circumstance or design, are unable to stay or have been forced from their country into the towns and cities the situation is little different. Either through covert and overt racism and economic circumstance most have been forced to the fringes of 'civilisation' into basic housing which is poorly serviced and overcrowded. The location of housing availability entrenches socio-economic disadvantage, poor health outcomes, poor employment prospects and impacts upon the ability to participate fully in education. The location of housing is affected by affordability which is dependent upon disposable income that is in turn a product of employment opportunities. The ability to gain employment depends upon the ability to access the source of employment, which is dependent upon transport, which is itself dependent upon disposable income. Compounding these factors are the health affects that come from overcrowding and poor nutrition that result from inappropriate and inadequate housing. These entrench the inability to seek, train for, gain and maintain meaningful employment which would allow access to more suitable and better located housing. Without measures to break this nexus there are unlikely to be any change in Indigenous Australian circumstance.

14.3.4 Health

The health outcomes for Indigenous Australians result from a combination of factors which make the causes difficult to address. As with the other social issues for Indigenous Australians these have their basis in past policies and practices, low socio-economic status, inappropriate and overcrowded housing, poor behaviour modelling and inappropriate or irrelevant health education. The policies of 'protection' and 'assimilation' have permeated the consciousness of Indigenous Australians. The very agencies that are charged with providing social services to Indigenous Australians were also the same one's responsible for devastating the Indigenous community through the forced removal of our children from their families. They were also the same one's that Indigenous Australians resisted and hated for their role in applying these mentally and physically devastating policies. In many instances there is a deep suspicion, active avoidance and lack of support for the activities of these agencies despite the benefits they can now offer. The effects of past policies have left many Indigenous Australians suffering from learned helplessness and hopelessness resulting in many forms of self harm such as substance abuse and suicide and susceptible to risky personal (inappropriate relationships involving family violence), physical (obesity, smoking, alcohol and drug use), sexual (underage and non-protected sexual activity) and criminal behaviour (vehicle theft and high speed chases with law enforcement). These are compounded by avoidance of many contemporary social service providers because of a continuing suspicion of their motives (however well founded) in applying current policy which is often seen as being culturally inappropriate. Indigenous Australian food gathering and preparation practices were supplanted by imposed diets that were nutritionally deficient resulting in a raft of health problems for those subject to state control and these poor practices have been inherited by the current generation of Indigenous Australians (resulting in high incidences of diabetes, renal disease and heart disease, etc.). The former control of every aspect of Indigenous Australians' lives by the state has left a population without the skills or modelled responses or in many cases the sense of personal agency to respond to their own health needs and that of their families and communities.

The other issues present in the social cycle, that is the Indigenous Australian experience, also impact negatively upon their health outcomes. Overcrowded, inappropriate and poorly located housing effect Indigenous Australian health in a number of ways to varying degrees. Overcrowding and lack of disposable income increase the prevalence of family violence, resulting in physical harm, family, education and employment disruption, and the possibility of criminal convictions, fines and incarceration (further impacting upon available disposable income and positive modelling of behaviour). Over crowded and inappropriate housing also increases the prevalence of infectious diseases and makes them difficult to eliminate. The poor socio-economic status decreases the ability of Indigenous Australians to purchase sufficient nutritious foods to enhance and maintain good health as well as limiting the ability to afford medical treatment and purchase medicines to treat illness. The location of affordable housing is often situated away from major hospitals and shopping centres forcing Indigenous Australians to rely upon public transport (less than one third of Indigenous Australians has access to personal transport) which is inconvenient (due to family responsibilities, financial constraints or incapacity) and further reduces their disposable income. The health services available locally are not as comprehensive or coordinated as those of the major hospitals leading to health issues being unrecognised and untreated. The fresh foods available at local shops are more expensive and often of poorer quality leading to reliance upon less nutritious and less healthy, but more easily transported processed packaged foods (tinned foods, packets of chips, biscuits and bottled drinks are easier to carry and store than fresh fruits, vegetables, flour and meat). The compounding affects of the poor health of Indigenous Australians entrenches them in the lower socio-economic strata of Australian society limiting their ability to access appropriate transport, access more centrally located and appropriate housing, undertake uninterrupted education and training and limits their employment readiness.

14.3.5 Education

As with the other areas of social cycle that Indigenous Australians experience, the impact of past policies have been pervasive. Access to education was not a priority prior to the granting of citizenship with the prevailing view that educational efforts be directed largely to equipping Indigenous Australians with the skills required to perform the more menial work tasks in Australian society (domestic, labouring and rural work on farms and stations and remunerated at rates much lower than Anglo-Europeans undertaking the same tasks). The effects of health problems, such as Otitis Media impact upon early schooling with intermittent hearing losses that have lasting impact on their abilities to learn and their inappropriate labelling as behaviourally or educationally deficient. Poor nutrition, lack of access to educational resources and inappropriate housing severely limit the ability to achieve positive outcomes in education. This reinforces and entrenches racist attitudes towards Indigenous Australians within the education system. Despite Indigenous Australians demonstrated abilities to excel in more academic endeavours (the accomplishment of the likes of David Unaipon were ignored when implementing education policy), the opportunities to participate fully in formal education were largely denied. This has resulted in many Indigenous Australians being denied the positive influences that derive from effective educational role modelling. For many Indigenous Australians there is a reduced expectation of educational outcomes affected by past racist treatment and a reduced belief in its efficacy. In higher education there is a deep suspicion of the theoretical development and implementation of policies and practises that relate to Indigenous Australians as these are often seen as counter intuitive to their own experiences. While many Indigenous Australians are now embracing higher education, the system does not allow them agency to influence their educational experience. They are forced to act in and conform to an academic environment that is reified into a patchwork of specialised academic areas. And while Indigenous Australians have moved from being the objects of academic research to becoming the researchers themselves they have been forced to adopt Anglo-European procedures and methodologies in order for their research to have credibility in the educational and social service environment – the same approach that has been failing in its application since it has been implemented. Those Indigenous Australians who have not been able to participate in, or have been denied or have rejected education due to the impact of other areas of the social cycle become further entrenched in the cycle of poor housing, low levels of employment, poor health outcomes and reduced disposable income.

14.3.6 Political Climate

The political climate is often overlooked as a factor in the poor performance of Indigenous Australians on social indicators. In a social environment where politicians deny any notion of responsibility for the effects of past policies upon Indigenous Australians, and refer to this as the 'Black Armband' view of history, it is difficult to raise awareness of the issues which are impacting the lives of the most socio-economically disadvantaged subsection of Australian society. The responses to past injustice and its continuing effects would be familiar to anyone who has worked in the areas of family and domestic violence. The perpetrators responses to reasons for their violence in family situations range from downright belligerence (it happened - who's going to do anything about it?) denial (it didn't happen, it resulted from the victims own actions), a justifiable response (they hit first), pre-emption (they were going to hit me so I hit first), provocation (the other's behaviour was such that violence was necessary), mutual (it clears the air for both parties, though the dominant party is seldom injured), accidental (it started as a small act that escalated out of control) through to partial responsibility (the other party gave as good as they got). These types of responses to the impact of past and present policies upon the lives of Indigenous Australians are all too familiar. There is belligerence (these things happen but its time to stop living in the past - draw a line under what happened and move on, people don't care about past history). There is denial (Indigenous Australians did not own this country – it was not stolen as they were too backward to have a system of land tenure; if the British hadn't settled Australia the French or Chinese would have – it was inevitable). There is the justifiable response argument (Indigenous Australians attacked heroic settlers/battlers who were forced to respond by punishing the perpetrators – a common response to the massacres which occurred). There are the pre-emptive response arguments (Indigenous Australians were a dying race who needed to be 'protected' for their own good by incarceration on missions and reserves). There are provocation responses (Indigenous Australians provoke government services to suspend or revoke funding through their inability to meet funding requirements). There are the mutual responses (public attacks upon Indigenous leaders helps to clear the air and resolve difference – after all they are critical of the services provided to them). There are the accidental type of responses (the significant fall in Indigenous Australians undertaking higher education is an accidental and unintended consequence of removing support mechanisms to achieve equality of education for all Australians) and finally there are the partial acceptance of responsibility (sure research resulting in policy that has adverse consequences for Indigenous Australians has been implemented but if they want to ameliorate this then they should engage more fully with the research process and be more clear about what they want). What is clear in domestic and family violence situations is overlooked or unrecognised in the political environment - the violence and abuse that occurs results from the unwanted domination of one party by another. The power to change or control the environment rests with the perpetrator. The victims can accept the situation, resist, avoid it (by leaving the situation) or attempt to change it. Indigenous Australians are responding to these options with their attendant consequences. Those who accept the situation, do so at the expense of their identity (the social environment is not accommodating of cultural difference). Those who resist either passively or actively are likely to become incarcerated (in correctional and mental institutions Indigenous Australians are disproportionately represented). Those who wish to avoid the current situation or leave have few options other than through self harming practices such as inappropriate eating (which leads to obesity with attendant health problems), substance abuse or suicide. Those who would attempt to change the situation must come to terms with the perpetrator and work with them to create an acceptable basis for interaction. This requires the recognition of the issues and the negotiation of acceptable responses and behaviour. A power imbalance will remain while the dominant party fails to recognise or address their role as key to resolving the issues. A first step in this process is the political acceptance of collective responsibility for past injustices - a necessary first step which is long overdue.

For Indigenous Australians the separate issues outlined above are not inseparable from each other. Each of these factors influences the other areas, affecting the ability to respond to initiatives that may bring improvement. In this environment any single initiative is unlikely to succeed as the preconditions for failure reside in areas over which there is no control. There is a clear need for a multi-perspective, cross-discipline approach to research and service provision involving Indigenous Australians.

14.4 THE APPROPRIATENESS OF RESPONSES TO INDIGENOUS SOCIAL OUTCOMES

The social research environment underpinning the provision of social services to Indigenous Australians appears to be driven by three imperatives, pragmatic, political and moral. While inseparable, the pragmatic and political imperatives are external drivers of individual action and these inform much of the implementation of initiatives to address the current inequalities experienced by Indigenous Australians. The pragmatic approaches are concerned with meeting goals (working within budget) and complying with state or federal legislation (Equal Opportunity or Anti-Racism policies etc). Political imperatives often result from expediency – the need to be responsive to issues of public awareness. These responses are not always consistent and require little or no personal commitment on behalf of researchers or service providers. Moral imperatives derive from personal determinations about right and wrong actions, and liability for praise and blame. As argued above there are two dimensions to moral considerations regarding Indigenous Australians – horizontal and vertical. The horizontal dimension involves resolving issues of equality without consideration of past influences upon the current situation. The response that this brings is the allocation of resources equally between all Australians and the maintenance of the *status quo*. Morally there is no personal commitment as liability for praise or blame rests with the institutional social service agencies - those charged with implementing these services are not complicit in causing unequal outcomes. However, the vertical dimension of morality creates a special moral imperative that extends the notion of individual responsibility to encompass the inherited responsibility for past injustice. The application of the conception of collective moral responsibility yields the simple conclusion that Anglo-European Australians have a special moral obligation in social research and the provision of services that is not shared by other social subgroups in Australia. Services to Indigenous Australians are appropriate areas for restitutional action and researchers and social service providers are uniquely placed to discharge a collective debt, even where this requires increased resources, effort and goodwill.

While the argument to act is compelling, the responses required to address Indigenous Australian disadvantage are less obvious and may be difficult to implement. Indigenous Australians are as diverse as their more than 200 languages and they practice their cultures with a diversity that reflects their environment and the language that encapsulates it. This diversity must be recognised and research and services tailored accordingly, with the role of cultural values and assumptions being understood. Despite views to the contrary the current social inequalities are not the result of a cultural inability to respond to changing circumstances. This is a continuation of discredited Darwinian notions of 'survival of the fittest' that underpinned the 'protection' and 'assimilation' policies that devastated Indigenous Australians in past centuries. Policies of 'self determination' have failed to achieve institutional outcomes as they are based on assumptions of commonality of perceptions that are not necessarily shared by both cultures. These will remain despite the removal of structural impediments to the allocation of services and resources. Nor can addressing racism and other forms of discrimination or educating Indigenous Australians overcome inherent cultural difference. The cultural predispositions that have underpinned past research and social service provision have failed Indigenous Australians and require examination.

14.4.1 Differing Cultural Predispositions

The basis of Indigenous Australians' cultural predisposition rests in the 'Dreaming' - a holistic construct of cosmological, ontological, epistemological and methodological beliefs about the nature of reality. The 'Dreaming' forms a conceptual framework of tools to evaluate reality and informs interactions in the world, orders the universe and defines personal relationships in a dynamic process incorporating all features of the environment, social, physical, mental and emotional within the present moment. Members of Indigenous Australian social structures are interconnected through family ties and 'Dreaming' affiliations that define their obligations and responsibilities. Implicit in this belief structure is the notion of shared identity where it is accepted that actions of individuals' are interconnected with attendant obligations and responsibilities on all who share affiliation. The concept of autonomous individuality, as understood by Anglo-European Australians, is counter intuitive to this understanding of reality. This worldview does not make philosophical distinctions between the ontology, the epistemology and the methods used for their evaluation. Indigenous Australians have a conception of the world that is largely holistic and incorporates all of the features of their environment.

In contrast the research environment and the social services derived from it are largely informed by mechanistic notions of 'reality'. The basis of research and social service provision in Australia is largely founded upon and dominated by a philosophy and practice derived from the hierarchical binary ordering of the ancient Greeks incorporating mechanistic notions of reality (the 'laws of nature'). The predominant feature of this system is narrow mechanistic reductionism (with the machine as its metaphor) focusing upon objective aspects of reality with empiricism as its methodological tool. Ontologically it results in a hierarchy of compartmentalized aspects of reality; notions of individual autonomy separate from context (society); and in notions of knowledge (information) separate from reality (context) – the reification of the whole into component parts that can be described, manipulated and acted upon individually free of its context. These assumptions about the structure of being, knowledge and method allow for the manipulation and control of nature however, advances in modern physics have shown that 'reality' is much more complex than that revealed by the narrow focus of the mechanistic model (and the conceptual absurdities that result from its application). Research based upon objective criteria discount important aspects such as quality of life, general well-being and cultural difference – that which cannot be objectively measured or easily described. Cultural assumptions are fundamental to interaction in the social environment but despite an awareness of this, the dominant fragmented approach of researchers has prevailed and with it the notion that there is a deficit in Indigenous Australian understanding of, and interaction with, 'reality'. This view is questionable and probably mistaken, given that environmental constraints alone would dictate the responsiveness of ontological and epistemological beliefs about the nature of 'reality'.

14.5 ETHICAL RESEARCHING OF INDIGENOUS AUSTRALIAN ISSUES

Indigenous Australians have been the subject of much research however, while researchers have built solid academic reputations, the circumstances for the most socially disadvantaged group in Australia has remained unchanged or deteriorated further. Research is not value free - it results from a particular cultural perspective and serves to reinforce those views. Research is seldom informed by or conducted from an Indigenous Australian perspective but the results are accepted as reflective of the situation under examination. The research produced is often narrowly focussed and often has little applicability in addressing the complexity of issues experienced by Indigenous Australians. This has resulted in an environment where Indigenous Australians are sceptical of the benefits of research and suspicious of the motives of researchers in general.

14.5.1 Indigenous Australian Research Concerns

The following concerns have been raised by Indigenous Australian communities and researchers about research practices:

14.5.1.1 Fair dealing with Indigenous participants and communities

Participants in research are seldom clearly notified that the information provided by them becomes the intellectual property of the researcher conducting the research or their organisation. In some cases the researchers and organisations concerned charge the community and participants for access to the results of the completed research. Indigenous Australians have lost the economic benefits of up to forty thousand years of research.

14.5.1.2 Plain language ethics procedures

The terms used in consent forms pertaining to research are often unclear and misinterpreted by Indigenous participants - many of whom have only limited English language proficiency (as many as thirty percent of Indigenous Australians are functionally illiterate).

14.5.1.3 Informed consent

In Indigenous Australian communities where English is a second language obtaining informed consent is problematic. Community members participating in research are reliant upon the skills of the researcher or those of the interpreter who may not be familiar with the terminology used. Trust in these situations has been misplaced in the past.

14.5.1.4 Ability of ethics committees to properly assess proposals

The ability to assess the value and applicability of research proposals is often outside the expertise of the members of ethics committees. While many ethics committees have competent Indigenous Australian members, the skill required may be outside their expertise. Membership of ethics committees is often determined by higher education institutions and other organisations irrespective of the members' knowledge of ethical practice. In universities there may also be a conflict of interest for Indigenous Australian members with resulting pressure to approve university research of dubious value.

14.5.1.5 Indigenous research by stealth

On many occasions research has been conducted in areas where there are large populations of Indigenous peoples (prisons and hospitals etc). Specific approval to research Indigenous Australians in some of these circumstances is not sought. This is dishonest where the results of such research are reported listing Indigenous Australians as a separately identified subgroup.

14.5.1.6 Ethics approval shopping

Most universities and many government and non-government organisation have their own ethics committees. Unscrupulous researchers can submit the same proposal to a number of committees (some without the skills to assess them properly) in order to obtain a favourable outcome for dubious research.

14.5.1.7 Lack of community representation on ethics committees

While there are many research ethics committees that have Indigenous Australian members few of these represent community interests. The legitimacy of research is questionable where approvals have been given for research without the community concerned having input through representation upon the research committee.

14.5.1.8 Differing concepts of ethical practice.

There have been different concepts of what is ethical behaviour and practice that is dependent upon the community being researched. Ethical research in these instances has failed to recognise and accept that there are different ethical spaces and that research practices and behaviours need to be negotiated accordingly.

14.5.1.9 Unethical research by 'independent' consultants.

There have been instances where ethical clearance process have been circumvented through the use of research consultants who have not been required to obtain ethics approval for the research conducted.

14.5.1.10 Lack of standardised ethical approval processes

While the National Health and Medical Research Centre (NH&MRC) ethical guidelines are recognised as the standard for most medical and health related research there is little standardisation in other areas. There is strong support by Indigenous Australian communities and researchers for standardisation of ethical clearance processes with clear guidelines as to which committees have the appropriate skills to assess particular research proposals.

These concerns have manifest in demands from Indigenous Australians that researchers who have been 'unethical' be 'white listed' to exclude them from conducting any Indigenous Australian research. While many of these researchers may have met the requirements of research ethics committees they may have failed to conduct their research in a manner deemed ethical to Indigenous Australians. The differing perceptions of what is considered ethical are often more than a mater of interpretation or simple miscommunication. There are cultural differences that underpin interaction between Indigenous and non-Indigenous Australians that remain largely unrecognised. While Indigenous Australians have been forced through their domination by Anglo-European Australians to accommodate the resulting ontological and epistemological assumptions that follow in their engagement in Anglo-European Australian contexts, a similar accommodation is seldom made for Indigenous Australians. These cross-cultural differences can be profound. One way to understand the complexities of trans-cultural and cross-cultural interactions is to consider the following. While one can stay in a non-English speaking community and hear a particular song that encapsulates the feelings and experiences engendered, these would be unlikely to engender the same feelings and experiences for a local familiar with the language. An understanding of language adds another layer to the meaning of the experience. The language and beliefs give enriched meaning to experiences and allow them to be interpreted in particular ways within a given context. Research of Indigenous Australians has been fraught with misunderstanding and the omission that result from shared experiences engendering differing meaning largely dependent upon cultural predisposition.

The questioning of the value of research by Indigenous Australians should not be understood as a rejection of Western science and the research that derives from it. It has however, resulted in demands that future research only be conducted where:

- it is community initiated
- conducted by Indigenous Australian researchers
- the intellectual property generated remains with the participants and
- there is a clearly defined benefit to the community

While this might assuage Indigenous Australian concerns about the practices of researchers it is a response that brings with it its own set of consequences. Indigenous Australian researchers, particularly those who have been inculcated with Anglo-European Australian beliefs, may be as unaccommodating of Indigenous Australian beliefs and practices as Anglo-European Australian researchers. They may also be prone to making the mistakes that their Anglo-European Australian colleagues make when they research Indigenous Australian communities which are not their own. This approach may also discount relevant similarities between Indigenous Australians in general that objective, non-community based research might establish. While reinforcing Indigenous Australians rights to their information there is a danger that responding in this way may open Indigenous knowledge and scholarship to the same types of criticisms that plague Anglo-European knowledge and scholarship – that it is exclusionary. While Anglo-European Australian research is dominated by a narrow empirical approach expressed in binary oppositions (true/false, good/bad dichotomies) there is the danger that the results of Indigenous Australian research will be discounted by Anglo-European Australians, even where Western scientific methodology is used. And if Indigenous Australians accept the same approaches from their researchers it may limit the applicability of the research generated. Good and appropriate research requires more than the representation of the issues from one particular cultural perspective however, the privileging of Indigenous Australian voices in research may well be required to redress their relative absence.

14.5.2 Addressing Cultural Issues in Research

For many non-Indigenous Australian researchers the issues raised above may appear insurmountable, as it requires greater consultation with other research disciplines and an expansion of the research paradigm to encompass the cultural predispositions of Indigenous Australians. However, research practises based upon a system of binary oppositions and approaches that have reified the social circumstances of Indigenous Australians into a patchwork of areas and issues for examination and action have failed. While there are distinct cultural differences these are not necessarily barriers to communication and research interaction. There are three things that can assist cross-cultural interaction in research and service environments: acceptance of a shared identity; acceptance of the limitations of language and an acceptance of the need to negotiate a space for difference.

14.5.2.1 Shared Identity

While ethnicity, language and skin colour may signify cultural difference; much of what make us human is shared irrespective of one's cultural identity. Gender identity defines us individually and underpins universal roles such as parenthood, childhood, sisterhood, brotherhood etc with many of these roles overlapping each other (one can be a parent, child, brother/sister and an uncle/auntie without contradiction). While there may be minor differences in how these roles manifest from one cultural group to another there is cross-cultural concurrence and the associated attributes are well understood. These are fundamental aspects of a shared humanity but there are also others which are less obvious but also inherent in human behaviour – group identity. Group identity is not discreet either with people being members of sporting clubs, organisations, social groupings, political groups and religious groups simultaneously. Membership of and identification with these groups in society is often a mater of personal choice with the aims and objectives of the group subsuming personal aims and objectives in that context. The challenge for researchers is to encompass the aspirations of Indigenous Australians in an extended identity that is not exclusionary on the basis of ethnicity, language or skin colour. The aspects of humanity that bind us are greater than those few differences that separate us.

14.5.2.2 Limitation of Language

While language issues can be a barrier to interaction in cross-cultural interactions much of what occurs in these contexts has little to do with signifiers used. Within Australian society the language in common usage often bears little concurrence with the language and terms used by academics (the term 'significant' is used to signify importance in common usage but in the research context it is used only to signify statistical difference). Indigenous Australians are no different and often use English terms in ways that differ from Standard English or that which is in common usage. Indigenous Australians may have a more contextual understanding and application of English that has been enriched with Indigenous language (the Indigenous Australian term 'gammin' can mean joking or untruthful and the English term 'deadly' has different meanings depending upon the context, ranging from sexy to well-done). These are signifiers that can be defined and the contexts in which they occur understood. But some concepts are more difficult to articulate and describe. An example of this type of issue is the concept of motherness. Most people have first hand experience of mothers and have developed concepts of what that entails. They can articulate values for the attribute motherness and conceptions of good and bad mothering abound. While there is some agreement that aspects of mothering can be easily defined and measured to support certain attributes such as caring, nurturing etc., aspects such as loving, compassion and empathy by a mother toward her child are less definable and elicit less agreement.

This is no less difficult to come to terms with than the different understandings and use of sexual terminology between Anglo-European Australian males and females. While males may understand these terms objectively and free of context females often understand them relationally, embedded in context. The response to 'Lets fuck' is very much dependent upon the context in which it is uttered and by whom, irrespective of the meaning attributed by the parties concerned. The term 'fuck' is often understood by males as a physical action devoid of emotion and attachment, whereas females would attribute an emotional component and expect some attachment even for casual relationships. Despite the differences that exist in an area that is so fundamental to human interaction, these are usually accommodated amicably with meaningful interaction (not necessarily long term) established between the sexes. The accommodations made are seldom articulated, for to do so may be counter productive to achieving a sense of shared understanding. Thus defining terminology may obscure the meaning for some parties rather than it being elucidated. Where there are no problems and there are procedures for allowing mutually agreed outcomes, then perhaps these incompatible understandings are better left un-stated. Where the measures used to describe interactions with Indigenous Australians are prone to such error then the researcher should recognise and accommodate these differences.

14.5.2.3 Negotiating a space for difference

What the above highlights is the need to recognise and accept that there are issues in research environments that may not be easily articulated and that these may impact upon the findings. There are concepts such as 'shame' and 'dreaming' which are often misunderstood or poorly understood. In research where concepts such as these are not a factor then the results can be recorded with a degree of certainty. However, where there are issues of disagreement about the reporting of unfamiliar concepts and terms then a space needs to be negotiated to leave issues that cannot be easily resolved unresolved.

14.5.3 Conducting Research with Indigenous Australians

The following suggestions arose from a workshop with Janet McIntyre on researching Indigenous Australian issues attended by non-Indigenous researchers in 2003. To establish good working relations that will enhance Indigenous Australian engagement and produce useful research outcomes it is important that:

14.5.3.1 Trust is established

In Indigenous Australian communities there is a climate of distrust regarding research. If commitments are made to participants and community then they must be kept. If research is conducted for a specific purpose, ensure the participants are aware of it. Where research is contingent upon funding ensure that when permission is being sought for community participation that this is clearly understood. There is nothing worse than raising expectations that an issue is going to be addressed only to have these dashed after those involved have made a personal commitment to the project.

14.5.3.2 The privileged position of the researcher is recognised

While the research outcomes result from the input of participants there is an inverse power relationship in favour of the researcher. This can be lessened through privileging the voices of the participants, engaging them in the processes through which their contribution is encoded in academic language. Allow participants to reconsider or adjust their contribution to keep their original voice if they wish. Map participants contributions according to the number of perceptions established and show which set of views seems most reflective to the researcher. Show that the researcher's views are informed by the contributors to the research.

14.5.3.3 Sufficient time is allowed

Given the unfamiliarity and distrust of research, sufficient time should be allowed to negotiate and conduct research with Indigenous Australians. While there may be time constraints associated with research funding, relationships with participants and communities require time to develop and maintain. Research processes involve a series of iterations and consensus building which needs to be factored into the time allowed to develop relationships with the participants.

14.5.3.4 The cultural bias and predisposition of the researcher is recognised

Research involves a number of perspectives, some of which are more relevant than others. The relevance of these perspectives needs to be justified including those of the researcher. While the researcher may have an overview of the process it may not be sufficient to explain what is under investigation where the context is different from their own. For example, one can observe and report what happens during a particle acceleration experiment but it would contribute little or nothing to theoretical knowledge unless the physicists present explained what had occurred in relation to the hypotheses being tested. Otherwise all that could be recorded from such observation is the whirring sound, the flash of lights and some lines on a piece of paper. Indigenous Australian participants should be given an overview of the research and the hypotheses being tested to inform their contribution. All research occurs within a context and in social research there are often hidden cultural predispositions that can distort the reporting of the situation under observation.

14.5.3.5 The issues raised by the consultation process are listed for consideration by participants

Indigenous Australian research participants seek to enrich their experience through reference to context - often viewed as counter to research aims which are to discount factors considered extraneous. It is important that participants be given access to all of the issues raised and be allowed to reflect and comment upon which issues should be given prominence for reporting,

14.5.3.6 The consequences of research conducted are defined for participants

While research is supposed to be objective it is conducted for a purpose. While the purpose may appear to be clear to the researcher there are often consequences that are not so clear to the participants. During the research process inform participants of the consequences for them, intended or unintended, of their contribution to research. Participants may change or withdraw their contribution where they perceive the consequences as being undesirable. Being frank, open and transparent will engender trust, assist engagement with the research process and create an environment conducive to further research.

14.5.4.7 Disagreement or differing perspectives among participants is reported

If there are a number of competing perspectives these should be articulated particularly where there are consequences for the parties involved. Indigenous Australians are not homogenous and where there are differing perspectives these should be reported with a rationale to support the reasons why one set of perspectives are more relevant to the research than another.

14.5.4.8 Recommendations from research where consensus has not been reached or suggested actions for which there is disagreement are avoided

Often research is conducted for a specific purpose which has funding and or service provision implications for Indigenous Australians. If there is no consensus in support of the recommendations flowing from the research then they should be avoided in favour of a series of options which attend to the various perspectives under consideration. These should be communicated to all participants with any known consequences signposted. While this may not satisfy the funding organisation, who may wish an unambiguous outcome, it is important to keep faith with the participants and maintain an environment conducive to further research. It is human nature that where there is disagreement about a particular issue, even if there is agreed support for the research project, it is the outsider who will be blamed for the problems which arise.

While the above principles pertain particularly to researching Indigenous Australians it is reflective of good social research practice. Observation and research is theory dependant and any research is better for this being articulated.

14.6 A NEW SERVICE PROVISION APPROACH

It is clear that the current paradigm and context in which social services are delivered has largely failed to address the social disadvantage of Indigenous Australians. It is also clear that a coordinated approach is required that recognises and encompasses Indigenous Australian perspectives and there would be many social service organisations that would argue that they are doing just this. The task facing service providers may appear insurmountable but these types of issues have been met and responded to in the past. There has been recognition of differing perspectives (albeit from within the dominant paradigms) with allowances made to accommodate the perspectives of children, women and the holders of divergent religious beliefs within the social service environment. However, while this has been occurring elsewhere, few changes have been made to accommodate the perspectives of Indigenous Australians. At a pragmatic level, service providers are charged with the social duty of delivering resources to those in greatest need and thereby ameliorating the differences in outcomes, but the duty to Indigenous Australians has yet to be met. While all Australians have a responsibility to address the issues of Indigenous Australians' poor social outcomes, it is social service providers who are uniquely situated to make a significant contribution and should receive appropriate support to do so. A *strategic praxis* approach offers a solution that can address service issues simultaneously but its application has been limited.

The measures that have been implemented to address Indigenous Australians social disadvantage have largely failed to recognise the all pervasiveness of one particular viewpoint and its impact upon those who do not share these views. This is ameliorated to some extent by social service organisations recruiting and training Indigenous Australian staff and the provision of cultural awareness training to those of their staff who are in contact with Indigenous Australians in the services environment. And while these measures are laudable they have had only limited impact upon Indigenous Australian disadvantage. Indigenous Australians working within Anglo-European organisations are presented with a cultural paradox. Either they are forced to reject their own cultural predispositions in favour of the dominant culture, as Anglo-European predispositions are based upon dichotomies that dictate that if one view is right (the dominant one) then the other view (Indigenous Australian) is wrong, or they are placed in a position where their views are discounted for not being 'objective'. Either way their ability to mediate the provision of services is limited unless they can find ways to accommodate the cognitive dissonance that results. The benefits of cultural awareness training for Anglo-European Australian service providers, though useful, may be of limited value. Awareness of relevant cultural differences does not necessarily imply that an understanding of the issues is attained, nor does it mean that the knowledge learned can be applied appropriately in the right context. The mental landscape into which this knowledge is accommodated has been shaped by the Anglo-European view of 'reality' and the context in which it is applied is a reflection of that 'reality'. In addition, the social service environment has been designed for and by Anglo-European Australians and favours those who are inculcated with those belief sets. Where cultural training results in cultural sensitivity this often amounts to little more than accommodating cultural affectations (such as clothing styles and food preferences) and avoiding embarrassing *faux pas* (such as paying proper respect to Elders) in cross-cultural contact situations. While useful and desirable in cross-cultural contexts cultural sensitivity does not necessarily allow meaningful engagement with Indigenous Australians, though it is a necessary precondition to cultural competence. Cultural competence, the awareness of another culture's predispositions and the ability to interact effectively and achieve outcomes for an organisation with relation to that particular cultural group, is the goal of cultural training. Cultural competency is desirable and sought after in service providers as it allows the provision of services in designated contexts irrespective of their cultural appropriateness. While there has been an emphasis upon cultural competency in the delivery of services to Indigenous Australians it is still premised upon the notion that the dominant approach informed by a particular cultural predisposition should prevail. And while those who are culturally competent are sensitive to the emotional and spiritual aspects that inform the Indigenous Australian social experience the environment in which the service provision occurs largely ignores or discounts its influence. The ability to interact, apply cultural knowledge, and achieve outcomes does not mediate the structure of the social service environment in which it is applied, irrespective of the degree of cultural competency.

Irrespective of the willingness to embrace cultural difference, the interactions that result will be determined by the dominant culture until such times as the social environment in which services are provided accords equality to Indigenous Australian cultural perspectives. Currently it is only in Indigenous Australian organisations and community run organisations that this is occurring to any degree – it is these that accommodate and mediate the provision of services. However, their ability to be effective in this is very much dependent upon the imposition of the dominant culture in this process (they provide the funding and dictate the terms of its use). In these Indigenous Australian organisations it is the role played by cultural brokers that largely determine the effectiveness of the provision of services. The role of cultural broker entails more than cultural competence. Cultural brokers have a deep understanding of both Indigenous and Anglo-European cultural predispositions, often have an understanding of language and its use and have often been inculcated in both sets of ontological beliefs. The impetus to examine cultural assumptions and predispositions is less likely to occur in situations where they do not clash with that of another culture. The assumptions of Anglo-European beliefs are seldom questioned as the environment in which they are acted upon supports their outcomes. However, Indigenous Australians interacting in Anglo-European Australian social service environments are uniquely placed to examine not only their own cultural predispositions but also those of the dominant culture. Likewise, Anglo-European Australians who live and work in Indigenous Australian organisations and communities are also more likely to have a deep understanding of cultural predispositions and the assumptions that underpin them. This understanding leads cultural brokers to dispense with constructs of 'otherness' and interact with different cultural groups according to the context. Where commitments to one's own cultural predispositions are mediated by an examination of their foundation, the cultural perspectives of others are more likely to be considered less judgementally and accorded greater equality. Those who are forced to examine their beliefs are more likely to understand the ontological and epistemological inconsistencies that result from assumptions applied outside the context in which they are applicable (such as the Anglo-European Australian preferences to treat time as an ontological construct as opposed to Indigenous Australians preference for epistemological constructs of time). They are also more likely to understand that the paradox that results from a commitment to an Anglo-European Australian cultural predisposition can be largely eliminated if it is dispensed with or replaced by those cultural predispositions that are appropriate to the context in which they are applied (this is not to argue that this can be articulated by them in the way it has been here, for such reflection is the domain of those who can indulge themselves with the time to do so – the academics).

14.6.1 Providing Services via Indigenous Organisations – The Strategic Praxis Approach

It follows that if the context in which services are provided must be cognoscente of the predispositions of Indigenous Australians' cultural beliefs and experiences then the appropriate forum for this to occur is through Indigenous Australian communities and organisations. Given that the issues that impact upon Indigenous Australians' poor performance on social indicators cannot be addressed effectively in isolation then processes will need to be implemented to address them simultaneously. The 'silo' approach to social service provision is coming under increasing scrutiny and the value of creating 'communities of partnerships' (COPs) to address the lack of communication and interaction between social service agencies is gaining greater acceptance. This approach has the potential to reduce duplication of services and the application of information technologies allow for greater integration and coordination of planning and service delivery through *strategic praxis*. In essence this is the theoretical construct of formulating a multi-disciplinary approach to addressing social issues. And COPs are one way of implementing *strategic praxis* whereby the social service agencies charged with providing a particular service negotiates the sharing of information and resources to be better able to deliver those services. Not only does this have the potential to improve the quality of service but there is also the possibility of significant reductions in the costs of service.

Where the COPs approach needs to vary from that which is applicable for Anglo-European Australians, is the context in which it is applied and the types of persons who will be charged with its implementation within Indigenous Australian communities. While the diversity and commitment of Indigenous Australians to their particular cultural beliefs are diverse and vary according to their exposure to Anglo-European belief systems and the strength of their identity, COPs are more likely to be effective where the services are provided in contexts that are not replications of the dominant structures. Essentially COPs will need to adhere to the following principles to be effective for Indigenous Australians:

- Be reflective of recipients cultural predispositions, practises and protocols
- Be based within an Indigenous community or organisation (these should be the nexus of COPs activities)
- Involve as many service organisations as required to address the issues of concern
- Be negotiated through cultural brokers who are accepted by and are reflective of the communities they represent
- Do not replace existing services
- Allow community representation and a measure of control over their implementation
- Be accepting of modes of delivery and service rationale that do not accord with those indicated by the dominant paradigm
- Use interpreters where the context dictates (in addition to cultural brokers)
- Share relevant information with partnership organisations
- Articulate all known options and their consequences to allow the community to make informed decisions regarding the services provided
- Be soundly based upon research that is reflective of Indigenous Australian perspectives

- Be conducted in an environment of goodwill that accords equality of respect for the options articulated

14.7 CONCLUDING REMARKS

Indigenous Australians have yet to benefit equally from the wealth that their resources have provided other Australians despite a moral and ethical duty to do so. This is not because there is an inherent inability within Indigenous Australians but because there are cultural predispositions that do not recognise or discounts Indigenous Australian perspectives. Researchers and social service providers have a crucial role in redressing the imbalance and privileging Indigenous Australian voices within the social research and social services environment. While offering a more appropriate approach to the issues of Indigenous Australian disadvantage in Australian society, COPs is not a quick fix solution (See McIntyre, 2002 a, b; McIntyre-Mills, 2003a,b, c). The problems experienced by Indigenous Australians are the product of more than two centuries of racist, inappropriate or misinformed policies. The effects of these will not be addressed in a political term or a generation, for some it will take a lifetime. The solutions to addressing these issues are not the sole domain of Indigenous Australians, but rests largely with Anglo-European Australian society as a whole which has an undischarged obligation to provide redress to Indigenous Australians. To do nothing about this is to doom succeeding generations to a moral and ethical obligation, with attendant blame, that this generation has a responsibility to discharge. To do nothing is not an option to be considered while Indigenous Australians continue to experience injustice that can and should be addressed by contemporary Anglo-European Australians.

Chapter 15

AN EXPLORATION AND EXTENSION OF CHURCHMAN'S INSIGHTS: TOWARDS THE TACKLING OF RACIAL DISCRIMINATION AS A WORLD PROBLEM

NORMA R.A. ROMM

Abstract: Churchman's argument concerning the link between human inquiry and action is explored in this chapter in the context of considering a variety of approaches to the study of "intergroup bias" in social relations. I argue that there are ways in which the study hereof might itself serve to perpetuate "group" thinking (and attendant practices) in the social fabric. Researchers should be more alert to this as a possible consequence of their inquiries, as such thinking may not be amenable to addressing the world problem of racial discrimination. Possibilities for making theoretical and practical provision for indeed tackling this world problem are explored in the chapter. The discussion highlights, extends and endorses Churchman's considerations regarding the importance of our embracing an ethic of (global) good, kindness and care into our way of both "knowing" and "being" so as not to become complicit in the perpetuation of racism.

Key words: global good, global kindness, global care, complicity in racism, (ab)use of collectivity constructions

1. INTRODUCTION

Porra comments that over the years, in his writings, Churchman has frequently talked about ethics in terms of "good", "kindness", and "care"

(2001, p. 20). She notes that he defines ethical behaviour as "purposefully good". But "good", "kind", and "care" have to be seen "from the perspective of all humanity". She elucidates that:

> "This means that narrowly defined interests such as those of a nation, profession, race, gender, political party or religious group *cannot* determine 'good', 'kind', or 'care' for all human kind." (2001, p. 20).

Mentioning her own series of dialogues with Churchman, Porra refers to Churchman's vision of what sort of century the 21st Century should be. She notes that Churchman expresses his hope that:

> "We can do something in terms of caring about people. That is what I want. That also means trying to figure out how to do that. That is not easy. It requires a lot of knowledge and a lot of sympathy … . We are interested in the global, the whole world not just those who have wealth … . Twentieth century will supply many lessons of what constitutes unethical management." (Churchman, cited in Porra, 2001, p. 22).

Churchman here points to the need for the creation of "a lot of knowledge" to address/rectify what he sees as the "unfairness" and "injustice" of the 20th Century. He also suggests that tackling these issues requires "a lot of sympathy". In this chapter, I concentrate on exploring some implications of Churchman's linking of knowledge constructions with global human caring. My aim is to provide an entry into discussion around ways of tackling racial discrimination as a world problem.[83]

Some authors have argued that the social discourse of "ethnicity" as a way of referring to people's identities, can be considered as an advance on the discourse of race, in that at least the former discourse offers people the potential to define themselves in terms of what is particular to them (some kind of cultural location) while the discourse of race does not allow for people to escape from (externally) defined characteristics. Thus Fenton (1999, p. 237) suggests that we can reflect on "the dynamic and hopeful dimensions of an emphasis on ethnicity rather than 'race'". He continues: "If we can construct ethnicities we can also deconstruct and reconstruct

[83] Fuenmayor indicates that Churchman's argument for adopting a holistic approach to our understanding of "the world" is at the same time a plea for us to focus on world problems such as "relationships between developed and undeveloped countries, racial discrimination, the so-called 'new international order', etc." (2001, p. 44).

them" (1999, p. 238). Other authors have argued that a reference to "ethnicity" in people's discourse often simply serves to make more palatable to those seeing the world in these terms their own propensity to regard "others" in terms of negative evaluations. As Anthia and Yuval-Davis note (1996, p. 12) "ethnocentricism occurs when one's own culture is taken for granted as natural". They suggest that this is "characteristic of all ethnicities to a greater of lesser extent" (1996, p. 12). In other words, as soon as people "identify" with some group, they tend to forget that its naturalness is a product of socialization into particular ways of life that are not universal: hence other ethnic groups become inferiorized in discourse and practice.

Churchman's way of thinking about our tendency to categorize people in terms of their (supposed) heritage/culture (and or other supposedly existing "characteristics"), can be gleaned from a number of comments he makes in the context of outlining his systems approach. In his discussion with Porra, he posits that in his view "the present policy [of a particular political party in the USA] towards immigrants ... is outrageously bad. [In the] first place, there should be no nations" (2001, p. 23)[84]. Churchman here shows that as far as he is concerned, the division of the world into nations is a division that can and should be questioned because of the consequences it may have for people in society. He points out that the idea of "letting the importance of your nation be one of the guiding principles or your life" is especially problematic (because of the way that this idea pans out in practice). He remarks that "The U.S. suffers from that disease [this way of thinking/being] more than any other country I can think of" (cited in Porra, 2001, p. 24).

Churchman goes on to note that as regards our understanding of "the human species", "we don't know enough about it. We hardly know ourselves as species at all" (cited in Porra, 2001, p. 25). This is not to say that gathering knowledge about the species can ever be regarded as a value-neutral enterprise – see Romm's discussion on Churchman's epistemology in this volume. Churchman emphasizes (1982, 1994) that "knowing" is an enterprise that is already informed by specific ways of asking questions, and seeking "answers" thereto. He mentions in his interview with Porra that when thinking about issues and developing "information" in regard to them,

[84] Short and Magana have studied the way in which people in the USA seem to respond differently to "immigrants" from different origins. In a survey that they undertook they found that attitudes of respondents to scenarios involving people from Latino backgrounds and those from Anglo-Canadian backgrounds (in which the person had supposedly received a parking ticket) differed significantly depending on the background of the "immigrant". They attribute the differences partly to the way in which the former group is stigmatized in the media (2002, p. 709).

we need to consider the implications of our thinking for our acting in society:

> "If it [information] reduces the probability of attaining the objective [good and kindness], then it is negative information. If it increases it, then it is positive. You can begin to measure the amount of information by the decreased probability of attaining your objective [good and kind], and if it does nothing, it's irrelevant … . I think that information (systems) theory should be a branch of ethics." (Churchman, cited in Porra, 2001, p. 23).

At first sight – especially if we adopt a realist-oriented epistemology (rather than a more pragmatically-oriented one) – this way of defining "information" may seem unreasonable and indeed anathema to the enterprise of truth-seeking (cf. Popper, 1994; Hammersley, 1995, 1997). But in terms of Churchman's epistemological position, the gathering of information is always beset by values/assumptions that need to be reflected upon. Considering any information regarding people's "regular" or "normal" way of interacting with others, the method of gathering the information needs to be carefully examined, so as not to restrict our chances of indeed "serving humanity" via our investigations. Churchman's view on what it means to be involved in serving humanity is linked to his view of caring for others, which for him implies "wanting to join" with them, rather than to create divisions (cited in Porra, 2001, p. 24).

In considering Churchman's epistemological and ethical position insofar as it relates to our humanity, Porra points out that:

> "Churchman sees that humanity is heading toward difficulty. The current emphasis on diversity whether related to nations, gender, race, politics or research, is a trend leading to division of human kind into smaller and smaller special interest groups. Churchman warns that this fractionization leads to erecting boundaries, which in turn [can] foster [destructive] conflict …" (2001, p. 26).

This argument is consistent with the deliberations of Brubaker and Cooper (2000), who point to the dangers (in terms of restricting options for political choices) of "deploying the language of bounded groupness" as if it refers to *pre-existing real entities* in social reality rather than to the "making of the groups it invokes" (2000, p. 33). They indeed criticize Young on this score for her vision of group-differentiated citizenship (even within nation-states) built on group representation. They cite Young's argument that "different social groups have different needs, cultures, histories, experiences, and perceptions of social relations". They note that according to her understanding of these groups, politics should not be aimed at trying to transcend (group) differences but should recognize them as "irreducible" (Brubaker and Cooper, 2000, p. 32).

But Brubaker and Cooper ask (of Young' argument): "what makes them groups rather than categories around which self- and other-identities may but certainly do not necessarily always crystallize?" (2000, p. 32). They point out that Young does not answer this question but simply construes heterogeneity as "a juxtaposition of internally homogeneous, externally bounded blocs" (2000, p. 32). In this way she hides difference at the level of the constituent "groups". This, as Brubaker and Cooper emphasize, in itself has implications for the subsequent choices that seem open to people to pursue in their political involvement in society. The identification of "groups" in a society – whether by (professional) analystis of social life or lay actors therein – is thus not neutral in its (political) consequences.

Of course, Young's vision of a society where provision is made for what she calls "effective representation and recognition of the distinct voices and perspectives of its constituent groups that are oppressed or disadvantaged within it" (1989, p. 261) could be argued to be clearly preferable to the practise of "ethnic cleansing/genocide" (cf. Mitroff, 1994, p. 96). Young's assumption that ours is now a group-differentiated society, and that it will continue to be so for some time to come" (1989, p. 261) could be argued to open up possibilities, at least, for "self-organization of group members so that they can attain a sense of collective empowerment" and "voicing a group's analysis of how social policy proposals affect them" (1989, p. 261).

One of the questions asked in this chapter is whether Young's assumption that groups of diverse collectivities *do* need to be identified (in and between nation-states) in order to generate a more just way of living in the 21st Century is consistent with Churchman's systemic thinking, or whether his thinking may lead in a different direction. My own belief is that his thinking leads more in the direction of supporting Handler's argument that the global culture of the present is in danger of fetishizing and reifying identified "groups" as indeed identifiable and stable rather than fractured and processual (1991, p. 70). Anselmi and Gouliamos too have pointed to the importance of exploring how the reification of identities takes place in society, in order at the same time to "formulate a counterhegemonic ethnoracial discourse which is politically and socially conscious of itself" (1998, p. 70). Such a discourse, they argue, can make provision for the subjects' specific lifeworlds to be properly taken into account (1998, p. 71). Likewise, Brubaker and Cooper suggest that the term self-understanding (rather than that of "identity") would allow us to consider people's own understandings of their affinities, connections, affiliations, and ways of defining relevant features of "themselves" – while admitting that people's understandings can be "overridden by overwhelmingly coercive external categorizations" (as defined by social categorizations) (2000, p. 18).

Brubaker and Cooper argue that framing the political options by talking about groups whose distinctiveness needs to be respected, can "miss the variety of forms [other than bounded groups] that affinity, commonality and connectedness can take" (2000, p. 33). Hence they propose a "more supple vocabulary" to "illuminate the range of [viable] options" (2000, p. 33). They argue that framing (political) claims in terms of group membership of bounded, distinctive solidary groups (even for certain defined purposes at hand) could serve to limit people's thinking and acting in relation to their human sociality (2000, p. 34).

In this chapter, in considering the debates about human options for living together in a fair and just fashion, I draw on Churchman's view that it is possible to both study and institute forms of caring/human sociality that allow people to revisit their allegiance to "group" thinking. I start by considering a number of "scientific" studies about group behavior – showing how they can be revisited to make provision for a way of thinking/living more in line with Churchman's views (as I interpret them).

2. "SCIENTIFIC" EXAMINATION OF GROUP BEHAVIOR: SOME EXPERIMENTAL STUDIES

Ashburn-Nardo *et al.* indicate (2001, p. 789) that many scientific studies of group behavior have been inspired by Allport's much quoted statement that:

> "The human mind must think with the aid of categories … . Once formed, categories are the basis for normal prejudgement. We cannot possibly avoid this process. Orderly living depends on it." (Allport, 1954, p. 19).

This statement of Allport has been the inspiration of many studies aimed at testing hypotheses about intergroup bias. As Ashburn-Nardo *et al.* point out:

> "In recent years, many prejudice researchers have used implicit measures to obtain evidence of bias even in persons who consciously reject prejudiced attitudes and beliefs. Findings of such hidden prejudices certainly seem to lend support to Allport's suggestion that intergroup bias is a normal, universal phenomenon." (2001, p. 789).

Ashburn-Nardo *et al.* remark that their own research in some sense supports the supposition of the "automatic nature of evaluative differentiation between in-groups and out-groups, suggesting that people are hardwired for intergroup bias" (2001, p. 797). This implies that "people who differ from us even in inconsequential ways may bear the brunt of intergroup bias of which we are not even aware" (2001, p. 797). But the authors make the point that even if it is taken that intergroup bias "runs deep" and that our minds are "fertile ground for the seeds of intergroup bias", this does not commit us to accepting that the process of creating intergroup bias in society is socially inevitable (2001, p. 798).

Ashburn-Nardo *et al.* cite two ways in which scientific studies can be done to approach the topic differently. The *first* kind of approach that they identify focuses on exploring/testing (via experimentation) the so-called "*common in-group identity model*" – which suggests that "recategorization, or the transformation of group members' mental representations of separate groups into one common in-group, can reduce intergroup bias" (2001, p. 798). The *second* kind of approach focuses (again via experimentation) on exploring people's propensities to "effectively *inhibit prejudiced responses*", insofar as they become alerted (by researchers) to biases in their mode of responding (2001, p. 798, my italics).

In Sections 2.1 to 2.3 below, I review these two kinds of approach to exploring the issue of intergroup bias – by undertaking a critical examination of their focus points. My deliberations around the studies referred to are aimed at highlighting – following Churchman – that the way information is sought by would-be scientists *has to be carefully considered at the very outset of an inquiry, because it constitutes a specific way of framing the inquiry.* My argument is that while there are aspects of the "scientific" work that appear compelling, we can revisit the mode of proceeding as well as the results developed. This in turn makes provision for a different way of seeing and addressing the issue of our social connectedness.

I begin with a detailed discussion of the so-called "Common Ingroup Identity Model" that Dovidio *et al.* drew on when conducting their (1997) experiment; and I consider what I regard to be lacunae in their way of conducting the inquiry. I show how alternative concerns (to the ones with which they were concerned) can be brought to bear in reconsidering the import of their work. I then follow a similar approach with a study conducted by Monteith *et al.* (2001) – Monteith being one of the authors of the Ashburn-Nardo *et al.* article (2001) cited above. In this latter case, I suggest that the assumptions of the authors about "normal" human behavior are less restrictive than Dovidio *et al.*'s conception and might provide more opportunities for re-opening debate around "ideals for the human species" (to use Churchman's terminology, cited in Porra, 2001, p. 25).

2.1 The Common Ingroup Identity Model

Dovidio *et al.*'s work (1997) in investigating scientifically the "Common Ingroup Identity Model" is explored in this section as an example of the design of experiments to test the benefits of what they call "recategorization" in intergroup relationships. My focus in the section is on reviewing their mode of experimentation – with the aim of locating possible consequences of their approach, that, I argue, have not been fully accounted for. (See also Romm, 2001, pp. 136-144.)

Dovidio *et al.* (1997) describe an experiment that was organized around manipulating people's "intergroup contact". They were interested in exploring the effects that might be created when members of different groups were induced to "conceive of themselves more as a single group rather than as two separate groups" (1997, p. 401). The manipulation amounted to the experimenters trying to instil an impression of people's belonging to one group (rather than two groups) in the experimental situation — so as to induce people to *recategorize* the situation (1997, p. 401). (The participants involved in the experiment were students studying Psychology at a University in the USA.)

Dovidio *et al.* state that the experiment was aimed at "investigating the effects of a manipulation designed to induce recategorization" (1997, p. 403). They indicate that they wished to examine the effects in terms of people's self-reported *evaluations* of others, and also in terms of their "forms of intergroup interaction" (1997, p. 404). Two types of interaction were considered. *Self-disclosure* was the first type. The researchers see this as "communicating a desire for intimacy and promoting closeness" (1997, p. 405). The second type of interaction investigated was *helping* orientations. This the researchers see as activity that is "co-ordinated to the needs of another" (1997, p. 404). Dovidio *et al.*'s expectation (or hypothesis) was that in people's *evaluations* of others, as well as in their *self-disclosure* and *helping* orientations, there would be lower bias scores (on average) for members from the situation where recategorization had been induced by the experimenters, as compared with the situation where it had not.

The experiment was designed in such as way that 24 sessions with sets of six participants in each case were arranged (1997, p. 406). In each session, participants met initially in three-person groups to which they were randomly assigned. Their group behavior was ostensibly being examined (so they were told) with a view to exploring how they manage to interact as they "attempt to reach consensus" on a problem set for them (1997, p. 407). The groups were all asked to choose their own names. The groups were given some time to work on the problem and then escorted to another set of rooms where they had been told a new interaction would take place (1997, p. 407). At this point, the manipulation of intergroup contact was undertaken. As Dovidio *et al.* note, "the goal of the manipulation was to vary the participants' cognitive representations of the six participants as either one group or as two separate groups" (1997, p. 407).

To induce the persons to have a *one-group* representation of the aggregate, instructors seated them in an "integrated seating pattern" and "emphasized their common identity as students at the same university". They were also instructed to wear the same color T-shirts (maroon), reflecting the school color. Furthermore, "an experimenter assigned the members of both groups a new single name to represent all six people and instructed them to use the new group name on all subsequent forms" (1997, p. 407). A TV recording of the session was arranged around the notion that the groups had merged to form the STARS (their group name).

In the *two-group* condition, a "segregated seating pattern" was arranged and each group wore a different color T-shirt – one group wore maroon, the other blue. Instructions "made no mention of their common university identity". The groups kept their separate three-person group names throughout the session. (The experimenter announced the session using the names of both groups, making no mention of any merger.) "Finally, each of the two TV cameras that recorded the session was aimed at a different three-person group" (1997, p. 407).

The task of the groups was again (as in the first task) to discuss and reach consensus on the same problem that they had been set earlier. They were given time for discussion and then each member was given questionnaires to fill in that asked about:

a) their conceptual representation of the group; and

b) their rating of each participant.

The first questionnaire asked whether it felt like one group or like two separate groups – and used a scale ranging from 1 (not at all) to 7 (very much) in order to establish the member's *conception of the aggregate*. The second questionnaire asked members to *evaluate each participant* with reference to how much they liked them, as well as how honest, cooperative, and valuable they found them. The rating again ranged from 1 (not at all) to 7 (very much), for each of the four items on the scale.[85]

Once the forms were completed, "an experimenter prepared the participants for the next phase of the study, which involved helping and self-disclosure tasks" (1997, p. 407). Participants were chosen from the original subgroups and assigned into groups of pairs. The way in which they were chosen, provided the opportunity to assess both intragroup and intergroup *self-disclosure* and intragroup and intergroup *helping* for each six-person session (1997, p. 408).

2.1.1 Results and Opportunities for Further Study as Presented by Dovidio et al.

In discussing the results of the experiment, Dovidio *et al.* note that, "consistent with the hypothesis derived from the Common Ingroup Identity Model, creating conditions that foster a stronger one-group representation significantly reduced intergroup bias in both evaluations and behavior" (1997, p. 410). They indicate that the Common Ingroup Identity Model (as previously developed) is based on a social categorization perspective of intergroup behavior that "recognizes the central role of social categorization in ameliorating as well as in creating intergroup bias" (1997, p. 402).

Part of the hypothesis that guided this experiment, was that "the effect of the Group Manipulation on evaluations and behaviors would be mediated by representations of the aggregate" (1997, p. 411). In other words, it was hypothesized that the effect of the group manipulation as organized by the researchers, would be mediated by participants' representation of the aggregate in the different conditions. Dovidio *et al.* summarize the overall results:

> "The manipulation of the intergroup contact situation that created stronger impressions of one group [rather than two groups] significantly reduced intergroup bias. Furthermore, ratings of the extent to which the six participants felt like one group mediated the reduction in bias." (1997, p. 413).

They note that what their study has left open for continued exploration is whether there may be more proximal mediators of interpersonal cooperation (1997, p. 415). For example, they suggest that it is possible that creating a

[85] These four items utilized in the questionnaire were considered as reflecting different dimensions of evaluative bias.

common group identity (through group manipulation) "produces higher levels of interpersonal trust, which, in turn, facilitates self-disclosure and other types of interpersonal cooperative behavior (Kramer and Brewer, 1984)" (1997, p. 415). They suggest that the degree to which the intergroup factors combine with interpersonal factors to facilitate such behaviors "may be a productive issue for further research" (1997, p. 415).

They suggest furthermore that a challenge for future research in regard to helping behavior is to:

> ... find direct mediation of this sense of we-ness and common ingroup identity on helping, to determine when and how common ingroup identity may relate to interpersonal processes ... and to understand how intergroup and interpersonal processes can combine to influence helping (1997, p. 416).

Generally, they propose that the advantage of the focus that they have put on:

> ... intergroup as well as interpersonal processes in bias articulated in the Common Ingroup Identity Model permits not only a parsimonious theoretical explanation for the effects of diverse contextual factors, but it can also inform practical interventions designed to reduce bias (1997, p. 416).

They refer specifically to the importance of the findings regarding recategorization as a "cognitive mechanism" to produce "a more inclusive, superordinate categorization of the membership" (1997, p. 416). Recategorization as a strategy encourages members to *employ a superordinate category to define their relationship with others* as part of some (created) group. Because of the way in which they come to see one another under the umbrella of an inclusive (group) category, their "bias" is now likely to be favorably inclined, allowing "personalized impressions" to "quickly develop" (1997, p. 417). As Dovidio *et al.* put it, "the newly formed positive bias is likely to encourage more open communication" (in which the *de*categorization processes of individuating and personalizing can set in) (1997, p. 417).[86] They suggest further research in this direction.

[86] Miller points out that decategorized responding requires an "awareness of the distinctiveness of individual members". But he goes on to indicate (in terms of his research) that it is likely that when intergroup cues (about the groups in question) are "highly salient", perceptions of typicality and atypicality "will become more extreme" (2002, pp. 400-401). In other words, on the one hand the "typicality" of a prototypical group member is exaggerated and on the other hand an atypical group member is regarded as even more atypical. This effect tends to be stronger for the out-group,

2.1.2 A Consideration of Implications for Practice

Dovidio *et al.*'s focus in their study (1997) was on testing whether an experimentally induced recategorization of people's group representations, in which they come to see themselves as belonging to one group rather than to two separate ones, might have an effect on their evaluative and behavioural responses to others. They present the experiment as having been designed to obtain evidence that would either support or disconfirm their proposed hypothesis and the theory of common ingroup identity from which it was derived. As it happens, they note that the findings that they obtained largely confirm their initial expectations (1997, p. 401). They point out that the theoretical model of ingroup identity, that guided the formation of their hypothesis, has thus received supported from their findings. As they note, the observed results, as they interpreted them, were "generally consistent with the framework [the Common Ingroup Identity Model] that guided the present research" (1997, p. 411).

But what Dovidio *et al.* seem *not* to have accounted for is that by presenting this model as "supported" by their "scientific" study, people (participants and other audiences) having access to the information made available could decide that it is normal to operate in terms of specific in- and out-group categorizations. They would now have evidence to believe this to be a "normal" response (in terms of Dovidio *et al.*'s statistical analysis of participant responses in the various situations) to their own experience (as they see it) of group situations. In other words, it is possible that people may decide to consider it normal that as long as they belong to *some* group (in this case a supergroup created through recategorizing themselves as part hereof) others in the said group are treated more positively by them (in terms of evaluations, intimacy, and helping behaviour).

In an earlier article, Dovidio (1993) argues that (negative) "bias" that people show in their evaluations of others (if they are seen as part of an out-group) could be seen to be "based on normal processes" (1993, p. 55). He explains that one of the "normal" processes implicated in aversive racism (which he defines as involving subtle racist feelings of which people seem not to be aware, because they suppress them) is the process of "categorization of people into in-groups, 'we', and out-groups, 'they'". He

---

especially as regard atypicality – that is, people who are seen as not fitting the (outgroup) category, tend to be regarded as very atypical exceptions. Miller points out that when intergroup aspects of the setting are heightened (in people's perceptions) in the sense that the given categories become highly "salient" as features of the setting, this often goes hand in hand with "perceptions of ingroup as well as outgroup homogeneity" (2002, p. 401).

explicates the link between what he takes to be the normal process of thinking in we/they terms and the negative evaluation of certain "minorities" (1993, p. 57):

> "To try to understand how these normal processes could account for some of the test results [around the operation of aversive racism], we ran the ... study over again, except that instead of presenting the words 'blacks' and 'whites' subliminally, we used the words 'we' and 'they'. The results were similar. Subjects associated more positive characteristics with 'we' than with 'they'. Subjects associated more negative characteristics with 'they' than with 'we'." (1993, p. 55).

Dovidio remarks that although the we/they distinction does not explain all racism as it manifests in society, the "normal" process in everyday life of utilizing this distinction can be seen as contributing to racist attitudes and behavior. As he puts it:

> "That's not to say that the we/they distinction explains all racism or bias. Racism is much more complicated than that. But it should be emphasized again that normal processes may contribute to subtle, unintentional forms of racial bias." (1993, p. 55).

Here Dovidio re-iterates that as far as he sees it, we cannot avoid cognitive processes of categorization into "we" and "they" (in- and out-groups) – even though they may contribute to forms of racism. All that we can do is try to generate different categories of representation – so that people can define themselves and others in terms of *different* categorizations (that are not set along race/ethnic lines).

His solution to the problem of racial bias (which he notes can occur at "cultural, individual and institutional levels") (1993, p. 57) is thus to educate people as well as try to train them to recognize *such* bias, so that they may act differently:

> Training and education are the next two pieces of the model. By educating people about subtle bias and training them to recognize it, we can help ensure fairness and avoid the damaging consequences of discrimination (1993, p. 157).

Dovidio's argument is that people can be educated and trained to reset their attitudes. When explaining more fully his view on aversive racism in the context of arguing for affirmative action, he emphasizes:

> "It is important for us to understand that although aversive racism may be unconscious, unintentional and subtle, it is neither inevitable nor immutable. Significant changes can occur in individuals and society. Expressed racial attitudes have changed dramatically since Congress enacted civil-rights legislation 30 years ago. The personal, social, and economic well-being of blacks, women, and other traditionally disadvantaged groups have improved since the advent of affirmative action." (1997, p. 60).

Dovidio argues, however, that there is still much work to be done in order to combat racism actively and self-consciously (1997, p. 60). In further work on the subject, he participated in research on how stereotyped attitudes associated with "skinheads" might serve to activate racist attitudes amongst respondents if they are "primed" in this direction. A study was done (in which he was involved) that tested the hypothesis that: "compared with participants who were not primed, participants primed with the skinhead category would respond more extremely on items reflecting negative intergroup attitudes but not on items related to other attitudes" (Kawakami *et al.,* 2003, p. 315). The idea was to test whether specific attitudes (negative evaluations) can be "activated" via category priming, that is, whether prejudicial attitudes associated with the skinhead category would appear more often after such priming (2003, p. 316).

The results of these experiments into the "priming" effect, were seen as supporting the contention that "social category priming can influence people's attitudes so that they become more consistent with attitudes associated with the primed groups" (2003, p. 318). Kawakami *et al.* point out that it is not only attitudes that can become activated via social priming, but also "behaviours that are congruent with the primed group" (2003, p. 318). That is, after priming, people can change attitudes as well as behaviour so as to be more in line associated with the group. The implication here is that people can "'socially tune' their responses to make them more consistent with their social environment" (2003, p. 318). Kawakami *et al.*'s conclusion is that their research offers "evidence of the power of social categories to shape personal responses and ultimately social interactions of people without their awareness or intention" (2003, p. 318). Because of the subliminal way in which categories (and activating representations of a category) may function, Kawakami *et al.* consider it all the more important to render these processes visible (via their research).

2.1.3 Revisiting the Experiments

There are a myriad of ways in which Dovidio *et al.*'s (1997) way of organizing and presenting their scientific work can be revisited. While one can appreciate that they try – via a rigorous scientific approach – to provide the "information" that will allow people as actors to re-look at their prejudices, they still *seem to operate in terms of the idea that some kind of group categorization into "we" and "they" as features of social interaction is to be regarded as "normal" (and can be scientifically presented as such).* In this section I show how this assumption can be questioned at root. Such questioning, in my view, is more consistent with Churchman's suggestion that as a human species we can direct ourselves to try to learn not only to apply new categorizations (via the new representations that allow us to recategorize ourselves as belonging to a different "in-group") but indeed to work beyond invoking "bounded groups" as the salient way in which we structure our thinking about ourselves and others.

One way to begin to question Dovidio *et al.*'s research (and the conclusions which they draw therefrom) is to refer to the so-called interpretive tradition of social inquiry. Those espousing an interpretivist-type of argument, suggest that the aim of social researchers is to try to gain an understanding of what Fielding calls the "symbolic world" in which people live.[87] From this perspective, Dovidio *et al.* could be accused of not utilizing the appropriate methodology to be able to obtain an understanding of the variety of ways in which people might give meaning to their interactions with others. For example, Dovidio *et al.*'s use of a questionnaire asking people to decide using a 7-point scale whether "during the interaction it felt like one group" or "during the interaction it felt like two groups", would be regarded as an inept way of developing an understanding of how people were experiencing the situation. Clearly, the different participants need not have understood the meaning of "group" in the same way as they answered the question whether "it felt like one group" or "it felt like two groups". Or again, the meaning of one-ness and two-ness might have differed across participants' conceptions. Dovidio *et al.*'s own concern with the differences between single-group and two-group situations makes it

[87] By "symbolic world" Fielding means "the meanings that people apply to their own experiences" (1993, p. 157). Authors who focus on the symbolic character of social experience make the point that symbols as drawn on by people to make sense of the world need not be seen as "containing" fixed meanings existing independently of their use in social discourse. This also implies that cultural "boundaries" (as spaces of meaning-making) may be considered as porous and permeable (cf. Parekh, 1997; Chui, 2003; Romm and Hsu, 2002).

difficult to determine whether and in what ways, if at all, the issues on which they were focusing are relevant to people.

Considering in more detail the example of this experiment from an interpretivist point of view, it can be noted that when organizing the group manipulation into a single-group condition, Dovidio *et al.* tried to induce participants to recognize certain features that they shared in common, such as their common student identity, a common color T-shirt (maroon), and a common name (the STARS). By doing so, they tried to create the impression of a group condition grounded in a sharing of certain characteristics. However, it could be argued that there are varied ways in which a sense of "we-ness" might be developed in social life, and that Dovidio *et al.* have failed to highlight this variation. From an interpretive position, an understanding of the phenomenon of "group identity" cannot be gained unless efforts are made to *examine the complex ways in which people might treat their relationship with others in different contexts*.

McIntyre notes when considering the way in which people might draw on a distinction between "us" and "them" that it may not be necessary to "see/experience" our relationship with others in terms of such a clear distinction (2004, p. 53). In any case, the issue arises as to how people might be encouraged to *reconstruct* their conceptions of their relationship with others without necessarily relying on "group" categories as the basis for their conceptions. Although Dovidio *et al.* try to show via their experiments that people can be induced to recategorize themselves as belonging to a superordinate group within which they might create a sense of "oneness" with others (in the group*), the question arises as to whether this is still inducing them to think in (rigid) group categories.*

From what is called a "critical theoretical" point of view, any supposedly existing "regularities" in social life as identified by social scientists must be considered, at best, as being only quasi-causal. That is, they should be considered as "existing" only insofar as people continue to act them out (Habermas, 1976, pp. 208-209). But Habermas warns in addition that the likelihood of people continuing to act them out is strengthened through the operation of a social science that treats them as "given". Thus when Dovidio *et al.* treat as given the fact that people's creation of a superordinate group categorization leads normally in the direction of being less biased toward members of the (superordinate) group, this statistical normality starts to become equated with a norm of conduct in social life. When Dovidio *et al.* "reveal" the fact that people's application of a superordinate group category renders their responses significantly different (in terms of a comparison of mean scores) from those not applying such a category, they simultaneously give this "fact" normative support.

In the case of Dovidio *et al.*'s (1997) study, critical theory would prefer to concentrate on the possibility of people's experiencing forms of human sociality that express the vitalization of human communicative capacities. And human communicative capacities for Habermas are linked to our considering ourselves, when interacting with others, as part of the human species, rather than as members of bounded "groups" (1982, 1993). Critical theory tries to encourage people to reflect upon possibilities for developing social integration around what Delanty calls the "principle of discursivity" (1998, paragraph 6.5). This idea (and practice of "science") would be more consistent with Churchman's proposal that we utilize science to look for ways in which people's sense of their connectedness (as part the "human species") is strengthened (cf. Churchman, cited in Porra, 2001, p. 24).[88]

Returning to Dovidio *et al.*'s research, although the authors suggest that the research itself was not to be guided by their own value-preferences it can be argued that their "finding" that creating a *superordinate category* is a way to generate harmonious interactions between those (now) identifying with the superordinate group, *expresses their own preference* for ways of organizing people's identification with groups. The idea of producing superordinate categories as an avenue to harmonious social relations is an expression of a specific preference, which is *obscured by its being presented as based on the "facts" of group existence*. It is this aspect of their argument that would be questioned by Churchman, on the grounds that it is overly restrictive. Lincoln remarks in the same vein that the feminist ethic involved in breaking the monopoly of scientifically created stories, is that of "making the space for the lifeways of others and creat[ing] relationships that are based ... on mutual respect, granting of dignity, and deep appreciation of the human condition" (Lincoln, 1995, p. 284).[89] Science – by its very pretence to offer value-free "information" about human co-existence – may well prohibit us from revisioning it. In summary, I would suggest that Dovidio *et al.*

88 Space in this chapter does not permit a full discussion of the possible difference between Churchman's vision of possibilities for human dialogue and those who argue that social communication across (incommensurable) languages invokes an impossible ideal of universality. See Flood and Romm (1996, pp. 46-58) for a discussion of ways of reworking the so-called "incommensurability" thesis.

89 Many authors who place themselves in the tradition of "feminist scholarship" argue (in line with Churchman) that "reason" and "feeling" should not be separated in our thinking about "justice" (cf. Noddings, 1984; Okin, 1989; Sunstein, 1989). Okin makes the point that although Rawls in his Theory of Justice (1971) argues that people need to operate as "rational mutually disinterested agents" in order to form the basis for a just society, it is impossible to imagine that people can "think from the position of everybody" without invoking some principle of benevolence (1989, p. 244). She notes: "especially for those accustomed by class position, race, and sex to privilege, wealth, and power, a real appreciation of the point of view of the worst off is likely to require considerable empathy and capacity to listen to others" (2002, p. 245).

have not given sufficient credence to arguments and concerns springing from alternative views on what experiments (including their own) can, and cannot achieve. This makes it difficult to challenge their way of setting up their experiments, or their recommendations springing from their research regarding the appropriate practical ways to address problems of racial discrimination. This is not to say that their research should be regarded as offering no input into the social discussion as to how to address the issue of discrimination. Indeed, as Ashburn-Nardo *et al.* mention, their work does point to options for cooperation via the cognitive mechanism of recategorization (2001, p. 798). Nevertheless, I would suggest that the "findings" of Dovidio *et al.*'s research still need to be regarded circumspectly, as the underlying assumption about people's propensity to "identify" with bounded groups have not been queried via the research. Possibilities for developing human sociality along lines not accounted for in their research still, it could be argued, need to be considered as part of our "human inquiries".

Ashburn-Nardo *et al.* offer a second way (other than that of Dovidio *et al.*) of studying intergroup relations characterized by prejudice – an approach which they see as aimed at providing some insight into the possible inhibition of prejudiced responses (2001, p. 798).[90] Before explaining their way of setting up their experiments, it is worth showing how Monteith and Winters set out their understanding of the issues – as this understanding is linked to the second approach mentioned by Ashburn-Nardo *et al.* in their article (of which Monteith is one of the authors).

2.2 (Re)considering the "Drive" to Categorize into "Us" and "Them" (Social Identity Theory)

Monteith and Winters note that research to date seems to support the idea that "the drive to completely and quickly divide the world into 'us' and 'them,' is so powerful that it must surely come from some deep-seated need" (2002, p. 48). They explain the argument offered by "social identity theory" to account for such divisions:

[90] This does not mean that they regard their approach as inconsistent with that of Dovidio et al. They indicate that their approach is simply another option for exploring racially-oriented biases (2001, p. 798). Monteith and Winters (2002, p. 50) indeed refer favorably to Dovidio's work on how people can be encouraged to "reset" their mental divisions so that race and ethnicity become (via recategorization) less salient in their social relationships.

"Given even the slenderist of criteria, we naturally split people into two groups – an 'in-group' and an 'out-group'. The categories can be of geopolitical importance – nationality, religion, race, language – or they can be as seemingly inconsequential as handedness, hair color or even height. ... Once the division is made, the inferences and projections begin to occur. For one, we tend to think more highly of people in the in-group than those in the out-group, a belief based only on group identity. Also, a person tends to feel that others in the in-group are similar to one's self in ways that – although stereotypical – may have little to do with the original criteria used to split the groups."

They remark that there appears to be "ample evidence" to back up the "science" behind identity theory (2002, p. 49). They go on to explain, however, that "not many of us will admit to having strong racist or xenophobic biases" (2002, p. 49). Now they ask the question: "If categorization and bias come so easily, are people doomed to xenophobia and racism?" (2002, p. 49). They explain their position in relation to this question:

> "It's pretty clear that we are susceptible to prejudice and that there is an unconscious desire to divide the world into 'us' and 'them'. Fortunately, however, new research also shows that prejudices are fluid and that when we become conscious of our biases we can take active – and successful – steps to combat them." (2002, p. 50).

They refer us to Monteith's previous work that has:

... found that people who are concerned about their prejudices have the power to correct them. In experiments she [Monteith] told subjects that they had performed poorly on tests that measured belief in stereotypes. She discovered that the worse a subject felt about her performance, the better she scored on subsequent tests. The guilt behind learning about their own prejudices made the subjects try harder not to be biased (Monteith and Winters, 2002, p. 87).

The practical conclusion that Monteith and Winters draw from the research to date is that:

> It is difficult to escape conventional wisdom and treat all people as individuals, rather than as members of a group. But that seems to be the best way to avoid the trap of dividing the world into two – and discriminating against one part of humanity (2002, p. 87).

Having offered this context for understanding the concerns of Monteith and her co-author(s), I now examine in some detail one of the studies to which she refers to offer some substantiation of her claim regarding the possibility of "overcoming" prejudicial attitudes and behaviors.

2.3 Taking a Look Underground: Detecting, Interpreting and Reacting to Implicit Racial Bias

Monteith *et al.* (2001) suggest that racial prejudice can be considered as having "gone underground" following certain historical events such as World War Two and the civil rights movement. People are less inclined to express racial prejudice as such in "blatant" form; and even to themselves they may find it difficult to admit that racial bias is affecting any of their choices in regard to their way of interacting with others. They cite favourably Dovidio's work (as well as that of other authors) concerning people's reticence to express prejudice openly to themselves and/or others (2001, p. 395).

Monteith *et al.* contend that empirical research around "prejudice" that may be operative outside of conscious awareness has thus far not focused sufficiently on:

> … people's interpretations of and reactions to experiences that could potentially implicate them as being prone to subtle forms of prejudice. In other words, when given the chance to look underground, are people willing and able to attribute their responses to prejudiced tendencies? (2001, p. 396).

They note that they consider this to be an important question to address because "awareness of subtle biases and willingness to attribute them to internal forces are critical for learning to control them" (2001, p. 396). It is for this reason they consider it worthwhile to offer "lay persons the opportunity to take a look underground at their own implicit racial biases" (2001, p. 397). One way of "looking underground" is via *racial Implicit Association Tests (IATs).* This refers to a process of testing that is designed to measure implicit prejudice. The test measures the strength of pleasant versus unpleasant associations in relation to, say, "Whites" versus "Blacks" at the automatic level (2001, p. 396). For example, automatic racial bias on the part of Whites (classed as such) is revealed "if Whites have an easier time pairing White names with pleasant stimuli and Black names with unpleasant stimuli than the reverse (i.e., Black names with pleasant stimuli and White names with unpleasant stimuli)" (2001, p. 397).[91]

In the research reported in their article they mention that several goals were operative:

1. Firstly, they wanted to examine the magnitude of relation between IAT performance – how (White) participants fared on the Implicit Association Test – and reports of "feeling" the IAT effect (that is, a sense on the part of participants that they had shown some hidden bias). The researchers were interested in knowing whether only those who have an especially strong IAT bias (as measured by the IAT test) were able to detect this, or whether such detection is more general (2001, p. 397).
2. Secondly, they were interested in "how people interpret and react to the detection of biased performance on the IAT" (2001, p. 397). They mention that it is of course possible that even if people "feel" the IAT effect, they may differ in their interpretations thereof.
3. Thirdly, they were interested in exploring the possible moderating effect of individual differences on IAT performance and reactions to such performance. They explored this in the light of previous findings regarding discrepancies between how participants report that they *should* respond in a variety of situations that involve Blacks, based on their beliefs about what is appropriate, and their reports about how they *would* respond in such situations. The

[91] Brendl et al. qualify the significance of IAT tests (done to assess people's biases) by noting that in their view, "whereas prejudice leads to an IAT effect, an IAT effect does not unambiguously indicate prejudice, because it can have multiple causes" (2001, p. 769). For instance, "the influence of familiarity [with names used in the test] compromises the IAT as an ambiguous marker … " (2001, p. 770). Also, when two groups are measured [by respondents] relative to one another, it is possible for one group to be preferred to another without the second group being evaluated negatively (2001, p. 770). The IAT thus appears to measure only "relative preferences" and not necessarily prejudice against others. Brendl et al. suggest that developing the IAT will require "extensive work". They also propose that one way of refining our understanding of attitudes is to "explore the relationship between these measures of attitude [such as IAT tests] and other behaviours" (2001, p. 771).

> "difference between participants' *should* and *would* ratings is taken for each scenario to index the extent to which people would respond with greater prejudice than they believe is appropriate (that is, the extent to which there is a discrepancy [in each case])". Monteith *et al.* indicate that they were considering the role that "self-reported proneness to discrepancies would play in people's interpretations of and reactions to their IAT performance" (2001, p, 399). They wondered whether people would "experience negative self-directed affect (e.g. guilt) when they realized they had engaged in prejudiced response" (2001, p. 399). They were also interested in the implications of raising awareness of discrepant responses for the subsequent self-regulation of prejudiced responses.

Monteith *et al.* set up their study (involving an experimental procedure) to examine these three areas of interest, using seventy-nine participants (studying Psychology in the University in the USA). They note interestingly that one participant who had initially been involved, withdrew because s/he was "opposed to categorizing people as African American versus White American" (2001, p. 400, footnote 1.) One might argue that this is indeed one of the dangers of employing (social) categories as if they refer to an acceptable grouping of people – especially if the aim is ultimately to encourage people to try to transcend their prejudicial thinking (and to treat "all people as individuals, rather than as members of a group", as explained by Monteith and Winters, 2002, p. 87). Use of the black/white categories in the research could possibly contribute to reifying the categories in people's consciousness – just by virtue of being invoked as if they refer to real entities. This is whether or not people specifically object to their use (as the one participant did). However, Monteith *et al.* would counterargue that a pre-requisite to our overcoming prejudicial responses might be to indeed recognize how, unconsciously, we tend to invoke stereotypical thinking – hence the need to set up the IAT tests in this case to "take a look underground".

The participants in the experiment proceeded to complete the IAT tests designed to detect "hidden" biases. (Some of the blocks of trials were for practice purposes to begin with.) After the participants completed the IAT fully, the experimenter provided them with a questionnaire. This was aimed at assessing their awareness of the extent to which they showed bias on the IAT, as well as their affective reactions to any biases they may have had (2001, p. 403).

As part of the questionnaire, participants were asked whether they thought they were particularly slow or particularly fast on some parts of the IAT test. They were asked to consider their slowness and fastness in regard to the classification of various presented stimuli:

- White and Black names
- Pleasant words and unpleasant words
- White names paired with pleasant words
- Black names paired with unpleasant words
- White names paired with unpleasant words
- Black names paired with pleasant words

Participants were also asked in the questionnaire (via an open-ended question) to speculate as to why they may have been especially slow for certain conditions (trials). Their speculations were later examined (via content-analysis) to consider the way in which they explained any perceived difficulties with the IAT. The questionnaire also asked people how they felt "as a consequence of the experimental procedures": They were asked to indicate how they felt about the task by circling a number between 1 (=does not apply) and 7 (=applies very much) on 32 affect items (2001, p. 404). This was designed to ascertain the existence and extent of any negative feelings about their own biases (as measured by the IAT tests) that people might have experienced when their scores were revealed to them by the researchers.

2.3.1 Monteith et al.'s *Discussion of Results Generated via the Research*

Considering the results regarding the IAT scores, Monteith *et al.* indicate that "only four participants had negative IAT scores, such that they responded faster on incongruent trials than on congruent trials" (2001, p. 405). (That is, they responded faster to Black American names paired with pleasant words and White American names paired with unpleasant words than to Black American names paired with unpleasant words and White American names paired with pleasant words.) The rest showed a bias in favor of Whites (by responding faster for congruent trials) – although there was a "good deal of variability in the extent of bias" shown by the different participants (2001, p. 405).

Monteith *et al.* examined statistically the relationship between participants' actual IAT scores and their own detection (recognition) of an IAT bias. A moderate-sized correlation was found. They interpret this as follows: "the fact that this effect is not stronger suggests that it is not just participants who have an especially strong IAT bias who can feel the IAT effect" (2001, p. 406).

The next question Monteith *et al.* explored was how people themselves react to and interpret a biased IAT performance (that is, how they react to their bias as measured by the IAT test). This was examined by considering the "affect [feelings] they reported in relation to the IAT task" (2001, p. 406). An examination of the affect indexes showed that, interestingly, the (statistical) relationship between the actual extent of IAT bias and negself [negative feelings about one's reactions] was not significant. This suggests that it is the person's "detection of IAT bias [that] drives negself feelings rather than the magnitude of IAT bias" (2001, p. 407).

Monteith *et al.* also content-analyzed participants' written explanations for why they thought they were particularly slow on certain trials – this was in order to determine how the participants interpreted their IAT performance. The responses were coded according to three categories: One category was used if people indicated that racial associations or stereotypes caused a slowing of response; another category was used if people indicated that color associations caused the slowing; and a third category was used when the response had nothing to do with race or color (e.g. if people mentioned the fact that "Black names are unfamiliar and harder, so they went better with the unpleasant words" or "I know this is a test on racial bias, but I felt that since the White names were paired with the pleasant things first, that set the pattern") (2001, p. 407). The results of the coding were that 37% of people did attribute their slowness to race or stereotypical reactions. And (as would be expected) "participants who interpreted their biased IAT performance as related to racial associations tended to report greater feelings of guilt than participants who made other attributions of their performance" (2001, p. 407-408).

Monteith *et al.* were further interested in establishing whether participants who self-report a discrepancy between their standards and their actual behaviour (as measured by *should/would* discrepancy scales) are "less prone to automatic associations revealed by the IAT than other participants" (2001, p. 408). In addition, they wished to establish whether discrepancy scores were related to participants' detection of their own IAT biases, as well as to their interpretations of and reactions to this detection. The findings indicate that:

> ... participants who reported that they were not prone to discrepancies showed just as much evidence of implicit racial bias as their discrepancy-prone counterparts and were just as likely to detect the IAT effect. However, they were less likely to feel guilty in relation to their IAT bias and were less likely to interpret the bias as reflecting race-related factors (p. 411).

Monteith *et al.*'s overall understanding is that completing the IAT and having the opportunity to detect a biased IAT performance can serve to "highlight implicit racial biases for some individuals" (2001, p. 411).[92] Furthermore, Monteith *et al.* find it worth noting that:

> The finding that many individuals (particularly those who are discrepancy-prone) recognized their implicit biases, and felt guilty in relation to them, increased our confidence that many people are willing to detect prejudiced responses when they occur and self-regulate in relation to them (2001, p. 412).

Monteith *et al.* conclude that tools such as the IAT may be useful devices to provide self-insight into racial bias and may indeed allow people to self-reflect on their way of interacting with others. For example, they indicate that:

> ... perpetrators of discrimination may well recognize that their group discriminates against others but, because of the unsettling and image-threatening effects of recognizing personal discrimination within oneself ... people may be motivated to deny it at the personal level (2001, pp. 413-414).

According to Monteith *et al.*, when faced with a biased IAT performance and the possibility of detecting and interpreting their score, this may lead people to be more *self-conscious* about their own way of responding in future behavior. However, Monteith *et al.* comment that one disturbing finding from the research is that low-prejudice participants with small discrepancy scores "were resistant to interpreting their IAT performance as having anything to do with racial biases, and they did not feel guilty about their biased performance" (2001, p. 415). This implies (for the researchers) that it is possible that people sometimes do respond in prejudiced ways, but do not recognize their responses as prejudiced or self-regulate in relation to them" (2001, p. 415). Monteith *et al.* continue (and make the final point) that:

> ... people cannot possibly gain control over a process or its effects if they are not aware that the process is occurring ... . This underscores the crucial need to encourage people to take a look underground at their implicit racial biases (2001, p. 415).

[92] Considering the potential argument that possibly individuals who "misattribute" their IAT bias to non-racial factors may actually be correct in assuming that bias has nothing to do with their performance, Monteith et al. contend that this is highly unlikely because of other research which shows that performance on the IAT is related to other measures for assessing racial bias – including to explicit self-reported attitudes (2001, p. 412).

2.3.2 Revisiting Monteith et al.'s *Experiments*

As noted above, Monteith *et al.* suggest that by taking a look underground, participants may be offered the devices to rethink their way of relating to others. The "looking" underground, they hope, can help to serve this purpose. The IAT tests (and experiments aimed to establish how people fare on such tests) are thus presented by Monteith *et al.* as a way of "finding out" about prejudice as well as a way of possibly providing people with options for re-thinking their prejudicial reactions.

Monteith *et al.* rely on the language of rigorous science to set up their experiments and to create findings – which they present as referring to the realities of people's responses during the Implicit Association Tests as well as their responses to how they fared on the tests. The question arises as to whether this way of doing the research and presenting their findings does justice to Monteith *et al.*'s intentions to encourage people to reconsider/reform their attitudes. One could argue that their experimental work could be strengthened insofar as the researchers (or others) were to follow up and elaborate upon the open-ended questions in which people were asked to comment on their experience of the experiments. Through further open-ended questioning, interviewing, and indeed active interviewing and action research[93] a *discourse can be opened in society around the utility of the "IAT tests" or other devices used as tools to encourage people to rethink prejudice.* This need not amount only to trying to "find out" if and under what conditions people experience guilt when they do not score as they hoped on the tests; but it also could be a process of working with people to *consider some of the ways in which their inconsistencies can be worked through.*

Experimental research (especially if it is couched simply in terms of being a "finding out" exercise) can be argued to restrict discursive exploration of the ambivalences that people might feel – ambivalences that can become thought through (and revised) in social discussion. Even experiments that allow people to first take part in an IAT test and, having heard their score and being given a chance to reflect on it, take part in another test, do not necessarily provide possibilities for people to *work*

[93] Romm (2001, pp. 231-232) offers an example of a way of using Holstein and Gubrium's (1995) style of "active interviewing" as an approach to interviewing that enables people to broaden their horizons. She also discusses an example (Weil, 1998) of a way of doing action research as an effort to activate people's capacity for "reflexive inquiry" in action (2001, pp. 259-279).

together, in natural settings, to reconsider and revise ways of dealing with situations.

Churchman argues (not necessarily in terms of this issue, but considering human psychology more broadly) that he became frustrated with logic early on in his student life because of the strong requirement of the discipline (of logic) for precision. This, he argues, excludes "such essential world features as how people reason psychologically or sociologically. I began to realize that the psychology of inconsistency is hardly captured by the symbolic logician's "p and not-*p* are both true" (1979, p. 10). Churchman notes that people seem to him to have the facility to render both *p* and *not-p* "true" (despite the inconsistency in logic) in their social interactions. They can, for instance, recognize the truth that people should not be discriminated against on grounds of certain characteristics (because they have no control over these characteristics) while also finding it "true" that there are reasons for finding people unsuitable for certain positions that demand certain qualities.[94] Mason takes Churchman's stance to imply that, as humans, we are called upon to reflect, with others, upon our way of organizing our responses (our discourse and our actions). Mason summarizes the importance to Churchman of the "examined life" – aimed at securing some practical improvement in terms of our commitment to humanity. And he notes that the only "guarantor" according to Churchman of a better human system is indeed the propensity of people to reflect, with others, on their way of seeing and acting (1994, pp. 71-72). It is not clear whether Monteith *et al.*'s way of setting up their experiments can make provision – via the research process – for this kind of reflection.

My argument in regard to Monteith *et al.*'s work is that, at the least, it needs to be extended so that opportunities become created through the research process – with participants and also with other audiences – to reconsider the meaning of the "findings". I would suggest that at some point during the research process, participants (at least those willing) could be brought into discussion with the researchers and with one another to think through their responses, including their inconsistencies and ways of interpreting them. This in turn might help them to reflect with others upon the considerations they brought to bear in explaining their IAT scores; and these reflections could be rendered available to other audiences, to in turn open up reflection on the areas discussed.

[94] Dovidio (1997, p. 60) gives an example of an employer who is "influenced by feelings of aversive racism" and might therefore subtly re-evaluate the most important qualifications of a job, depending on the race of different applicants: "If, say, a white applicant had broader experience and a black applicant had more up-to-date training, the employer would decide that experience was more important; if the white applicant had more recent training and the black more experience, the employer would decide that experience was less important".

While Monteith *et al.*'s research seems to offer scope for us to recognize (and work with) human prejudice as fluid rather than as static (cf. Monteith and Winters, 2002, p. 50), I would suggest that they (and/or others) can capitalize more on this aspect of their inquiries – allowing lay people to indeed *explore the fluidity in their attitudes and thus to recreate/reform their way of seeing and acting*.

The next section shows how one may approach the question of our human sociality from a more action-oriented approach, with the expressed intention of using the research process to help people to work across "boundaries" that might have been operative as part of their "mapping" of the situation.

3. ENRICHING TRANSCULTURAL UNDERSTANDING VIA ACTION RESEARCH: AN ALTERNATIVE WAY OF "DOING SCIENCE"

The example discussed in this section is used to illustrate the possibility of developing an action research approach to what McIntyre-Mills (drawing explicitly on Churchman's systemic work to inspire her thinking and practice) calls "working the boundaries" – which included conceptual boundaries used by people in their "narrations", as well as social, political and spatial boundaries (2003, p. 5). In terms of the action research approach described here, the "scientific" inquiry is not done or presented as an exercise of "finding out" about (aspects of) reality, but is seen as an exercise in people forming together ways of thinking and acting in relation to issues of (practical) concern.

McIntyre-Mills's example is set in Alice Springs, where, she remarks, she tried to utilize "systemic thinking, practice and policy" in order to "challenge boundaries through attempting to co-create shared meaning". She presents her own story of this effort as "an attempt to present narratives as multilayered, multitextured perceptions of reality" (2003, p. 5). In other words, she does not present her own story as if it refers to "findings" to be appropriated by others – but rather she presents it as an opportunity for others to consider what to make of her text, which does not pretend to offer a clear-cut account of "the situation" in which she was involved.

McIntyre-Mills's work in this case (as I read it) involved working with so-called Indigenous people, whose culture included a history of resistance to what McIntyre-Mills summarizes as "an intolerant environment [that] created a strong cultural response" (2003, p. 4). McIntyre-Mills suggests that the Indigenous "social, cultural, political and economic concerns" have been expressed in various ways, for example, through "Native Title Claims, The Land Act 1992, Amendments to the Pastoral Land Act, 1992, Aboriginal

Land Rights Act 1976, Sacred Sites Act 1989, and Aboriginal Heritage Act 1984, now being amended" (2003, p. 4). She suggests that in working through *the issues of concern*, one does not have to treat "culture" as referring to repositories of fixed meanings, but rather as spaces where issues can be and are discussed as relevant features of the lifeworld. Defined in this way, "culture" refers to a space for meaning-making, rather than to a definite set of meaning-patterns apparently "possessed" by a group.[95] But she also notes that in considering the issues of concern "in" the culture, one needs to sweep in the colonial history that generates these issues of concern to the people involved. This is not to say that the issues are seen in the same way by all people in "the community". McIntyre Mills's argument is in this respect tallies with that of Bouchard (2004), who argues that despite the "collective" character of the rights to land, etc. of Indigenous people, one should recognize that the community consists of people who may have different understandings of the issues to be debated, and may "bring different perspectives" to bear on discussions with others. She prefers to use the word "community" rather than the word "group" to refer to Indigenous communities, and argues that the word "group" already invokes an "us/them" relationship between (groupings of) people. (She also argues that who indeed may be considered as part of "the community" is itself not clear-cut.)

McIntyre-Mills stresses that it is crucial for those working with Indigenous leaders and others in the community not to expect a homogenous response on everyone's part to what are taken to be the issues at stake (or even a homogeneous way of seeing the issues) (2003, p. 229). She notes in particular the ambivalence around the "successful Indigenous capitalists in Alice Springs" – ambivalence expressed by both "Indigenous and non-Indigenous informants" (2003, p. 219). Different perceptions of "commodities" and the "nature of work" were at issue here, as the (capitalist) way of life might seem threatening to certain people. She notes that in such a context:

> "To achieve civic pride through generative communication requires a quantum leap, based on trust and real participation across a range of sectors and members of the community in participating in 'mainstream' as well as Indigenous community organizations." (2003, p. 229).

[95] This definition of "culture" is consistent with that of Laitin who considers it as, "points of concern to be debated" rather than as "values [or ways of life] to be upheld" (1988, p. 589). There is no univocal solution offered to address problems of concern to people participating in a cultural arena. By speaking of cultural arenas (rather than cultural repositories of meaning) and of debates taking place as part of "the culture" (rather than culture offering fixed meaning patterns) we can open the possibility for debate not only within cultures but also across "cultural arenas".

She continues, though, that as far as the provision of services are concerned, "there is less enthusiasm for generic services when there are great differences in culture and life chances of citizens as far as service providers and users is concerned" (2003, p. 229). Here she suggests that people's perceptions that their specific concerns (arising from their "cultural" positioning) are not being provided for, contributed to distrust.

Despite the expressed attempt of service organizations in Alice Springs to cater for "socio-economic development outcomes", such outcomes have not been achieved for many Indigenous people – as "evidenced by the statistics" (2003, p. 229). It is for this reason that McIntyre-Mills argues that more is involved in generating "outcomes" than had currently been envisaged. For her, "involvement in decision-making is a more positive goal for development geared to enhancing the life chances of citizens" (2003, p. 229). She emphasizes that involvement in decision-making in this case implied that the involved people needed to commit to the value of (some form of) democracy. From her own experience in the situation she states that:

> Much could be learned about the ways of seeing and perceiving the world through engaging in a dialogue. Listening and learning from one another not only contributes to reconciliation, but also enriches transcultural understanding and enhances our understanding of cultural systems (2003, p. 225).

Here McIntyre-Mills points to the potential of people (from her reading of people's interactions) to indeed move beyond seeing "culture" as having some fixed boundary that precludes creation of shared concerns and ways of addressing them across (originally perceived) boundaries. Through her own work with those involved, McIntyre-Mills tried to instil and cultivate the idea that: "The valuing of people and the environment as resources need to be supported because the opportunity costs of neglecting them are disastrous in social, political and economic terms" (2003, p. 224).

But McIntyre-Mills avers that due to the history of their marginalization, many people have neither the confidence nor the will to participate in defining (co-creating) a future. She notes: "this is an area that needs particular work as a result of the history of marginalization and intergenerational welfare dependency" (2003, p. 227). Yet she points out that "non-engagement can allow others to determine one's life chances and one's future" (2003, p. 227). Hence she herself pleads for (and tries at the same time to introduce) a more participative decision-making process. She explains her position: "Participation based on co-creating a future is not however about winning or losing a debate, or a pool of funds. It is about achieving social and environmental harmony and justice" (2003, pp. 227-228). It is with this aim in mind that she proceeded, as a matter of her own "ethical responsibility" (2003, p. 228). But of course she notes that the values of those involved have to be taken into account, as these values are "essential determinants of the way things are and the way in which things could be done". Following the "soft" tradition in systems thinking, she suggests that "soft systems mapping of ideas, perceptions and values can be of assistance in understanding the different ways in which reality is perceived" (2003, p. 229).

As stated earlier, for McIntyre-Mills the aim of mapping different experiences of reality is to continue the (action research) project of developing co-created shared meanings across perspectives and apparent barriers to communication. Her book as a whole tries to highlight possibilities for people to indeed work across erstwhile boundaries and barriers, while not denying the challenges involved. Her language, like that of Churchman, grapples with the tension between offering a vision of what she believes is possible for humankind (through offering vignettes to illustrate such possibilities), and recognizing that her vision is also a looking forward to a different (more ethical) future (cf. Churchman, cited in Van Gigch *et al*., 1997, p. 744).

I have used the example of McIntyre-Mills's work in Alice Springs to try to draw out certain prospects for researchers/actors to become involved in reworking conceptual and social boundaries that might have been taken for granted. I have concentrated on aspects of her narrative concerning the way in which those defined as Indigenous people may decide to indeed self-identify with this category, while also working beyond it as a way of developing transcultural understanding with others across areas of concern. In the following section, I draw on an(other) example of people trying to work past fixed categories that were experienced as coercive in a particular institutional context. I use the example to illustrate the dilemmas that may be involved in using apparently taken-for-granted framing principles, while trying to avoid their ossification. I use the example also to point to the emotional upset that can become created, leading to people trying to challenge the fabric of dehumanizing social relationships in which they find themselves enmeshed. Because I myself was involved in the case (as an "active" participant observer) I became close to the emotional experience of those who were trying to find a way to activate the institution in question to at least take seriously their recounting of their experiences of "unequal treatment" within the institution. (I became involved because those alleging "inequality" turned to me as one of the staff whom they trusted to help get their concerns raised within the institution.)

4. STRIKING AT THE FOUNDATIONS OF, ETHNORACIAL CATEGORIZATIONS: A CASE STUDY BASED ON "PARTICIPANT OBSERVATION"

The case recounted in this section was set in a University in the United Kingdom – in relation to a Masters class consisting of students from diverse backgrounds (whose meaning itself came to be defined in a particular way). My narration of events is structured around offering my deliberations regarding the political choices made by certain actors in situ, to (in terms of my understanding) strike at applied categorizations of "belonging" (to ethnoracially defined groups). As the story unfolds, my account of "what might have been" (in terms of an experience of alternative human relationships) had those (apparently) reifying the categories been able to review their status, is elaborated. But first some contextual background is supplied.

Briefly stated, the Masters class at stake in this example consisted of students from many countries all over the world, including many from the Far East. The (two) home students were conscious of the proportion in the cohort of "the Chinese" – whom they later came to define (in writing) as "the numerous Chinese population". Meanwhile, during the course, many of the students (from a range of countries) and certain lecturers considered these home students to have a hostile attitude towards in particular "the Chinese" (as well as to the other "foreign" students).

At a certain point all of the students from the Far East signed a petition stating that their class representatives (one of them being home student here called "Ms. X" and the other being someone they regarded as led by Ms. X) had failed to keep harmony in the group (that is, in the class as a whole). They asked on these grounds for new representatives to be elected. However, the re-election of representatives was held up for more than a month and in the meantime the experienced aggression increased. The (Chinese) students arranged for an appointment to see the Dean of the Business School to express their concerns with the attitude and behavior that they had experienced. Two of them were at the meeting (with the others on standby). They pointed out that the two home students had in their view been insulting them and indeed had presumed to judge them as inferior on various counts: academically, ethically and culturally. Below are some quotations from their conversation, which was recorded. (The "… not clear" indicates points on the tape where the voices were unclear.):

J.Y.:– [one of the students representing the concerns of "the Chinese" in the meeting[96]]:

... And also in the Research module because Ms. X and Mr. Z. [Mr. Z being the other erstwhile class representative] were speaking ... (not clear) not only the Chinese but the Indian students as well. And they think they have, you know, a high position they can judge academic and ethic and cultural things. So always when the students want to contribute they, they ... funnily enough we got a meeting with Dr. B [the Director of post-graduate programs] ... (not clear) they said we didn't want to contribute, but in every class ... (not clear), just like what the English students do ... (not clear).

Dean of School:-mmmm

J.Y.:– ... (not clear) they don't know if there are some standards to select the students, to give the offer to the students in the Business School, they don't know. So in the Research module they think they are better and they just ... (not clear).

And further on during the conversation J.Y. continues:

J.Y.:– [They say] we cannot stay here. We have no ...

Norma Romm:– Capability.

J.Y.:– Capability, capability to study here ... (not clear). And then Dr. B [Director of post-graduate programs] goes to the class on the Friday ... (not clear) example ... (not clear) employer ... (not clear) employee ... (not clear) conflict ... (not clear) according to English law, English law these kind of things happen ... (not clear). Quite Serious.

Dean of School:– mmmm

J.Y.:– And [Dr. B implied that] it is English culture, the students have the right to do these things ... (not clear). So they have not stopped the trouble. And in the meeting they said the Chinese students do not contribute. During the leadership module, it is Ms. L who is our lecturer. I always give a contribution there ... (not clear) to discuss

[96] Mr. J.Y. (who later followed up the case) was happy to have his name revealed in my reporting of the saga – which he experienced as "horrible" throughout, and which he experienced as increasingly frustrating as the institution failed on several occasions to follow Regulations in dealing with complaints raised.

> ... (not clear). I am not finished my sentence and I am interrupted by Ms. X and Mr. Z ... (not clear). I don't understand! The same things happened when we had the meeting with Dr. B. Some students used very rude words. ... (not clear) an English student [Ms. Y] used "f***" during the meeting when I was trying to ... (not clear).

At the end of this meeting the Dean promised the students that the issues would be fully investigated. He admitted (earlier in the meeting) that he had understood that the issue at at stake for the students was having to cope with feeling aggressed against; and he acknowledged that they had spoken with "passion" in recounting their experiences. I would like to add here that J.Y. had spoken to me too with extreme upset (as had certain other of the students from Chinese background) at what they experienced as antagonistic treatment leveled against them due to their nationality (as they put it), as well as against some other students from other countries (such as India). It was not one incident but several that brought J.Y. and fellow students to the edge to take the step of lodging with the Dean their account of their experiences – hoping that the matter could be looked into within the University. Unfortunately, the ensuing "investigation" never resulted in any written statements being taken (from witnesses) by the person supposedly investigating the issue – thus it is not possible to establish what people might have said if questioned about the concerns complained about.[97]

Returning to the chronology, after a meeting with the person asked to investigate the matter, namely Dr. B, J.Y. and a fellow student (who had been present in the initial meeting with the Dean too) sent an email to the Dean summarizing some points raised in the meeting with Dr. B and emphasizing that "we just want equal treatment". Yet when the Dean

[97] Actually, some witness statements were obtained by me more or less accidentally (but were never elicited by the University). One case in point is a statement from an exchange student from France who volunteered her views (when matters were being investigated). She wrote that "Ms. X does not make enough effort to understand people from other countries who she thought were not as good as her", and that Ms. X created "an upsetting atmosphere in the class". She pointed out that she had "never seen such a case before". Another student (from Nigeria) also was keen to offer a statement expressing her "deep anger and bitterness" over the "matter which I [she] felt should have been dealt with properly and adequately during the MSc session ... ." She continued that "Ms. X, Ms Y., and Mr. Z should have been disciplined and taught how to behave to foreigners". She indicated (to me) that she had never been questioned by anyone about the attitude of Ms. X and Ms. Y to the other students. Meanwhile, an (external) lecturer who specifically stated (both verbally and in a written report sent to Dr. B) that a "small cohort of the Caucasians seemed to be openly hostile to the Chinese majority" was not on the list of witnesses (even as an anonymized name) that the Dean eventually offered to J.Y.

eventually (nearly a year late in terms of University Regulations) reported to Jin Yi about what he saw as the issues, and what witnesses had been used and what evidence collected, no mention was made of any references of the complainants to unequal treatment, to their concerns that they had experienced others as treating them as inferior (culturally, ethically and academically), or to their expressed anxiety that their capabilities to study in the University had been called into question, etc.

Interestingly, Jin Yi would (so he remarked to me much later) have preferred not to talk of the two home students as "the English" and not to have to define as "the Chinese" himself and the other students on the course whose countries of origin was the Far East. Yet had he not used this terminology when addressing the Dean of the School and had he not made comparisons showing that "the Chinese" were feeling inferiorized, he would have been left (in all likelihood) in an even more vulnerable position. The University officers could have then stood more easily by their claim that issues of discrimination were never put on the table by the complainants.[98] This claim was indeed made in a letter sent to Jin Yi by the Vice-Chancellor, after Jin Yi had asked for the handling of the "investigation" within the Business School to be looked into.[99] In one of his responses to Jin Yi, the Vice-Chancellor stated that had the issues been about (alleged) racial discrimination, the University policy on the prevention of harassment would have been activated (admitting that the University never did address the issue as involving such an allegation).

Now – taking into account the kinds of responses briefly mentioned above – we can ask the question whether by Jin Yi initially referring to "we" (the group of students whom he noted had felt affronted by those who seemingly posed as superior), he was unnecessarily essentializing the group classifications. Is it necessary to have used the category for the purposes of drawing attention to its coercive application by Ms. X and Ms. Y who saw "them" indeed as a group to which "the Chinese" belonged, and referred to

[98] It is worth noting in this context that two days before the "investigator" (Dr. B) handed his report to the Dean, he told myself and a colleague that: "What I am saying to you … and you have been at it for a long time. You know as well as I do that allegations of racism are a) very, very serious and b) take a lot of substantiating".

[99] Jin Yi had mentioned when he lodged his complaint to the Vice-Chancellor that although the Dean had promised in writing (subsequent to the "investigation") that action against those who had acted inappropriately and offensively would be taken, no-one of whom he knew was informed of anything that had occurred; nor had a report (stating the substance of the issues and how they were examined by the investigator) been provided. To this the Vice Chancellor replied to Jin Yi that "the response of the Business School was not within the strict guidelines prescribed by University procedures [for dealing with student complaints in general]", while at the same time he stated that the matter was handled informally in the Business School "(Section 9 'Informal Resolution')" – apparently unbeknownst to the complainants.

"them" as "the Chinese population"? Furthermore, while drawing on a conception of an imaginery collective that had been imagined to exist as "the Chinese" on the course, could "the Chinese" try at the same time to subvert the status of the "collective" categorization? Had a dialogue opened between the parties (which unfortunately never did occur) the parties may well have been able to speak about, around and past the categories and they might have generated a different conception of what was involved in drawing on categories of collectivities. The collectivity constructions then might have become re-visioned as not referring to ontological givens, but indeed as potentially opening up a dialogue around central issues of concern to the concerned parties. As it stood, "the English" appeared to inferiorize "the silent Chinese" without acknowledging that one could interpret the cultural agenda of "the Chinese" as not predisposing people to speak in public unless specifically encouraged and unless the social atmosphere is made conducive to this. For "the Chinese" on the course (at least for those who still undertook to enter class discussions), being interrupted when trying to speak was prohibitive of meaningful involvement in the conversation. In this context, use of the collectivity construction "the Chinese" served to occlude their involvement rather than to point to potentially culturally distinct meanings and styles of interaction that could be drawn upon.

Bauman argues that "politically, ethically, socially, it's a very important experience – to be attracted by otherness, to be inclined in some sense to know something you didn't know before, go where you were not before …" (2003, p. 215). He goes on to say that meeting "the other" may be easier by "staying at home" than by being a tourist abroad.

"Paradoxically, the chance of meeting the other (I mean genuinely meeting, not mis-meeting) may be greater when you stay at home in the big cities than when you go a thousand miles away in order to land up in a Holiday Inn. When I ponder the prospects of humanity, I derive more hope in this 'globalization coming home to roost'." (2003, p. 216).

In my view, there were many missed opportunities for people in the Masters class referred to in my cited example to indeed meet one another – missed primarily because of the orientation on the part of the two home students towards "the other" as something that could not hold anything new from which to learn. Instead of being curious about what "the other" might have held if they were prepared to enter a dialogue (which I would suggest certain students from the Far East did try to open, by for example, organizing a party to which they invited all students, and by making efforts to contribute input in the class), the dialogue was not pursued – and this had traumatic consequences for those involved (as reported to me and interpreted by me).

The dilemma for all those concerned about the coercive use of the overgeneralized collectivity construction used to identify "them" ("the Chinese"), was how to work past this construction while not thereby enabling those tasked with investigating their complaint to pass off the matter as simply "tension [in the cohort] and lack of respect for individuals".[100]

I have tried to show that one strategic device was to indeed utilize the group construction (the construction of "the Chinese") considered to have been used to treat this (defined) group unequally, but with an important proviso: that the invoked construction should not imply the existence of a collectivity of people all with the same experience of the(ir) cultural framework (that facilitated their expressions). By putting in place this proviso regarding the status of the collectivity construction, space might be opened (in principle) for a genuine meeting from which all parties could learn. But although this was possible in principle, the various actors' communicative propensities and the necessary activation of the institutional fabric that would make this practically feasible still needed to be accomplished (which was lacking in this case).[101]

In terms of what I call a "trusting constructivist" position (Romm, 2001, 2002a,b,c) when people fail to take responsibility for the possible social consequences of the use of their constructions, they necessarily become complicit in the furthering of a less-than-ethical social fabric (see also Shotter, 1993, p. 52-61). Trusting constructivism as a position highlights possibilities for trust creation by encouraging people's recognition (via social discourse) of their complicity, through their way of using constructions, in the way in which social outcomes might unfold. This position implies that people are brought to account – through processes of social discourse in which a variety of arguments and concerns are brought to bear – for their way of employing all constructions.

[100] This is how Dr. B phrased matters in his report to the Dean, although he had said to myself and a colleague verbally (two days before he submitted his report) that from what he could ascertain "it is still very much a two-groups' culture. And one group is intolerant of the other. Perhaps Ms. X's group is far more intolerant".

[101] As advised by a legal advisor from the Commission for Racial Equality in the UK, efforts at holding the University to account are still (at the time of writing) being made.

5. CONCLUSION

In this chapter I have suggested (following Brubaker *et al.*, 2004, p. 45) that the hold of "groupism" on our way of seeing and relating to "others" can be rendered less tenuous in practice once we are prepared to recognize that our focus on groups can be shifted to the focus on our own *grouping activities* such as classification, categorization and identification. With this shift, we can imagine possibilities for developing social relationships across so-called "bounded groups".

Bauman argues that this capacity to indeed move across borders that we may have drawn in our imagination is part of our humanity. As he states: "We are 'transcending beings' which constantly look forward beyond the border they have drawn, beyond the limits they set, and we need this propensity for transcending today ..." (2003, p. 215). He continues: "either we will teach each other and learn from each other or we will live unhappily ever after" (2003, p. 215). He cites our human need (that he posits) to "top up the proximity of otherness with recognition of shared humanity and enrichment of its contents" (2003, p. 214). But he also notes that, unfortunately, our capacities to "transcend our reciprocal otherness" keeps being "used up, diverted, channeled away, squandered by ... commercialized pseudo-multiculturalism ... in lieu of genuine conversation or a real attempt to get insight into the other's life and thought" (2003, p. 215).

In considering the prospects of genuine dialogue, McIntyre-Mills points out that such prospects need to be created as a process of people involved trying to "find interpolations across maps [of experienced realities] that will create shared lines of reasoning and shared understanding (if not shared narratives)" – in relation to issues raised on the agenda for discussion as issues of concern (2003, p. 129). For this "reasoning" to be workable (in Churchman's sense of being "humane"), our landscapes for communication need to be spaces where we (try to) "join with others". As Porra explains, according to Churchman: "In 'caring', 'I want' is replaced by 'we share'" (2001, p. 27).

In whatever way one inquires and acts in the social world (the two being seen as inseparable by Churchman) an appreciation of the need to combine "reason" with a "caring for humanity" seems to be an important requisite for addressing the world problem of ethnoracial discrimination.

I used an example set in a University in the United Kingdom to highlight the dilemmas that may be operative for those faced with the upsetting (and frustrating) effects of the invocation of ethnoracial categorizations, and wishing to input an alternative way of addressing the human relationships involved. My recounting of the example may touch audiences of this chapter who can "identify" with the story in some way (as it resonates with some of their own or others' recounted experiences, which may indeed point to more disturbing sagas). The example serves to illustrate some of the challenges of individuals (and collectivities) trying to "take on" the fabric of social and institutional power that seems to block the development of accountable, humane responses.

Considering this example (along with the discussion in the rest of the chapter), it appears that "putting brakes" on the world problem of racial discrimination requires that "we" (those interested) make concerted efforts to "do something in terms of caring about people". This, I would suggest, is the import of Churchman's appeal to us to consider a response at the global level, so that we can institute some kind of "ethical management" towards activating (more) "caring" as a global enterprise (Churchman, cited in Porra, 2001, p. 22).

One last point I would like to make in this conclusion is that I have clearly not offered in this chapter a (theoretical) account of the debate around the analytic power of the concept of "*class*" as a way of helping to cast light on issues of ethnoracial discrimination as experienced by people and as reproduced within institutional structures. Marxist-oriented authors who emphasize the importance of undertaking a Marxist analysis of the causes of exploitation in capitalist social formations, suggest that we need to be careful of treating "class" as just one among the many ways in which patterns of oppression develop in society (along with, say, race/ethnicity, gender, and disability). They suggest that the analytic power of class does not just point to another form of oppression, but to the way in which *capitalism structurally produces and reproduces poverty* (rather than reducing it) and *simultaneously creates ethnoracial patterns of social inequality*. According to this argument, it is capitalist social relations that constitute the basic form of exploitation (the exploitation of human labor), which itself contributes to the configuration of different forms of prejudice (including race and ethnic prejudice). The practical consequence of this (Marxist-inspired) way of arguing is that, as McLaren puts it: "to defeat racism we must defeat capitalism". Or, put differently, "capitalist social relations need to be challenged" (2004).[102]

When asked (in an interview with Pozo, 2003) about the relevance of post-modern theory with its focus on different epistemological stances (ways of "knowing") in society, Mclaren notes that he does not want to deny "the role of language in the ordering and regulation and reproduction of power". But he is concerned that:

[102] Churchman too concurs with the sentiments of Marx – whom he considers to be "one of the true good philosophers", who "stood for" the idea of giving "more power to working class". Churchman avers: "[this] still should happen" (Churchman, cited in Porra, 2001, p. 23).

> Many of these contributions by postmodern theory and its perfumed vocabulary of 'difference' hasn't been up to the task of exploring adequately how differences are shaped by the historical shifts within the globalization of capitalism that are currently devastating the entire globe (2003, p. 19).

McLaren's argument is that alongside a concern with respecting difference in our way of addressing "others" so as not to become complicit in inferiorizing what is "other", we require an understanding of the dynamics of global capitalism – which creates a lack of concern with others' poverty. For McLaren, it is only an anti-capitalist movement that can adequately address the "ruthless totalizing aspect of capitalism" (2003, p. 25). As he notes (commenting on the U.S.):

> We rarely talk about the war against the poor by the rich and the odious practices of the state to keep the poor in a condition of powerlessness. Here in the U.S. there exists an implosive reduction of the central antagonism of labor and capital to a single, uniform, denial of structural class conflict (2003, p. 25).

To address the points raised by McLaren (and others wishing to focus our attention on the structural inequality built into the capitalist system), would require that we explore a variety of theoretical debates. These are, for instance, debates about whether and how we can presume to locate "structures" existing outside of the human experience thereof (cf. Solomos and Back, 1996: 7-11; Romm, 2001: 115-116); whether genuine social transformation should be viewed as grounded primarily in altering *social structures*, out of which better (more ethical) human relationships can follow, or as grounded in the development of our *ethical consciousness* as the basis of progress towards a better society (cf. Gouldner, 1980; McKay and Romm, 1992; Von Foerster and Poerksen, 2002); and whether the postmodern location (and urging) of "difference" can be reconciled with the critical theoretical plea for us to develop our communicative capacities across "difference" (cf. Habermas, 1993; Flood and Romm, 1996; Simpson, 2000; Alleyne, 2002). As Churchman notes, in regard to our understanding of the human species, "we don't know enough about it. We hardly know ourselves as species at all" (cited in Porra, 2001, p. 25). In this chapter, I have not tried to address (or find ways of re-viewing) the myriad of debates around what it might mean to transform human relationships that are built on what McLaren considers as ruthless noncaring. I have attempted only to show how Churchman's concerns (including his insistence that our very way of knowing/understanding itself is not neutral in its practical consequences) might help to cast some light on the question of how we might direct our attention towards tackling racial discrimination as a world problem by combining "reason" with "caring" from the start.

REFERENCES

Alleyne, B.W., 2002, *Radicals Against Race*, Berg, Oxford.

Allport, G., 1954, *The Nature of Prejudice*, Beacon Press, Boston.

Anselmi, W. and Gouliamos, K., 1998, *Elusive Margins*, Guernica, Toronto.

Anthias, F. and Yuval-Davis, N., 1996, *Racialized Boundaries*, Routledge, London.

Ashburn-Nardo, L., Voils, C.I. and Monteith, M.J., 2001, Implicit associations as the seeds of intergroup bias: how easily do they take root? *Journal of Personality and Social Psychology*, 81(5):789-799.

Bauman, Z., 2003, The tourist syndrome (an interview with Zygmunt Bauman by A. Franklin), *Tourist Studies*, 3(2):205-217.

Bouchard, M., 2004, Panel presentation on "cultural copy", presented at the *Fourth International Conference on Diversity in Organizations, Communities and Nations*, UCLA, Los Angeles (July 2004).

Brendl, M.C., Markman, A.B. and Messner, C., 2001, How do indirect measures of evaluation work? Evaluating the inference of prejudice in the implicit association test, *Journal of Personality and Social Psychology*, 81(5):760-773.

Brubaker, R. and Cooper, F., 2000, Beyond "identity", *Theory and Society*, 29:1-47.

Brubaker, R., Loveman, M. and Stamotov, P., 2004, Ethnicity as cognition, *Theory and Society*, 33:31-64.

Chui, E., 2003, Unmasking the "naturalness" of "community eclipse": the case of Hong Kong, *Community Development Journal*, 38(2):151-163.

Churchman, C.W., 1979, *The Systems Approach and Its Enemies,* Basic Books Inc., Publishers, New York.

Churchman, C.W., 1982, *Thought and Wisdom,* Intersystems Publications, USA.

Churchman, C.W., 1994, What is philosophy of science? (Guest Editorial), *Philosophy of Science*, 61(1):132-141.

Delanty, G., 1998, Social theory and european transformation: is there a European society?, *Sociological Research Online,* 3, <http://www.socresonline.org.uk/socresonline/3/1/1html>

Dovidio, J., 1993, The subtlety of racism, *Training & Development*, 47(4):50-57).

Dovidio, J., 1997, "Aversive racism" and the need for affirmative action, *Chronicle of Higher Education*, 43:46.

Dovidio, J.F., Gaertner, S.L., Validzic, A., Matoka, K., Johnson, B. and Frazier, S., 1997, Extending the benefits of recategorization: evaluations, self-disclosure, and helping, *Journal of Experimental Social Psychology*, 33:401-420.

Fielding, N., 1993, Ethnography, in: *Researching Social Life*, N. Gilbert, ed., Sage, London.

Fenton, S., 1999, *Ethnicity: Racism, Class and Culture,* Macmillan Press Ltd., Basingstoke.

Flood, R.L. and Romm, N.R.A., 1996, *Diversity Management,* Wiley, Chichester.

Fuenmayor, R., 2001, The oblivion of Churchman's plea for systems approach to world problems. II. The rise of the modern constellation, *Systems Practice and Action Research*, 14(1):29-45.

Gadamer, H-G., 1975, *Truth and Method,* Seabury Press, New York.

Gouldner, A.W., 1980, *The Two Marxisms*, Macmillan, London.

Habermas, J., 1976, A positivistically bisected rationalism, in: *The Positivist Dispute in German Sociology*, T.W. Adorno, H. Albert, R. Dahrendorf, J. Habermas, H. Pilot and K.R. Popper, eds., Heinemann, London.

Habermas, J., 1982, A reply to my critics, in: *Habermas: Critical Debates*, J. Thompson and D. Held, eds., Macmillan, London.

Habermas, J., 1993, Justification and Application: Remarks on Discourse Ethics, Polity Press, Cambridge.

Hammersley, M., 1995, *The Politics of Social Research*, Sage, London.

Hammersley, M., 1997, A reply to Humphries, *Sociological Research Online*, 2, <http://www.socresonline.org.uk/socresonline/2/4/6.html>

Handler, R., 1991, Who owns the past?, in: *The Politics of Culture*, B. Williams, ed., Smithsonian Institution Press, Washington.

Holstein, J.A. and Gubrium, J.F., 1995, *The Active Interview*, Sage, London.

Kawakami, K., Dovidio, J.F. and Dijksterhuis, A., 2003, Effect of social category priming on personal attitudes, *Psychological Science*, 14(4):315-319.

Kramer, R.M. and Brewer, M.B., 1984, Effects of group identity on resource use in a simulated commons dilemma, *Journal of Personality and Social Psychology*, 46:1044-1057.

Laitin, D.D., 1988, Political culture and political preferences, *The American Political Science Review*, 82(2):589-593.

Lincoln, Y.S., 1995, Emerging criteria for qualitative and interpretive research. *Qualitative Inquiry*, 1:275-289.

Mason, R.O., 1994, Securing: one man's quest for the meaning of therefore, *Interfaces*, 24:67-72.

McIntyre-Mills, J., 2003, *Critical Systemic Praxis for Social and Environmental Justice: Participatory Policy Design and Governance for a Global Age*, Kluwer Academic/Plenum Publishers, New York.

McIntyre, J., 2004, Facilitating critical systemic praxis (CSP) by means of experiential learning and conceptual tools, *Systems Research and Behavioral Science*, 21:37-61.

McKay, V.I. and Romm, N.R.A., 1992, *People's Education in Theoretical Perspective*, Longman, Cape Town.

McLaren, P., 2003, Towards a critical revolutionary pedagogy: an interview with Peter McLaren, *St. John's University Humanities Review*, 2(1):1-26.

McLaren, P., 2004, Plenary session presentation, presented at the *Fourth International Conference on Diversity in Organizations, Communities and Nations*, UCLA, Los Angeles (July 2004).

Miller, N., 2002, Personalization and the promise of contact theory, *Journal of Social Issues*, 58(2):387-410.

Mitroff, I.I., 1994, The cruel science of world mismanagement: an essay in honor of C. West Churchman, *Interfaces*, 24:94-98.

Monteith, M.J., Voils, I.C. and Ashburn-Nardo, L., 2001, Taking a look underground: detecting, interpreting, and reacting to implicit racial biases, *Social Cognition*, 19(4):395-417.

Monteith, M.J. and Winters, J., 2002, Why we hate, *Psychology Today*, 35(3):44-50, 87.

Noddings, N., 1984, *Caring: A Feminine Approach to Ethics and Moral Education.* University of California Press, Berkeley.

Okin, S.M., 1989, Reason and feeling in thinking about justice, *Ethics*, 99(2):229-249.

Parekh, B., 1997, Dilemmas of a multicultural theory of citizenship. *Constellations*, 4(1):54-62.

Popper, K.R., 1994, The Myth of the Framework: In Defence of Science and Rationality, Routledge, London.

Porra, J., 2001, A dialogue with C. West Churchman, *Information Systems Frontiers*. 3(1):19-27.

Rawls, J., 1971, *A Theory of Justice*, Harvard University Press, Cambridge, Mass.

Romm, N.R.A., 2001, *Accountability in Social Research: Issues and Debates*, Kluwer Academic/Plenum Publishers, New York.

Romm, N.R.A., 2002a, A trusting constructivist approach to systemic inquiry: exploring accountability, *Systems Research and Behavioral Science*, 19(5):455-467.

Romm, N.R.A., 2002b, Responsible knowing: a better basis for management science, *Reason in Practice* 2(1):59-77.

Romm, N.R.A., 2002c, A trusting constructivist view of systems thinking in the knowledge age, in: *Systems Thinking and Practice in the Knowledge Era*, G. Ragsdell, D. West and J. Wilby, eds., Kluwer Academic/Plenum Publishers, New York.

Romm, N.R.A. and Hsu, C-Y., 2002, Reconsidering the exploration of power distance: an active case study approach, *Omega* 30(6):403-414.

Short, R. and Magana, L., 2002, Political rhetoric, immigration attitudes, and contemporary prejudice: a Mexican American dilemma, *Journal of Social Psychology*, 142(6):701-712.

Shotter, J., 1993, *Cultural Politics of Everyday Life*, Open University Press, Buckingham.

Simpson, L., 2000, Communication and the politics of difference, *Constellations*, 7(3):430-442.

Solomos, J. and Back, L., 1996, *Racism and Society*, Palgrave Macmillan, Basingstoke.

Sunstein, C.R., 1989, Notes on feminist political thought, *Ethics*, 99(2):219-228.

Van Gigch, J.P., Koenigsberg, E. and Dean, B., 1997, In search of an ethical science: an interview with C. West Churchman, an 80th birthday celebration, *Journal of Business Ethics*, 16(7):731-744.

Von Foerster, H. and Poerksen, B., 2002, *Understanding Systems: Conversations on Epistemology and Ethics*, Kluwer Academic/Plenum Publishers, New York.

Weil, S., 1998, Rhetorics and realities in public service organizations: systemic practice and organizational learning as critically reflexive action research (CRAR). *Systemic Practice and Action Research*, 11: 37-62.

Young, I.M., 1989, Polity and group difference, *Ethics*, 99(2):250-274.

Chapter 16

DESIGNING A REPLACEMENT FOR THE UN

RUSSELL L. ACKOFF AND JAMSHID GHARAJEDAGHI

Ackoff Centre for Advanced Systems Approaches, University of Pennsylvania, USA.

FOREWORD

West was always greatly concerned about the plight of people who are marginalized or disadvantaged wherever they lived and regardless of their culture religion, race, nationality, or gender. He made whatever effort he could to address related problems and he encouraged and supported our efforts to do the same. Neither of us would be so bold as to suggest we have had anything more than some local and limited successes. We dedicate to his memory this effort to get something larger going. I am sure he would have said, "Amen."

West was also a strong advocate of discussion. Our intention in writing this article is to provoke widespread discussion of all of its features. We invite suggested ways of improving the design.

16.1 INTRODUCTION

The world is in a mess. This statement hardly needs documentation. The terrible state of world affairs is disheartening, particularly to those of us who hoped the UN would be able to bring some order and sense into these affairs. Unfortunately, the UN has turned out to provide little more than the defunct League of Nations had. It is clear that it is only through effective, decisive and strong international action that any improvement in the state of the world can be brought about. A unilateral approach by any super power not only places a great and unfair burden on its own people but it does not produce desirable results. In fact, up to now, such interventions have only exacerbated the mess. The only workable solution requires the world community to accept its collective responsibility and act with authority to create a world order that is capable of managing national interactions. Therefore, it seemed to the two of us that a design of a potentially effective multinational organization might minimally encourage some productive changes in the UN. We have chosen to go about producing such a design by employing *idealized design.* That is, we start with a clean sheet; and then go ahead with no constraints except technological feasibility — no science fiction. The system we design does not even need to be implementable but, clearly, it should be able to improve the world if it were implemented. We do not expect our design to be implemented, but we wish it could be. However, we believe an approximation to it may be possible either as a replacement for the UN or as a new multi-national venture.

First we develop the idealized design and then consider what aspects of, or approximations to, it are feasible. It will become apparent that all the obstructions to implementation of even the idealized design lie in the minds of people, not "out there." But these are the hardest kinds of obstructions to remove. Fortunately, in the past, mobilizing ideas and visions have produced massive mental changes and brought about social revolutions of huge proportions.

> "...man(sic) has been able to grow enthusiastic over his vision of ...unconvincing enterprises. He has put himself to work for the sake of an idea, seeking by magnificent exertions to arrive at the incredible. And in the end, he has arrived there. Beyond all doubt it is one of the vital sources of man's power, to be thus able to kindle enthusiasm from the mere glimmer of something improbable, difficult, remote." (José Ortéga y Gassett, 1966, p. 1.)

Yogi Berra is reported to have said something like, "*If you don't know where you are going, you won't know where you are when you get there.*" Idealized design produces a vision of where you will be when you get there.

It provides continuous direction to ones activities. It reveals opportunities to make changes that take one closer to the state envisioned. In effect, it converts problems that arise along the way into opportunities. It puts meaning into both long and the short pursuits. We hope this ideal design will stimulate discussion that yields improvements in the way world affairs are handled while generating consensus as to what is most desirable. In other words, we hope it will raise the level of discussion of global issues to a higher level than that at which such discussion currently takes place.

Current discussion tends to accept the global system that exists and ask how parts of it or their behaviors can be changed so as to reduce the mess. Such discussion although pervasive is based on an assumption that one can dissolve a mess by patching or tackling its parts separately. Improving the performance of parts of a system may not improve the performance of the system of which they are part. In fact, improvement of the parts can destroy a system. For example, replacing the motor in, say, a ford Taurus with a more powerful motor, say from a Rolls Royce, will not improve it — even if it could be fitted in — because the other parts of the car are not designed to work with such a motor. The performance of a system depends on how its parts *interact,* not on how they perform when considered separately. In some cases it is possible to make a system perform better by making one or more of its parts perform worse. For example, when a store offers loss leaders (items sold at a loss) it does so with the intention of increasing total sales.

16.2 SPECIFICATION OF DESIRABLE PROPERTIES

There are certain important properties that we want to design into the organization that we envision. These must provide nations with a good reason for joining; that is, there must be a significant advantage to membership in the organization.

1. It should be a voluntary association of autonomous nations committed to democracy, pluralism, the protection of human rights, and the promotion of development of its member nations and those that are external to it. Because of this we call this organization the *Union of Democratic Nations.*
2. It should have the ability to dissolve or manage conflicts (a) between its member nations, (b) between its members and nonmembers, and (c) between nonmember nations when invited to do so or when such conflicts endanger one or more of its members.
3. It should provide security for all member nations by protecting them against aggression from any internal or external source.

4. It should facilitate unrestricted economic, cultural, educational and technological exchanges between its members.
5. It should make possible completely free communication and movement of individuals, goods and capital among member nations.
6. It should prevent exploitation within member nations of any group that is based on discrimination and/or segregation by race, nationality, religion, language, sex or age.
7. It should help eradicate poverty, illiteracy, genocide, tyranny, health hazards and environmental deterioration all over the world.
8. It should assist nations that are not eligible for membership but want to join the Union in becoming eligible for it.

16.3 GETTING INTO THE UNION

- Membership in the Union would be open to all nations irrespective of their location, race, religion and size as long as they met the following conditions:
    - Members must have been an autonomous pluralistic democracy for at least 20 years with at least two peaceful transitions of administrations; that is, transfers of power.

> *By a "democracy" we mean a nation within which (1) everyone who is affected by a decision has an opportunity to participate either directly or indirectly (through elected representatives) in making it, (2) anyone in a position of authority over others individually is subject to their collective authority, and (3) anyone can do anything they want to providing that it has no effect on anyone else's ability to do the same, otherwise they require the agreement of those affected.*
>
> *By "pluralism" we mean a society that permits any and every point of view on any issue to be expressed and acted on unless it restrictsthe ability of anyone else to do the same thing.*
>
> *By "autonomous" we mean not subject to any external source of control.*
>
> *By "peaceful transition of an administration" we mean the transferof authority at the top of the nation's government from one elected person or set of persons to another person or set of persons.*

- Application for membership must be approved by a majority of its eligible voters.
- Members must accept and fully meet requirements of the universal

- declaration of human rights without any exceptions, including allowance of no discrimination or segregation based on race, sex, religion, national origin, language or age.
- Members must accept the role of the Union and commit to making the changes necessary to conform to the transfer of responsibilities from the nation to the Union.
- Members must not be at war with any other nation, nor be experiencing a civil war, and not be engaged in any territorial disputes.
- Members must not belong to any other association of nations, obligations to which take precedence over those to this Union.
- Members must not harbor or support terrorism in any form.
- Members must abolish all forms of capital and physical punishment.
- Members must provide all nationals (except prisoners and those awaiting trial) with the freedom to emigrate.
- Members must provide free public education to all residents for a minimum of eight years, also universal healthcare and unemployment insurance along with incentives to work.

16.4 SEPARATION FROM THE UNION

- Membership would be suspended if any of the criteria for membership were violated. Members would lose their membership with three violation and or refusal to honor the Union resolution.
- Any member nation of the union could be expelled from the union by a two-thirds vote of both governing houses.
- Any member nation could secede from the Union with support of a majority of its eligible voters.
- Any part of a member nation may secede from that nation by a majority voted of its population. Once separated it would have to go through normal procedures to enter the Union.

16.5 CONDITIONS OF CONTINUED MEMBERSHIP

- Members would maintain no national military force. The Union would provide a military force to protect all members against aggression from within or without.
- Each member would pay dues annually consisting of a per capita assessment that would be proportional to average per capita income in that nation.

- Members would provide complete freedom of movement within the Union.
- Members would facilitate free trade within the Union.
- Members would limit nonrenewable terms of all elected officials to six years.
- The relevant government would finance all campaigns for public office and limit them to three months.
- The Union would adopt an official language that would be taught in all schools in member nations.
- No member of the Union could ship into any other member of the union goods or services outlawed in the receiving member — for example, guns or drugs.

16.6 SERVICES PROVIDED BY THE UNION

In addition to military forces provided and maintained by the union, it would :

- Operate a Food and Drug Administration and an Environmental Control Service.
- Establish and maintain a Center for Disease Control directed at prevention or control of epidemics originating both within and outside the Union.
- License products ad services and establish and administer safety standards and grading systems applicable to these in all member nations.
- Operate a conflict resolution service available both internally and
- externally to the Union.
- Establish and maintain a Commission to monitor all nations in and out of the Union for presence and development of weapons of mass destruction.
- Provide a patent office, registration of trademarks, and copyright facilities.
- Provide a stock market and a Security and Exchange Commission.
- Provide custom and immigration services.
- Establish, maintain and operate a fund and consultation service to encourage and facilitate development of nations both within and outside
- the Union.
- The Union would establish and support (but exercise no control over) a news agency that would publish a daily newspaper and would broadcast the news over radio and television. The output of this agency would be available in all member nations and, on request, by nations outside the Union.

16.7 GOVERNANCE OF UNION

- There would be two policy-making bodies One, *the Council of Heads of State* ,would consist of the elected Presidents or Prime Ministers of all member states with their deputies (e.g., Vice Presidents or Deputy Prime Ministers) serving as alternates. The other, the *Chamber of Representatives*, would consist of representatives of each member nation, the number proportional to their population and their GNP per capita.
- The Union would have an Executive Director elected for a nonrenewable term of six years. At least three persons would be nominated by the Chamber of Representatives, and be elected by a 2/3-majority of the Council of Heads of State.
    - Any elected executive of the Union could be dismissed from office by 2/3 vote of both the Council of Heads of State and the Chamber of Representatives.
- The Union would have a court whose function is to handle issues between member nations and between member and nonmember nations. Members of the court would be nominated by the Council of Heads of State and selected by majority vote of the Chamber of Representatives.
    - The court would consist of nine members each of whom serves one nineyear term. One member of the court would leave each year and be replaced. Nominations would be made by the Executive Director of the Union but selected by a 2/3 vote of the Council of Heads of State. Nominees would be drawn from former heads of State or national Supreme Courts.
- The Executive Director would have a cabinet consisting of the head of each division of the Union, these appointed by the Executive Director, but requiring approval of 2/3-majority of the Council of Heads of State.

16.8 THE RIGHTS OF MEMBER STATES

- We make no effort to make all the member states behave in the same way. We have tried to minimize the amount of autonomy that members in the Union are required to relinquish.

REFERENCE

Ortéga y Gasset, José, 1956, *Mission of the University*, W.W. Norton, New York.

Chapter 17

RESCUING THE ENLIGHTENMENT FROM ITSELF:
Implications for Re-Working Democracy and International Relations

JANET MCINTYRE-MILLS
Flinders Institute of Public Policy and Management

Abstract: All those at the receiving end of a decision should be party to the decision making process, so that power, knowledge, lived experience and creativity are shared at all stages of the democratic process. The capacity to improve governance and international relations is based on openness and an ability to work with, rather than within closed frameworks. I draw on West Churchman and many of those who have read and incorporated his ideas whilst developing Critical Systemic Praxis (CSP). The chapter works with Churchman's 'Design of Inquiring Systems' to address the need to bridge different knowledges, in the interests of improving governance and international relations.

Key words: Enlightenment, hegemony, democracy, Critical Systemic Praxis

17.1 INTRODUCTION

In order to rescue the Enlightenment from itself, we need not only idealism, but also an appreciation of different perceptions and different notions of what constitutes an issue based on different experiences of power. This chapter celebrates the contribution of West Churchman by working with some of his ideas developed in '*The Design of Inquiring Systems*'

(1971), and draws on Churchman's works such as: *'The Systems Approach and its Enemies'* (1979*) and Thought and Wisdom* (1982), as well as ideas from a range of thinkers. Churchman argues that human beings cannot comprehend total systems. Our comprehension of ultimate patterns will always be incomplete. When we think we have understood, it is quite likely that we will have a view filtered by time, place and our lived experience of it. Nevertheless we need to strive for a better understanding of power, communication and participation, in order to address the most appropriate forms of governance and international relations (Pierre 2000).

Human will is a vital consideration for those concerned about achieving better governance. But it cannot be tempered without greater mindfulness. Unlike Nietzsche, I argue that mindfulness can be encouraged when people realize that thinking systemically and critically is vital for sustainable futures and thus it is both idealistic and in their pragmatic interests. Critical Systemic Praxis (CSP) is a means to bridge the opposition between the ideal and the reality. Good governance (that supports democracy and good international relations) is about praxis that can lead to a better balance of social, cultural, political, economic and environmental concerns. Thinking idealistically in this manner can ensure a sustainable future for ourselves and future generations. Surely this is also pragmatic? Thinking about the self and ignoring the interests of others and the impact our decisions could have on the environment leads to the high risk of creating war and wastelands. Moving from pragmatism to idealism is not difficult once we appreciate not only 'the boomerang affect' (Beck 1992) but also the continuity across states, regions and the implications for nationality, race, culture, citizenship and democracy (Beck, 1998; McIntyre-Mills, 2000; Grugel, 1999).

Idealism and pragmatism are one. This is the *'low road to morality'* to use a phrase from Alinsky's (1972) *'Rules for Radicals' (*but at least people would be traveling in a sustainable direction!). Taking 'the low road' could of course be aided by means of laws that require accounting and accountability to be systemic and mindful of multiple concerns. The United Nations Aarhus convention of 2001, for example is a useful step in this direction (Florini, 2003), because it advocates broad public participation in vigilance on matters that impact on both social and environmental justice and transparency. Critical and systemic thinking needs to be "given teeth" by means of international law. This could help to redress the imbalances of pollution and dumping in Third World countries by the First World. For example, the so-called 'Pacific Solution' provided a short-term answer for the policy of Australian 'border control'. Testing and dumping nuclear waste outside national boundaries has provided short-term solutions for international powers.

Critical systemic thinking and practice (CSP)[103] enables boundaries to be drawn and redrawn on the basis of iterative dialogue that includes those who are to be affected by the decisions. The challenge is to ask questions that will enable sustainable social and environmental outcomes to be achieved from the point of view of those affected by the decision.

The chapter strives to address (if not answer) the following questions:

- What is democracy these days?
- To what extent does the so-called 'Enlightenment'[104] cast a shadow of hegemony over the powerless?
- What is good governance? Can it help create a sustainable future?
- Why should humanity care if the powerful choose to ignore and exclude the powerless?
- How can we recognize 'the enemies within' (all values based) that include: "politics, religion, morality and aesthetics", to use West Churchman's (1979 a, page 23)concepts?
- We need to examine the human potential for hubris in our individual, (social, cultural, political and economic) lives and its systemic impact on the environment and all those sentient beings affected by our decisions. West Churchman stresses that we should always be mindful of criticism and the opposite point of view. The closest we can get to truth is through dialogue. The ongoing process (together with the structures that support this open process) is directly relevant to governance. Maintaining civil liberties and justice requires that rights and responsibilities be respected and maintained. The process of maintaining rights and responsibilities holds within it the potential for abuse.

The Enlightenment and democracy need to be seen not so much as a static universal law, but as a dynamic structure and process[105]. The West faces the challenge of preserving this openness and trust and redressing the

[103] Critical systemic thinking and practice uses a complementary approach to theory and methodology and is open to many ideas. At the heart of CSP is the belief that social inquiry is not merely about observation, but also about intervention (Midgley, 2001).

[104] Western thinking has tended to emphasize thinking in compartments. We need to think about why the powerful who do the planning would want to listen to the powerless as Midgley (1996: 44, citing Romm, 1994, 1995a,b) stresses. The tendency today is to emphasize only some of the legacy of the early Greek philosophers in Enlightenment thinking[104]. Some Greek philosophers tended to categorize. But other Greek philosophers developed knowledge through dialogue based on questioning (see Banathy, 1996, 2000).

[105] In the introduction I stressed that the central challenge for democracy is balancing the eternal paradox that: On the one hand, openness to debate and to other ideas and possibilities is the basis for both the Enlightenment process of testing and for democracy. On the other hand, for openness to occur there has to be some trust that voicing new ideas will not lead to subtle or overt marginalisation of oneself or one's associates.

imbalances in wealth and power caused by centuries of colonization, modernization and globalization that are based on the single bottom line (see Elkington's 1997 critique) of economic profits for competitiveness in markets that support hegemony (rather than on a multiple dynamic awareness of socio-cultural, political, economic and environmental factors) that when considered together support a sustainable future.

West Churchman's work can be said to help rescue the Enlightenment from its own failings through mindfulness. Critical reflection and conversation could be the basis for rescuing the Enlightenment agenda from itself. For transformation that enhances democracy we need both mindfulness and conversation. The use of stories, metaphors and organic maps (preferably iterative and dynamic) are vital for understanding and making sense of events. This is because conversation and stories enable the participants to hold in mind multiple variables simultaneously. The challenge for good international relations and good governance is to make meaning from information through finding and drawing connections across variables. A respectful conversation using metaphors and heartfelt stories can make all the difference to our ability to understand complexity.

If we are drawn out of our fixed categories by a metaphor, a story or a conversation we are able to comprehend the dynamic and fuzzy nature of sets. What is the connection across these snippets for governance and international relations?

> 'I heard on the news today'[106] (to misquote the Beatles) that the probe to Mars has shown that there was once water on Mars and 'the oracle' suggests that life could have once existed there.

> I heard last night a local government official saying: 'Recycling is not cost effective. I do not want to 'put the mockers' on this, but we have found that it is not cost effective. We cannot mulch green waste and make it available for reducing evaporation in South Australia'.

South Australia is the driest state in the driest continent and recycling was listed with four other indicators of a sustainable green community[107] as the most important factors to consider in a strategic planning exercise on behalf of members of the public.

[106] March 1 Australian Broadcasting Service, 6 am news.

[107] Namely: 1. conservation of rainwater by individual households,re-use of grey water for irrigation and thus the use of organic cleaners, 2. development of urban hubs that prevent urban sprawl and celebrate diversity and 3. the conservation of green spaces within the urban area.

The will and power to make a difference for this generation and those that follow, is based on an understanding that idealism and pragmatism are one. The binary opposition between the ideal and the reality can and must be bridged when planning and implementing sustainable futures. I make a case for participatory design based on CSP to address social and environmental justice. Much of Western knowledge is in the area of narrow areas of technical ability. Economics has been rethought by considering more variables. There is plenty of room for transforming accounting and accountability to address the dynamic interconnections across social, cultural, economic and environmental factors (Elkington, 1997; Gallhofer and Chew, 2000).

Conversations with diverse stakeholders are vital for democracy. I explore some of his key concepts[108], adapt and apply his '*Design for Inquiring Systems'* to international relations and governance[109]. In so doing I

[108] In 1971 he wrote The Design of Inquiring Systems and developed tools for theoretical and methodological literacy. In 1979 he wrote The Systems Approach and its Enemies and developed reflexive thinking based on the dialectic of unfolding values and sweeping in multiple variables. We need to recognize 'the enemies within' and without- these include: "politics, religion, morality and aesthetics", to use West Churchman's (1979a, 23) concepts, so as to examine the human potential for hubris, which means thinking we can behave like a God who has all the answers for other people. Instead we have to accept that we all have potential to make contributions. West Churchman argues that we can never find a total or absolute solution and for that matter we are unlikely to ever know the full extent of any system, but nevertheless we need to try to see or appreciate wider horizons (Vickers, 1983). The greater the use of critical reflection and participatory design for governance and international relations, the better the local and international policy outcomes will be. In 1982 he wrote Thought and Wisdom in which he suggests that the system of interest should in fact start with 'unfolding' values and 'sweeping in' (in the sense used by Churchman 1979 a,b, 1982, who drew on Singer) multiple variables (social, cultural, economic, political and environmental) within the context of the inquiry. Dialogue and conversation to achieve resonance and radiance of shared meaning was a focus for this work. Critical reflection is the basis for testing out ideas and finding out how they are perceived, because the closest we can get to truth is through dialogue (McIntyre-Mills, 2000, 2003 2004).

[109] "The answer lies in the concept of governance. It is not in opposition to the state-centric perspective; rather it encompasses and extends beyond the governance sustained by states. By locating rule systems at the heart of our theoretical formulations, we can trace and assess the processes of governance wherever they may occur. That is, through focusing on rule systems we will not be confined to the work of states and will be empowered to explore issues and processes in terms of the way in which authority is created, dispersed, consolidated, or otherwise employed to exercise control with respect to the numerous issues and processes that states are unable or unwilling to address. Put differently, a governance perspective will enable us to observe and compare the situations within and between the multi-centric and state-centric worlds, thereby allowing us to probe the challenges embedding globalization and its corollary, localization, and to be sensitive to the fragmegrative dialectics whereby globalising and localizing forces are inextricably linked." (Rosenau in Pierre, 2000: 188).

draw on the work of others[110] to make a contribution towards working conceptual and geographical boundaries. This reflexive, discursive conversation needs to be built into 'Enlightenment' processes of reviewing ideas. The viewpoints of multiple stakeholders are needed, in order to rescue the Enlightenment from itself. Openness will help to prevent the hubris of top down, one-way communication. Discursive conversation /dialogue is a way to promote better understanding.

Appreciation (in the sense used by Vickers 1983) of multiple points of view can help us to work towards bridging Foundationalism, anti-Foundationalism and non-Foundationalism (Romm, pers comm. 2004). Rescuing the Enlightenment agenda from its failings, building bridges of understanding and open debate to achieve transparency ought to be the ideal for democracy. This is a far cry from the present situation of war, cover-ups and toadying to the powerful -nationally and internationally -within the so-called coalition of the willing warring groups. All those at the receiving end of a decision should be party to the decision making process, so that power, knowledge, lived experience and creativity are shared at all stages of the redeemed democratic process.

17.2 DISCURSIVE STORIES AND METAPHORS

Discursive stories and metaphors can help to develop a sense of human and environmental rights and responsibility. What are the connections across the ships (known as theTampa and SIEV X) carrying asylum seekers, the ship called the Cormo Express[111] (carrying rejected, diseased sheep to the

[110] Bhaba (1994), Gaventa (2001), Haraway (1991) and Borradori (2003) in conversation with Habermas and Derrida. CSP uses theoretical and methodological literacy (drawing on and adapting Banathy 1996, 2000, 2001a,b, c, 2003, Bausch 2001, Flood and Carson, 1993; Jackson, 2000; Midgley, 1991, 1997, 2000, 2001; Romm, 2000, pers. comm. 2004) to explore paradoxes and is based on the assumptions of: Enlightenment truth and emancipation linked with the process of dialogue, Complementary theory Complementary methodology and Social/cultural, political, economic and environmental awareness. The influence of the following are also central to better governance:

Gaventa, J & Valderrama, C., 1999, Participation, citizenship and local governance: Background note for workshop on "Strengthening participation in local governance", Institute of Development Studies, June.

Gaventa, J., 2001, Towards participatory local governance: six propositions for discussion. Paper presented to the Ford Foundation, LOGO Program with the Institute of Development Studies, June.

Gaventa, J. and Cornwell, A., 2001, Power and Knowledge Handbook of Action Research, Reason, P. And Bradbury, H. eds., Sage, London.

[111] "...Australians agricultural heritage imposed itself on the international stage. Few noticed when a livestock transport vessel I, the MV Cormo Express, sailed from Fremantle,

Middle East and dumped in Eritrea) with unnamed ships exporting nuclear waste?

In all three cases an attempt was made to dump the unwanted and commodified - whether they were labeled asylum seekers (as opposed to refugees with rights), live meat (as opposed to sentient beings) or nuclear waste. Drawing the boundaries between what is considered culturally and politically acceptable or unacceptable[112] is based on values.

In his work *'Democracy without Enemies'* (Beck 1998) discusses the way in which people are marginalized so that democracies only give rights to the citizen. 'The other' is rendered without rights and outside the boundary – or beyond the pale. The so-called 'boomerang affect' (Beck, 1992) of pollution does not respect national or class boundaries. Nor can governance and international relations afford to ignore the *pay back* of social and environmental injustice.

Western Australia in 6 August 2003, It made a far more dramatic public impression later that months once it arrived at its destination in Jeddah. Authorities in the Saudi Arabian port city rejected the ships cargo of 57000 sheep, citing unacceptable levels of scabby mouth, a viral infection common in Australian flocks. Australian on board vetinarians denied the charge, but the Saudis were uncompromising and refused the shipment. Mostly older wethers, the sheep were intended for slaughter and had already been through an arduous journey. The livestock trade …represents around 1 billion in exports annually. Seeking to protect this trading reputation, Australian officials set off a frantic diplomatic maelstrom as they scoured the planet in search of a new port that would take the stranded animals. In all Australia approached fifty-seven countries. There were fleeting rumours that Kenya, Kuwait, Oman, Bangladesh, or even Iraq might take the sheep as a form of live food aid…In Iraq, despite initial confidence an option had been found, the prospect collapsed when the British forces in control of the Southern region realised the daunting logistical task of such a mass slaughter, coupled with the prospect of civil unrest. Reports of diseased sheep arriving provoked local fears of a plot to spread pox among Iraqis. Domestic pressure mounted in Australia to expedite a humane outcome, with some suggesting that the shipment be returned home and others urging a butchering at sea. The farm lobby vehemently opposed repatriation, fearing the introduction of foreign pests, which the animals might have picked up during their unhappy voyage. The option of slaughter on the open sea was also fraught with problems. In the end Eritrea accepted the shipment nearly two months later. In mid-October Australia donated the sheep to this small, chronically poor country on the Horn of Africa, along with 3000 tonnes of feed and $ 1 million to meet transportation and abattoir costs. More than 5500 sheep, or 9.6 percent of the original cargo, had died from stress, heat and fatigue. The Australian taxpayer copped a bill estimated well in advance of $10 dollars. …." (Longo 2004:239). The international context of the so-called war on terror, suspicion, and the poor relations with the Middle East exacerbated by the apparent attempts to initially cover up the rejection from the first port of call (Long 2004: 239) added to the problem of the commodification of animals.

112 See Douglas, 1978 for a discussion on the sacred and profance and Midgley 2001, for a discussion on boundaries and pollution.

The war on terror needs to be seen as a *continuation of* and *fall out from* cold war relations (as suggested by Derrida, 2003, in his conversation with Borradori). The building the wall in Jerusalem is a response to a war quite literally about olive trees and access to the groves of one's mothers and forefathers (to draw on Friedman, 2000). But, according to him, it is also about access to the Lexus (a car metaphor which stands for materialism, market and mobility) and world opportunities. The challenge is to find sustainable industries. The building of the wall between Israel and Palestine is an example of reductionist thinking and practice. Separating people formally by means of a wall, to prevent violence is short sighted. For some, this wall (just like the Berlin wall) prevents integration and access to places of work and to markets. For some, it has lead to loss of land and access to family. For some, it could mean protection from what they perceive as 'terrorists', whilst for others it could be a rationale for becoming 'freedom fighters'.But in the long term boundaries are unhelpful when they limit the extent that we are able to appreciate problems. Issues of conflict are complex and cannot be solved simplistically by means of attempts at reducing them or *'cutting off'* other areas of concern. We need critical and systemic praxis to redress reductionism and to *'sweep in'* social, cultural, political and economic variables when doing governance locally, nationally and internationally that appreciates multidimensional viewpoints in an ongoing dialogue.

Throughout Friedman's (2000) work '*The Lexus and the Olive Tree'* he sounds very much like an advocate for economic determinism and rationalism, until his last chapter, when he acknowledges that human creativity could remake or redesign the world. His choice symbols[113] for culture and the market provide a very useful starting point for conversations on governance and international relations. These dimensions are vital and the challenge for the future is to shape the market through our values and not to allow the so-called "golden straightjacket" of global markets (Friedman, 2000:39) to govern us.

Critical and systemic mindfulness would include more emphasis on the perceptions of multiple players.[114] Friedman does not rest his analysis on

[113] The Lexus is shorthand for mobile 'knowledge worker's and for the aspirations for development. The Olive tree is the organic metaphor that roots us in the soil of tradition, family, culture and the local space or sense of place.

[114] Friedman, for example refers to the role of credit assessments by Moore and Standard and Poor's rating agency in positive terms, linking high ratings with stable societies and low ratings with unstable societies. He does not engage in a discussion of power and the rationale for the striving for nuclear power by India. Instead the thrust of his argument is only that once they had tested a nuclear bomb that they were downgraded as a credit risk. Friedman veers strongly towards economic determinism and only right at the end of his book does he develop the argument that it is human beings who decide on how to use the

human creativity. It is an afterthought in his book. We can make cultural choices, despite the limitations of structures on our life chances. Our analyses can stress not only the categories (or bits of information) but also the continuums that make meaning. We can be digital (Negroponte, 1995) and we can emerge from his 'golden straightjacket' when we realize the dynamic connection across our concepts of material atoms, computing bits (based on either or thinking) and telephonic waves or logons (based on continuums of electronic pulses or energy) (See Friedman 2000). Ideas that are open to debate within context (and in ongoing and iterative cycles) are better for democracy than ideas that are simplistic and imposed. If knowledge is the basis of governance, what knowledge counts?

We are the designers and we can cut ourselves free from the determinism of economic rationalism and reconceptualise it, so that we transcend it to embrace social, cultural, economic and environmental factors that take into account the needs of people within the global context. This means that re-conceptualization of economics and international relations need to be underpinned by questioning [115] and critical , systemic thinking[116].

Conversation needs to be elevated for systemic governance purposes. The unstructured exploring of ideas and experiences is essential for good design for governance[117]. Through conversation we could realize that the projections of self onto the other is reciprocated. Bhaba (talks 1994) of the way the colonizer is mimicked and Greer (2004) talks about the way the colonizers mimic the colonized; whilst Baruma and Margalit (2004) explore the way in which 'the west is seen in the eyes of its enemies' and **the way both the east and the west 'project their perspectives' onto 'the other'.**

---

new technology of communication and that our culture and values shape the future and the market, although this does not come out strongly.

115 Such as: What is knowledge? Does it represent or does it explore dimensions of what we perceive as reality? Who decides what should be included or excluded? How can knowledge sharing be facilitated in a fair and open way? Is this possible given the commodified price tag on commodified knowledge today? Some, not all knowledge is valued. How does silenced knowledge become appreciated?

116 A heuristic example, can explain how we could be more democratic and creative and how we could address power and praxis. We could have a 'brain wave' about how to design a social inclusion project that could redress the closure for example of a car factory in South Australia, by coming up with innovations. We could suggest a post-Fordist ecotourism project that uses the factory as a hostel and staging post for tours to the wine lands and then on to the Flinders Ranges. We could also use critical thinking to encourage the two multinationals to stay and help develop the new ecotourism project. We could make this inspirational idea a reality and share the ideas by email with government and the ideas will be encoded in bits and travel along a phone line and be decoded as our ideas at the other end or phone and then reshaped through the co-created ideas of a critical friend (Reason and Bradbury 2001).

117 The work of Banathy (1996, 2000) is complementary to this agenda, as is the work of Gaventa (2001) on power and knowledge.

Thus stereotyped Orientalist perceptions (that the east is made up of religious fundamentalists) are reciprocated by the stereotyped Occidentalist perception (that the west is made up by people driven by market fundamentalism and greed). The point to remember is that we construct ourselves through consciousness of self-other and the environment.

Moving from pragmatism to idealism in governance is less challenging once we appreciate the 'boomerang affect' (Beck, 1992) of social and environmental injustice **on our own and future generations**. Can sustainable idealism and pragmatism be seen as merely different dimensions? Perhaps we need idealism (at best) and pragmatism (at worst) to be open to diverse knowledges to support sustainable futures?

Many Indigenous philosophies internationally stress the need for harmony across self, other and the environment, thus acknowledging the importance of sustainability and not denying the connections across multiple variables.

The United Nations and the World Health Organisation also stress sustainability in their health and development policies, for example the Brundtland Report (1987), the WHO (1986) Ottawa Health Charter and UN policy (1992) called Agenda 21)[118] recognize the wisdom of making connections. The seven principles of sustainable development outlined by Agenda 21 that can be applied to enhancing management[119] and to the wider process of governance are:

[118] 1972 Stockholm Conference on the Human Environment, 1976 Health for All conference in Alma Ata, 1985 Healthy Cities Conference all for health approach, 1986 Ottawa Health Charter, the principles underpinning Healthy Cities: a) healthy public policy, b) supportive environments, c) community action, d) personal skills and e) reorienting services. The Healthy City/Environment Movement follows the Ottawa Health Charter of 1986. It emphasizes an integrated multidisciplinary and multisectoral approach to community development, public health and planning. This means that many disciplines (social work, community development, education, public health, psychology, sociology, etc) need to address issues together, because one discipline alone cannot address interrelated social issues. It also means co-operation by representatives across many sectors, such as education, health, youth affairs, policing, the legal sector who need to work together not in isolation. Specialisation needs to be balanced by integration. An ecological approach to planning health and development is recommended. The Ottawa Charter underlined that health is determined by the quality of the living environment. 1987 Our Common Future by Dr Gro Brundtland was the basis of the sustainable development movement. 1994 the International Council for Environmental Development launched Agenda 21. Brundtland has stressed the links across health, development and peace studies as a basis for international relations.

[119] International Council for Environmental Initiatives 1998. Our community our future: a guide to Local Agenda 21.
International Council for Local Environmental Initiatives 1998. Model Communities Programme. International Council for Local Environmental Initiatives, Canada.

- Partnerships
- Participation and transparency
- Systemic approach
- Concern for the future
- Accountability[120]
- Equity and Justice (social and environmental).

17.3 FROM EXPERT COMPARTMENTALISED KNOWLEDGE TO SYSTEMIC GOVERNANCE

Dialogue for democracy needs to be ongoing and iterative. The cycles of history are driven by the dialectic of thesis, antithesis and synthesis. Once a version of democracy is attained it can and is undermined (Fukuyama, 1992). Some knowledge is silenced merely by virtue of age, gender, culture, nationality, and criminal or refugee status. The ongoing *development* of trust needs to continue to balance *the erosion* of trust – thus the process remains cyclical and needs to be based on dialectical dialogue[121]. This process of openness is the foundation of both democracy and the Enlightenment. Participation processes need to be mindful of the potential for power and willfulness to swamp reflexivity and to limit our ability to see things from the point of view of other people.

> "If it is true that the historical process rests on the twin pillars of rational desire and rational recognition, and that modern liberal democracy is the political system that best satisfies the two in some kind of balance, then it would seem that the chief threat to democracy would be our own confusion about what is really at stake." (Fukuyama 1992; 337)

Perhaps the real confusion is not about these twin pillars at all, it is about expanding our notion of what democracy is and ought to be. The starting point is rescuing democracy (and the Enlightenment) by examining the 'enemies within'. It is realizing that perhaps the definition of rationality has to be expanded and that this means allowing for more co-creation or co-

---

International Council for Local Environmental Initiatives.1996. The Local Agenda 21 Planning Guide: an introduction to sustainable development. International Council for Local Environmental Initiatives.

120 Answers and more iterative questioning needs to address consciousness, archetypes, culture, ethnocentrism and gendered knowledge. Vignettes and case studies provide a means to explore ideas.

121 This means engaging in conversation that explores an argument and counterarguments, in order to create a new synthesis through respectful communication.

design[122] through dialogue. The closest we can get to truth is through dialogue. Thus truth is a process, just as democracy is a process supported by social structures (according to Warren, 1999) that ought to strive for the ideal of ensuring 'frank and fearless' participation[123], the so-called mantra of the public service- that is currently being ignored in many Western democracies today.

A systemic process for design is based on avoiding decisions that 'cut off' (Churchman 1982) options by:

- Achieving resonance through 'unfolding' values and 'sweeping in' the social, cultural, political, economic and environmental aspects.
- Understanding that we need to consider our own relationship to others and the way this affects systemic intervention.
- Recognizing that we can never know the entire system that we strive to appreciate or understand, so systems models will always be partial and or flawed.

Refusing to engage in systemic dialogue occurs, because people do not understand that it is in our mutual interests to engage in accountable processes. The risk that we face is a world of conflict and pollution that will undermine the quality of life for all in this generation and the ones that follow. We can make progress if we strive to 'own the enemies' of 'politics, religion, morality and aesthetics' and not blame them on others on whom we project our own failings[124]. We label people and the actions of others:

[122] See the work of Peter Reason (1988), Reason and Bradbury (2001)and Bela Banathy(1996) , respectively.

[123] This is a good starting point for rescuing the Enlightenment agenda from its own 'enemies within' and taking note of the concerns raised by both Habermas and Derrida (2003) in their discussion of terrorism and how to address it. Habermas stresses the need to improve communication, whilst Derrida stressed the need to widen our frame of reference beyond what we take for granted (as 'right and true'). Midgley's (2001) approach to critical systems thinking draws on West Churchman but emphasizes the role of multiple stakeholders in the process of co-creating a 'Design of Inquiring Systems'.

[124] Fukuyama (1992, 1995) argues that democracy is still the only option and that despite the troubles faced by democracy it needs to be accepted as the route for history. But the concept democracy is meaningless, unless it is based on a process that enables participation by all those who are to be at the receiving end of decisions. Whilst Habermas and Derrida agree (in their conversation with Borradori 2003), that despite the poverty of democracy it is still worth pursuing via the Enlightenment (albeit revised) agenda Habermas advocates a normative approach based on testing out ideas. Whilst Derrida reframes issues in terms of wider domains, in order to avoid cutting of areas of understanding (and thus having simplistic responses to international concerns). Thus in West Churchman's terms 'where to make the cut' remains the central challenge. Another essential dimension is realizing power inequalities in everyday life undermine the ideals put forward by Habermas. We can strive towards the ideal, but we cannot deny the everyday challenges as policy makers and managers.

'refugees who enter unlawfully into the countries of others, terrorists who engage in terrorism, infidels who torture and commit acts of mindless violence'. These labels need to be balanced by considering to what extent our own thinking and actions and those of our elected governments constitute forms of terrorism. If we can try to see ourselves as others see us, we may be able to redress the archetypal Jungian[125] shadow within that we cast onto others.

17.4. SYSTEMIC CHALLENGES FOR DESIGNERS

The systemic connections across thinking and practice and the way we construct ourselves and others need to be given more attention in international relations.

Fukuyama (1992: 55) cites Nietzsche *"The Use and Abuse of History"*. In his chapter entitled *"An Idea for a Universal History"* Fukuyama (1992) begins with an extensive quotation from Nietzsche, who subscribes to Hegel's view that history is driven by the passions of humanity striving to satisfy their own meanings and values on their terms:

> "The historical imagination has never flown so far, even in a dream; for now the history of man is merely the continuation of that of animals and plants; the universal historian finds traces of himself even in the utter depths of the sea, in the living slime...he stands proudly on the pyramid of the world-process; and while he lays the final stone of his knowledge, he seems to cry aloud to listening Nature: " We are at the top, we are at the top; we are the completion of nature!"

The West's notion of democracy and Enlightenment needs to be open to change (this is by definition a characteristic of both) and I disagree with the cynicism of Nietzsche and Huntington (1996). The former and the latter deny the possibility of creativity and re-design. Huntington proposes an inevitable clash of civilizations. But they are after all constructed by people who can make choices. Cultures and groups of cultures or civilizations can be made and remade in response to changing circumstances. Sustainable democracy needs to be open to diverse ideas and to re-work itself to cope with diverse interest groups who must be able to take decisions if they are to be at the receiving end of them! Huntington argues that culture and civilization are more important than the nation state. But the nation state is

[125] Jung, C.J. 1972 Translated by Hull, R. Mandala symbolism from the collected works of Jung. C.G. Bollingen Series. Princeton University Press.

also only a construct. Culture and politics are linked constructs that *can and do* change![126]

We need to rework democracy and governance culture through redesigning forms of wider systemic governance along the lines suggested by Peter Singer (2002). We need to think about the so-called 'Enlightenment' and democracy and the implications closed thinking or practice has on power. Western democracy and the Enlightenment have become an excuse for hegemony (Chomsky, 2003; Pilger, 2002)[127].

Human beings need to move from the hubris of certainty that is tied to a culture or to a favoured scientific paradigm to embrace an appreciation of diversity and the value of criticism. Perhaps this openness will help to rescue democracy and the Enlightenment from hegemony? The potential of democracy could be restored through realizing the value of diversity for creativity. Diversity should be valued to the extent that it does not undermine the rights of others.

"The West won the world not by the superiority of its ideas or values or religion (to which few members of other civilizations were converted) but rather by its superiority in applying organized violence. Westerners often forget this fact; non-Westerners never do" (Huntington, 1996:51).

Huntington (1996, 54 draws on Mortimer) and argues that the Western separation of religion and politics (Treaty of Westphalia) is not acceptable by non-Western civilizations. But in fact the separation of religion and politics is not so clear in the West today. Furthermore, if the market is the new religion, then it could be argued that fundamentalism is not the preserve of any one civilization. Huntington (1996) suggests that the basic unit of

[126] Consciousness is based on connections across matter, mind and emotion in context (Greenfield 2000). Consciousness is a continuum. The less connections we make, the less conscious we are! The geographical continuity of 'one world' (Singer 2002) is an alternative to conceptualizing separate states and regions and the implications for nationality, race, culture, citizenship and democracy (Beck 1992, 1998; McIntyre-Mills, 2000; Grugel, 1999). Understanding the 'boomerang affect' or the social and environmental 'payback' is pragmatic. Idealistically, the more we are able to hold in mind diverse viewpoints (West Churchman, 1979 a, b, 1982) the better we will be able to appreciate complexity, risk and what accountable entails.

[127] For example, though those who were oppressed by the authoritarian Taliban regime could consider regime change desirable. But the invasion of Afghanistan needs to be seen as a result of the cold war empowerment of the Taliban in a proxy war between America and Moscow. Similarly the invasion of Iraq is a product of the armament of the Iraqi régime by America, empowered paradoxically by weaponry for widespread destruction, supplied by America to Iraq who was at that time was a convenient bulwark against Iran (Chomsky, 2003; Pilger, 2002).

investigation does not need to be states[128]. Instead he suggests that it could be civilisations in the post cold war era. Huntington's (1996) case that civilisation is the most basic variable for explaining international relations is both bleak and simplistic. It denies the ability of human beings to make cultural governance choices (Singer 2002) and to design things differently (Banathy 2000).

17.5 "TURTLES ALL THE WAY DOWN[129]": INTERNATIONAL RELATIONS, POLITICS AND PUBLIC POLICY

"There is an Indian story- at least I heard it as an Indian story- about an Englishman who, having been told that the world rested on a platform which rested on the back of an elephant which rested in turn on the back of a turtle, asked (perhaps he was an ethnographer; it is the way they behave), what did the turtle rest on? Another turtle. And that turtle? Ah, Sahib, after that it is turtles all the way down."(Geertz, 1972:29).

Asking questions and then critiquing them in terms of our own limitations would be yet another iteration to the turtles.

International relations (IR), politics and public policy cannot be couched simplistically in terms of any one set of variables, such as culture/religion or economics acting in a simplistic billiard ball relationship of cause and effect to other variables (Kaplan 1957 in Flood and Carson 1993, Huntington). Nor can fractals - or patterns within patterns - explain the situation[130], albeit non-reductionist approaches to iterations can provide an opening up of what were deemed to be deterministic patterns.

[128] Huntington makes an argument that the widest cultural entity with which a group of nations can identify can be called a civilisation and that it is this level of organization that needs to be studied, not merely states.

[129] Clifford Geertz (1973) in his first chapter of 'Interpretation of Cultures', stresses that culture is contextual meaning . Although models can be helpful as a conclusion to 'thick description', they cannot be regarded as anything more that a starting point for thinking about other situations.

[130] Fractals can be defined as follows:"Fractals are...complex by virtue of their infinite detail and unique mathematical properties (no two fractals are the same), yet they're simple, because they can be generated through successive applications of simple iterations.... it's a new brand of reductionism...utterly unlike the old reductionism, which sees complexity as built up out of simple forms, as an intricate building is made out of a few simple shapes or bricks. Here the simple iterations in effect liberate the complexity hidden within it, giving access to creative potential. The equation isn't the plot of a shape as in Euclid. Rather, the equation provides the starting point for evolving feedback (Wheatley, 1999:106, cites Briggs and Peat, 1989).

Prigogine and Stengers (in Bausch, 2001; Wheatley, 1999:78) stress that we need to take into account change and process and not focus too much on structure. Their concept 'Dissipative structures' is a parallel concept to the social science concept of the dialectic (Bausch, 2001). The dialectic is the basis of history. Fukuyama (1992) draws on Hegelian and Marxian versions of the dialectic. Marx inverts Hegel's idea of the dialectic, meaning that instead of the individual driving the dialectic of change, Marx saw social structures (the products of relations of production) as driving history. Flood and Carson (1993:141) stress that international relations theory has been used to emphasise structure, but they argue for the need to introduce process into the analysis.

Flood and Carson (1993:145) argue that a systemic study of international relations does not need to start with states. It can start with people and their potential to rethink and reconceptualize possibilities. But as Chomsky (2003) and Pilger (2002) argue, unfortunately the power of the market has given America a position of hegemony. The billiard ball impact of power is a dimension of IR, but unfortunately IR has been only functionalist and structuralist oriented. The potential of people to use social movements for change and to reshape meanings is possible (McIntyre-Mills, 2000). If we were to consider the basic grid of Burrell and Morgan (in Jackson, 2000:23), it is clear that the systemic thinking has been limited to functionalism and that there is scope for expanding systemic thinking to enable it to be more critical and to encompass change and subjective ideas to balance out the existing models that tend to be regulatory, reductionist and objective:

Table 1: A Systemic view of governance and international relations derived from Burrell and Morgan's Grid (1979)

	Systemic change		
Subjective-ideographic	Radical humanist approaches to redesign our future	Radical structuralist approaches to redesign the future.	Objective - nomothetic
	Interpretive approaches that emphasise culture as the basis for maintaining the status quo	Functionalist – billiard ball approaches	
	Systemic regulation		

Fukuyama (1992: 337) stresses that culture and meaning making are perhaps one of the most important dimensions of humanity and argues following Plato that humanity is driven by reason, desire and self-esteem.

Nietzsche he argues believed only self-esteem and our own meanings and values drive us. Kaplan (is cited by Flood and Carson, 1993:142) as introducing cybernetic concepts. He argued that states could be seen in terms of 6 basic patterns[131] of relationships. The work of Kaplan and Rosencrance (cited by Flood and Carson, 1992:145) is also functionalist, takes a deterministic view and uses only positivist methods. The simulations and models used in functionalist terms do not take into account human nature, emotions and meaning.

Huntington comes up with four patterns of relationships, namely a) universal harmony, b) multipolar billiard ball of cause and effect, c) cold war bipolar and d) chaos where mutual destruction is possible (as a result of the nuclear weapons of mass destruction owned by hegemonic states and coalitions). He does not consider the possibility of systemic governance in the interests of social and environmental sustainability that spans international, regional and national boundaries in a new kind of federalism based on subsidiarity as suggested by Singer (2002). Subsidiarity is based on the idea that those at the receiving end of an decision need to be party to the decision making process. People should be free to make decisions as they wish to the extent that they do not undermine the freedoms of others.

Flood and Carson (op cit) review a number of works (Kaplan, Rosencrance, Modelski and Brecher, Weltman) that conclude that stability is functionally linked with a situation in which there are few variables with which to deal. Weltman (according to Flood and Carson, 1993:144) concludes that systems ideas are merely functionalist. But as they point out (Flood and Carson, 1993:145) the system of interest does not need to be states.

Flood and Carson (1992:147) go on to quote the work of Deutsche '*The Nerves of Government*' (1963) which although also functionalist in their opinion is more organic and introduces politics as a more dynamic recognition of power plays and the role that people have in making and remaking their world. They also discuss other writers such as Burton who begin the shift away from the objective and regulatory and introduce the subjective dimensions of power and its implications for change. Flood and Carson stress that if these early writers had looked at the work of West Churchman they would have found a useful source of ideas. But they did not.

[131] "Multipolar: a balance of power arrangement; literally a billiard ball model;
Loose bipolar: two main opposing spheres with satellites of varying degrees of adherence to either side;
Tight bipolar: like 2, but with clear allegiances to one sphere;
Universal: a confederation; all groups controlled by one government;
Hierarchical; significant grouping that are functional rather than territorial; a federation;
Unit vet: each group can destroy all others, there is mutually assured destruction (MAD)."

Flood and Carson suggest that system metaphors[132] can help us to think across the grid of options. Later to the original set of metaphors Jackson (2000:133) adds additional metaphors, in order to take into account power and domination and also flux and emergence for transformation. The latter metaphors are useful to assist understanding, as is the work of West Churchman on whom Flood and Carson (1993) and Jackson (2000) draw. Churchman made it very clear that he did not believe it possible to comprehend a total system. Flood and Jackson's term 'TSI or total systems intervention' should not be understood literally to mean striving for total solutions. It needs to be seen as a practical process for working with many different ideas and striving to appreciate the wider system and as a means to inspire creativity.

17.6 ITERATIVE QUESTIONING AS GOVERNANCE

- What strategies can be used to bolster good governance and good international relations? The role of alternative media, the Internet and social movements can be vital for making a case and lobbying from a position of strength. Access to multiple arenas is essential.
- What roles are there in the future for individuals, groups, organizations such as the UN and a high court and social movements for social and environmental justice?

[132] Machine – closed system; Organism –open system; Brain-learning system; Culture – norms and values; Team -unitary political system; Coalition -pluralist political system; Prison –coercive political system

[133] Metaphors that can be applied to international relations
Machines- parts are related to wholes and human beings are no different from other parts
Organisms- organizations are open and part of a wider whole
Brains-organisations are like neural networks
Cultures- people can make and remake organisations
Political systems – power plays dominate organisations
Psychic prison- our thinking is a product of psychology- authoritarian, anxious, fear of change
Flux and Transformation –structure and process define life
Instrument of domination- Marx and Braverman – organizations extract surplus value

Table 2: Checklist for working the boundaries of International relations and governance

Who currently is/ought to get what, when, why, how and to what affect? Have we considered the following variables:
Socio-demographic – age, gender, language/culture/disability.
Spatial: micro/macro.
Time: past/present/future.
Conceptual: values of diverse groups.
Arenas for strategic action: state, market, community, networks, and household.

Source: Inspired [134] by Ulrich's is/ought framework (See 1983, 2001)

People perceive reality in many different ways and instead of merely dividing the way people see things into right and wrong. We need to use communication to explore what people mean. Political dialogue can also help to appreciate diverse ways of understanding the world (Romm, 2001) and it can help to co-create meaning (Reason, 1988). Borradori (2003) in her recent conversation with both Habermas and Derrida (in the work called "*Philosophy in a Time of Terror*") addresses the need to explore differences in culture and to do this by dialogue. The process of appreciation of the fuller picture is part of the process of greater understanding that is vital for good governance that supports well-being and the reduction of poverty (Schneider, 1999). The essence of good governance is about fostering partnerships across the elected members of government, the corporate sectors and members of the public (Dean Jaensch Dec 4th 2003 - public lecture) at the local, national and international level as nested systems (White, 2001, 2002) that ensure that those who are at the receiving end of a decision are party to the decision making process. Thus decision making should be made at the lowest governance level possible. This is the principle of subsidiarity.

One of the challenges that government faces in Australia (and elsewhere) is to act as an enabler for sustainable local development. This could be measured in terms of multiple (Gallhofer and Chew 2000) or dynamically interconnected variables. 'Triple bottom line accounting' (TBA)(Elkington,

[134] Midgley (2001:639) explores a way to address either/or thinking by drawing on the notion of difference (as per Derrida), by suggesting that dialogue and questioning the boundaries (drawing on Ulrich, 1983) is useful. This is the intersubjective domain of developing arguments through unfolding values and sweeping in (in the sense used by Churchman 1979a, b, 1982 who drew on Singer) to appreciate the social, cultural, economic and environmental contextual variables.

1997) addresses social, economic and environmental factors, not just economic factors. It is a useful step towards the reworking of accounting and accountability at the local level. But further works is needed to build in more interaction across the social, cultural, economic and environmental factors. Software such as the cogniscope (See chapter by Christakis) could be useful in workshops for accountable, integrated decision-making. But space for difference needs to be balanced with attempts at integration (See chapters by Morgan, Romm and McIntyre-Mills).

The ability to think critically needs to be brought to bear in the development of relationships, where trust is earned over time through co-creation of meanings. Fostering relationships is usually discussed in terms of trust and networks of trust in the more traditional literature (Putman, 1995). But as Bourdieu (1986) stressed, social networks can be either positive or negative. In other words, in the interests of the powerful, not necessarily the powerless. Also the higher the level of education, the less likely people are to simply trust one another (Jupp, 2003). Tolerance however does go hand-in- hand with higher levels of education (ALGA and National Economics, 2002,2003). Respectful[135] communication and the willingness to reframe boundaries is the basis for better governance and international relations[136].

[135] For this reason I draw on the work of Habermas (1984 and his conversation with Borradori 2003), but not to strive for truth in such definitive terms. Communication tests (as per Habermas 1984) are useful up to a point (as suggested by Midgley 1992) provided we also engage with so called irrational discourses, in order to understand social history, policy and ourselves. Networks of relationships and webs of variables need to be considered in terms of the historical, conceptual and geographical space/time continuum when appreciating the nature of a problem, such as local and international terrorism and the way it is conceptualised.

[136] Habermas (1984) has stressed that of the three areas of human knowledge – technical (leading to objective knowledge), strategic (based on subjective interpretations) and communicative (based on shared or intersubjective knowledge), the latter is the least well developed. Some critical systems thinking strives to salvage the Enlightenment Agenda and others tend to criticize it. The arguments for the Enlightenment agenda are along the lines that truth is the basis for science, democracy and the rule of law that supports human rights. Those who critique this agenda argue that criticizing the boundaries is the only way we can approach truth and that truth is dialogue based on working with boundaries. We need to strive for the elusive goal of truth, but remain mindful that the striving for representation is the most important part. It is based on the recognition that definitions are always contextual and that the dialectic is the means to find the contradictions and to try to address them (not in order to try to find the hierarchical system that will explain all) but perhaps to find the iterative patterns of inquiry (see Hofstadter, 1979: figure 13.5, 690 of two hands drawing each other, drawn by M.C. Escher)[136]. We need to consider :

- how we create ourselves,
- our conceptions of god and the way in which self and other define and co-define each other.

17.7 DESIGN FOR SYSTEMIC INQUIRY BASED ON WEST CHURCHMAN AND CRITICAL SYSTEMS THINKING AND PRACTICE (PRAXIS)

Understanding requires addressing a complex array of variables. The following checklist for praxis and process (to be used in combination with a conceptual tool kit)[137] can help to assess the extent to which a problem has been appreciated systemically:

Table 2. Praxis tests for problem appreciation (drawn and adapted from West Churchman 1979 and Van Gigch 2003)

Have we considered the following:	
Subjective knowledge that may or may not be acknowledged by the powerful –socialization and experience- assumptions and values within context	Bhabha (1994[140]) on self and cultural other, Foucault[141] (knowledge and power) and Derrida[142] (postcolonial knowledge),
Intersubjective knowledge based on dialogue in a range of contexts (local, national and international)	Fonow and Cook (women and knowledge), Michelle Fine (1994) talks of race, class, culture and gender and how it constructs life chances, gender relations and analysis (March et al 1999, Moser, 1993).
Engaging in respectful dialogue will only occur in a context of trust built up over time as a participant observer or as an ethnographer/auto ethnographer138	Banathy (1996, 2000) stressed the need for public spaces in geographical and cyberspace to support designing for the future generations.
Using the mind and logic based on more than one cultural perspective. Western logic is linear (deductive and inductive)[139], but the logic of Indigenous cultures	Logic is the tool for analyzing arguments that are made up of premises and conclusions (Salmon 1973 1-4).

- that as we draw 'the other', they are drawing us.Being able to observe the self and other co-constructing each other is helpful for governance and international relations, because it helps us to be mindful of our own values and how they influence the way we see others.The working of boundaries (of what is acceptable and not acceptable within specific contexts) is at the heart of dialogue for governance. It is based on the realization that the closest we can get to truth is through dialogue.

[137] McIntyre, J., 2004, Facilitating Critical Systemic Praxis (CSP) by means of experiential learning and conceptual tools. This toolkit is expanded in Working and Reworking the conceptual and Geographical boundaries of international relations and governance (McIntyre –Mills forthcoming) to include tools to address orientalism and Occidentalism as well as other projections that undermine good governance and international relations.

[138] Ellis C. and Bochner, A., 2000, Autoethnography. personal narrative, reflexivity: researcher as subject, Handbook of Qualitative Research, 2nd ed., Denzin, N. and Lincoln, Y., Sage, London.

[139] Bhabha, H., 1994, The Location of Culture, Routledge. London.

[140] Foucault, M. and Gordon, C., eds., 1980, Power / Knowledge: Selected interviews and other writings 1972-1977, Harvester, Brighton.

Foucault, M., 1967, Madness and Civilization: A history of insanity in the age of reason, Routlege, London.

Foucault, M., 1979, Disciplinc and Punishment: The birth of the prison, Vintage, New York.

Foucault, 2001, in Tunisia In: Young, R., Postcolonialism: an historical introduction, Blackwell, Oxford.

[141] Derrida, 2001, in Algeria In: Young, R.. Postcolonialism: an historical introduction, Blackwell, Oxford.

Philosophy in a time of terror dialogues with Jurgen Habermas and Jacques Derrida interviewed by Borradori, Giovanna, University of Chicago Press, Chicago.

is more weblike, follows different rules and can help us to make (abductive) leaps and to find more patterns associated with power and privilege (retroductive logic).[143]	The statements of evidence may or may not be correct, but the purpose of logic is not to assess the evidence but the relationship of the premises to the conclusion using deduction (see Sarantakos [144], Salmon's logic[145] and West Churchman citing Leibniz's western logic nets that provide hooks of interconnecting reasoning 1979).
5. Empirical research based on the senses and testing falsification not only with reference to the facts (Popper and Habermas), but also in so far as those who are at the receiving end of policy decisions should have a say in the expressing views on the consequences. It is never falsification in terms of one set of facts, as presented by the powerful, but how multiple view points can be addressed through dialogue, co-creation and design.	External validity through comparisons across multiple stakeholders requires empirical research.
6. Dialogue based on the dialectic that appreciates the context of systems as part of the wider universe of which it is a part. For example: Dialectic (Hegel) 'Guarantors' of knowledge (West Churchman), inter-subjective communication tests for truth (Habermas)	Respectful communication and tests of significance – using the facilitation tools for working with many stakeholders using a tool kit (McIntyre 2004)[146].
7.Co-created (Reason and Bradbury 2001) or intersubjective knowledge based on praxis tools	For example, E. A Singer cited by, Churchman (1979a,b, 1982) for *sweeping in* the social, political, economic and environmental context and *unfolding* the values of all the stakeholders who are to be impacted by a decision and developing resonance.
8.Participatory action research that uses complementary approaches	For example, Reason (1988,1991,1994,2002), Flood and Romm (1996), Wadsworth (2001). Sight, sound, touch, hearing and feelings of many participants are considered. All the senses of multiple stakeholders are given attention in systemic approaches.
9. Empowered participatory governance "EPEPG "is part of a broader collaboration to discover and imagine democratic institutions that are at once more participatory and effective than the familiar	For example, Fung and Wright (2003), Edgar (2001) argue that stakeholders need to form a " virtuous circle of reciprocal reinforcement' (Fung and Wright (2003: 15), but a CSP view is that all circular thinking

[142] Deduction and induction (Salmon 1973:14)

If all the premises are true, the conclusion must be true

All of the information or factual content in the conclusion was already contained, at least implicitly, in the premises

All of the premises are true, the conclusion is probably true but not necessarily true

The conclusion contains information not present, even implicitly in the premises.

[143] Abduction and Retroduction

Abduction or "Leaping beyond" (Banathy, 1996) is possible when we design a new way of seeing. It helps to be able to enhance our creativity by listening to diverse voices.

Retroduction (Blaikie 1993) is about finding patterns that underpin or help us understand why things are the way they are. Listening to many voices can help this.

[144] Sarantakos, S., 1993, Social Research, Macmillan, South Yarra.

[145] Salmon, W., 1973, Logic. Foundations of Philosophy Series. Prentice Hall. New Jersey.

[146] McIntyre, J., 2004, Facilitating Critical Systemic Praxis by means of Experiential Learning and Conceptual Tools. To enhance our capacity to design systemic policy and praxis that promotes understanding through better governance, Systems Research and Behavioural Science, 21.37-61.

configuration of political representation and bureaucratic administration" (Fung and Wright, 2003: 15)	needs to be open to so-called good and bad news, if it is to avoid social entropy and the dissipation of energy for change (Flood 1998).
10. Participatory democracy	Christakis (2004)[147] and McIntyre (2004)[148] drawing on wisdom of many cultures for designing praxis that addresses the boundaries within and across self, other and the environment.
11. Transformation and emergence by means of accounting and accountability that supports a revised triple bottom accounting method (derived from Elkington, 1997; Gallhofer and Chew, 2000)[149] that is based on discursive democracy, compassion and sustainable decision making for social and environmental justice	Working the hyphens across self-other Fine in Denzin and Lincoln 1994) and the environment (McIntyre 2004 forthcoming) The rights of all sentient beings should also be taken into account. They have become unnecessarily commodified [150] in a wide range of social and political contexts. The work of Delueze and Guattari is useful on desire and the motivation for molar (fixed) and molecular (fluid) identity.

17.8. CONCLUSION: DESIGNING FROM MULTIPLE VIEWPOINTS AND THE VALUE OF CONVERSATION AND STORY TELLING FOR GOVERNANCE

> "Interpret Leibniz's 'apperception' to mean the ability of the system designer to design the system from many points of view- to design science as a management system or design physics as a psychology. To the extent that a system fails in its apperception, it is less than God, i.e., it is an imperfect monad." (Churchman, 1979:75).

Our challenge is to realize our imperfections, but to keep striving towards the elusive whole through never ending cycles of conversation that can weave together multiple views and to find out to what extent ideas can be improved upon. CSP is based on the premise that reality is best understood as comprising domains of knowledge and following Habermas (1984) these can be called: the subjective, objective and intersubjective domains. Knowledge is not so much a truth as multiple, multisemic (McCLung Lee, 1988), hyperlinked layers and dimensions of reality. Both openness and a process of iterative inquiry and structure are the basis of democracy, the Enlightenment and life[151]. It is vital to sustain that openness in times of terrorism and not to undermine civil liberties. Dialogue with many stakeholders on matters of social and environmental justice needs to be seen in terms of a complex web of international relations within specific contexts.

[147] Christakis, A., 2003, Wisdom of the People. President's Speech at the International Systems Sciences Conference, Herinossus, Crete.

[148] McIntyre, J., Participatory Democracy: Drawing on C.West Churchman's Thinking when Making Public Policy, Systems Research and Behavioural Science, 20:489-498.

[149] Elkington, J., 1997, Cannibals with Forks, Capstone, Oxford.

[150] Financial Mail, 1-2 November, Sheep Trade: no room for woolly thinking. September to October.

[151] Prigogine, I. and Stengers, I., 1984, Order out of Chaos, Bantam Books, New York. Bausch (2001) and Wheatley (1999) provided a way into this argument.

Philosophy and communication as a means to share thinking and to co-create responses, are perhaps more important than ever as the basis for peace studies and social policy. Unfortunately the paradox is that openness and sharing ideas (the idealist version of democracy and transformed Enlightenment approach) requires trust. Trust in turn requires openness and sharing ideas. The leap of faith required to develop resonance through respectful communication is where spirituality and science touch fingers across the abyss. Trust provides the first move towards embracing the dialectic. Better communication processes can only enhance democracy and Enlightenment through governance. In this way representation and co-creation can be addressed. It is for this reason that it is worthwhile to address or (try to bridge) the divides across ontological and epistemological understanding (See Romm, 2004, pers comm., Romm, 2001 a, b, 2002).

REFERENCES

Ashby, W.R., 1956, *An introduction to Cybernetics*, Chapman and Hall, London.

Alinsky, S.D., 1972, *Rules for Radicals: A Practical Primer for Realistic Radicals*, Vintage Books, New York.

Bhabha, H., 1994, *The location of culture,* Routledge, London.

Banathy, B., 1996, *Designing Social Systems in a Changing World*, Plenum, New York.

Banathy, B., 2000, *Guided Evolution of Society: A Systems View,* Kluwer/Plenum, London.

Banathy, B., 2001a, Self-guided evolution of society: the challenge: the self guided evolution of ISSS 45th *International Conference for the Systems Sciences*.. July 8th-13th.

Banathy, B., 2001b, The Agora project: self guided social and societal evolution: the new agoras of the 21st Century, 45th *International Conference International Society for the Systems Sciences*, July 8th-13th.

Banathy, B., 2003, Self-Guiding Evolution of Civilisation, *Systems Research and Behavioural Science*, 20:4309-323.

Baruma, I. and Margalit, A., 2004. *Occidentalism: The West in the Eyes of its Enemies*, Penguin, New York.

Bausch, K., 2001, *The Emerging Consensus in Social Systems Theory,* Kluwer/Plenum.

Beck, U., 1992, *Risk Society Towards a New Modernity*, Sage, London.

Beck, U., 1998, *Democracy without Enemies*, Polity, Cambridge.

Bhabha, H., 1994, *The location of Culture*, Routledge, London.

Burrel, G. and Morgan, G, Sociological Paradigms and Organizational Analysis. Heinemann, London.

Foucault, M. and Gordon, C. eds., 1980, *Power/Knowledge: Selected Interviews and Other Writings 1972-1977*, Harvester, Brighton.

Capra, F, Steindl-Rast, D. with Matus, T., 1992, *Belonging to the Universe. New Thinking about God and Nature*. Penguin, London.

Capra, F., 1982, *TheTturning Point: Science, Society and the Rising Culture*, Flamingo, Fontana.

Capra, F., 1996, *The Web of Life: A New Synthesis of Mind and Matter*, Harper Collins.

Chomsky, N., 2003, *Hegemony or Survival: America's Quest for Global Dominance*, Allen and Unwin, Crows Nest, NSW.

Churchman, C. West., 1971, *The Design of Inquiring Systems. Basic Concepts of Systems and Organization*, Basic Books, New York.

Churchman, C. West, 1979a, *The Systems Approach,* Delta, New York.

Churchman, C. West, 1979b, *The Systems Approach and its Enemies*, Basic Books, New York.

Churchman, C. West, 1982, *Thought and Wisdom*, Intersystems Publications, California.

Douglas, M., 1978, *Purity and Danger: An Analysis of the Concepts of Pollution and Taboo*, Routledge and Kegan Paul, London.

Edwards. and Gaventa, J., 2001, *Global Citizen Action*, Colorado, Boulder.

Ellis, C. and Bochner, A., 2000, Autoethnography. Personal narrative, reflexivity: researcher as subject, *Handbook of Qualitative Research,* 2nd ed., N. Denzin and Y. Lincoln, Sage, London.

Flood, R. and Carson, E., 1998, *Dealing with Complexity: An Introduction to the Theory and Application of Systems Science*, 2nd ed., Plenum, London.

Flood, R. and Romm, N., 1996, *Diversity Management: Triple Loop Learning,* John Wiley & Sons, New York.

Florini, Ann, 2003, *The Coming Democracy: New Rules for Running a New World*, Island Press.

Friedman, T., 2000, *The Lexus and the Olive Tree*, Random House, New York.

Fung, A. and Wright, E.O., 2003, *Deepening Democracy. Institutional innovations in Empowered Participatory Governance. The Real Utopias Project 1V.* Verso, London.

Fukuyama, F., 1992, *The End of History and the Last Man*, Free Press, Simon and Schuster.

Gallhofer, S. and Chew, A., 2000, Introduction: accounting and indigenous peoples, *Accounting, Auditing and Accountability Journal*, 13(3): 256-267.

Gaventa, J. and Valderrama, C., 1999, Participation, citizenship and local governance: background note for workshop on *"Strengthening Participation in Local Governance"*, Institute of Development Studies, June.

Gaventa, J. 2001, Towards participatory local governance: six propositions for discussion. Paper presented to the *Ford Foundation*, LOGO Program with the Institute of Development Studies, June.

Gaventa, J. and Cornwell, A. 2001, Power and Knowledge, *Handbook of Action Research,* P. Reason and H. Bradbury, eds., Sage, London.

Gibbons, M., Limoges, C., Nowotny, H., Schwartzman, S., Scott, P. and Trow, M., 1994, *The New Productions of Knowledge: The Dynamics of Science and Research in Contemporary Societies*, Sage, London.

Gouldner, A.W., 1971, *The Coming Crisis of Western* Sociology, Heinemann, London.

Geertz, C., 1973,*The Interpretation of Cultures, Basic Books*, New York.

Greer, G., 2004, Whitefella Jump up. The Shortest Way to Nationhood, Profile Books, London.

Grugel, J., 1999, Democracy Without Borders. Transnationalization and Conditionality in New Democracies, Routledge, London.

Habermas, J., 1984,*The theory of communicative action: reason and the rationalization of society.* Beacon: Boston.

Habermas, J., 1995, Reconciliation through the public use of reason: remarks on John Rawls's political liberalism, *The Journal of Philosophy*, XC11 (3): 109-131.

Habermas, J., Derrida, J and Giovanna, B., 2003, *Philosophy in a Time of Terror Dialogues with Jurgen Habermas and Jacques Derrida interviewed by Borradori, Giovanna*, University of Chicago Press, Chicago.

Haraway, D., 1991, *Simians, Cyborgs, and Women. The Reinvention of Nature.* Free Association Books. London.

Hofstadter, D., 1979, *Gödel, Escher, Bach an Eternal Golden Braid*, Basic Books, New York.

Huntington, S.P., 1996, *The Clash of Civilisations and the Remaking of World Order*, Simon and Schuster, New York.

Jung, C.J., 1972, Translated by Hull, R., *Mandala Symbolism from the Collected Works of Jung*, C.G. Bollingen Series, Princeton University Press.

Long, M, 2002, Beyond Traditional Boundaries: Government in the Information Age, *Australian Journal of Public Administration.* Vol. 61 no.1: 3-12.

Longo, M, 2004 Hostile receptions: Dilemmas of democracy: legitimacy and Supranational law. Australian Journal of Politics and History. Vol. 50. No 2. 211-228.

Reason, P. and Bradbury, H.2001 *Handbook of Action Research: Participative Inquiry and Practice*. London. Sage.

March, C., Smythe, I. and Mukhopadhyay, 1999, *A Guide to Gender-Analysis Frameworks*, Oxfam, Oxford.

McIntyre-Mills, J., 2000, *Global Citizenship and Social Movements: Creating Transcultural Webs of Meaning for the New Millennium*, Harwood. Netherlands.

McIntyre-Mills, J., 2003, *Critical Systemic Practice for Social and Environmental Justice*, Kluwer, New York.

McLung, L.A., 1988, *Sociology for People: Toward a Caring Profession*, Syracuse University Press, New York.

Midgley, G., 1992, Pluralism and the legitimation of systems science, *Systems Practice*, 5(2): 147-172.

Midgley, G. and Ochoa-Arias, A., 2001, Unfolding a theory of systemic intervention, *Systemic Praxis and Action Research*, 14(5): 615-649.

Midgley, G., 1997, Dealing with coercion: critical systems heuristics and beyond, *Systems Practice*, 10(1): 37-57.

Moser, C.O., 1993, *Gender Planning and Development: Theory, Practice and Training*, Routledge, London.

National Economic/ALGA 2002, *State of the Regions Report*, by P. Brain, National Economic Director, and C. Shepl, ALGA, published by Australian Local Government Association.

National Economic/ALGA, 2003, *State of the Regions Report*, published by Australian Local Government Association.

Negroponte, N., 1995, *Being Digital*, Vintage, New York.

Pierre, J. ed., 2000, *Debating Governance*: authority,steering and democracy, University Press, Oxford.

Pilger, J., 2002, *The New Rulers of the World*, Verso, London.

Reason, P., 1988, *Human Inquiry in Action*, Sage, London.

Reason, P. and Bradbury, H., 2001, *Handbook of Action Research: Participative Inquiry and Practice*, Sage, London.

Romm, N., 2001a, Our responsibilities as systemic thinkers, *45th International Conference of International Society for the Systems Sciences,* USA, Asilomar.

Romm, N., 2001b, *Accountability in Social Research: Issues and Debates*. Romm, N., 1996, Inquiry-and intervention in Systems planning: probing methodological rationalities, *World Futures,* 47:25-36.

Romm, N., 2002, A trusting constructivist view of systems thinking in a knowledge age, in: *Systems Theory and Practice in a Knowledge Age*, Ragsdell et al, Plenum, New York.

Ulrich, W., 1983, *Critical Heuristics of Social Planning: a new approach to practical philosophy*. Wiley. New York.

Ulrich, W. 2001, The quest for competence in systemic research and practice, *Systems Research and Behavioural Science*, 18:3-28.

Uslaner, E., 1999, Democracy and social capital, in: *Democracy and Trust,* M. Warren, Cambridge University Press, Cambridge.

Van Gigch, J., 2003, *Metadecisions: Rehabilitating Epistemology*, Kluwer Plenum.

Vickers, G., 1983, *Human Systems are Different*, Harper and Row Publishers, London.

Wadsworth, Y., 1991, *Everyday Evaluation on the Run*, Action Research Issues Associated.

Warren, E., 1999, *Democratic Theory and Trust*, Cambridge University Press, Cambridge.

Wheatley, M., *Leadership and the New Science: Discovering Order in a Chaotic world*, Berret-Koehler, San Francisco.

White, L., 2001, 'Effective Governance' through complexity thinking and management science, *Systems Research and Behavioural Science*, 18(3): 241-257.

White, L., 2002, Connection matters: exploring the implications of social capital and social networks for social policy, *Systems Research and Behavioural Science*, 19(2): 255-270.

INDEX

Page numbers followed by t indicate tables.